Analysis of Sport Motion: Anatomic and Biomechanic Perspectives

Third Edition

Analysis of Sport Motion: Anatomic and Biomechanic Perspectives

John W. Northrip

Gene A. Logan

Wayne C. McKinney

wcb
Wm. C. Brown Company Publishers
Dubuque, Iowa

CONSULTING EDITOR

Physical Education

Aileene Lockhart
Texas Woman's University

Formerly entitled *Introduction to Biomechanic Analysis of Sport*

Copyright © 1974, 1979 by Wm. C. Brown Company Publishers

Copyright © 1983 by Wm. C. Brown Company Publishers. All rights reserved

Library of Congress Catalog Card Number: 82–83663

ISBN 0–697–07206–1

Printed in the United States of America

2   07206   01

Contents

Physics of Sport 161

3

Theory to Practice: Techniques of Sport Motion Analysis 315

4

This book is designed to introduce the undergraduate physical education major to the exercise sciences known as anatomic kinesiology and biomechanics. One objective of the book is to help students understand the cause and effect relationships between these two academic areas as they are utilized in analyzing sport motions. Since quality teaching-coaching of neuromuscular skills in physical education should always be preceded by an analytic process where the professional physical educator synthesizes observations and theory from scientific and technical perspectives, this book is especially concerned with the teaching-coaching of individual skills in sport, athletics, and dance. It must be remembered that the teaching of physical education is an art with its bases in science. It is a major aim of this book to help the student learn how to apply concepts and principles from anatomic kinesiology and biomechanics in the actual practice of conducting daily teaching-coaching assignments. There is a strong emphasis on helping the student bridge the gap from theory to practice. Therein lies one of the unique features of this book.

Now in its third edition, the book, known in its first two editions as *Introduction to Biomechanic Analysis of Sport,* has been titled *Analysis of Sport Motion: Anatomic and Biomechanic Perspectives* better to reflect its amplified content in the anatomic area while also adding to the content in biomechanics. This change in emphasis came about, in part, due to the recommendations of the Kinesiology Academy of the AAHPERD for undergraduate professional preparation of physical education majors in the general field of kinesiology. The subject matter of this book meets those recommended knowledge objectives.

The book is divided into four parts. These were written in progressive fashion from the fundamental theory of anatomic kinesiology and biomechanics to the application of that theory in teaching-coaching situations. Numerous examples and illustrations from sport are included in all parts of the book. Part 1 is "Introduction to Kinesiology." Since the study of kinesiology includes both anatomic and biomechanic perspectives, the student is given a brief overview of the subject. The importance of analysis of sport motion is emphasized as the starting point for teaching neuromuscular skills, and a hierarchy for human motion analysis is outlined.

Part 2 is "Basic Concepts of Anatomic Kinesiology." This introduces the student to anatomic kinesiology, or it can be used as a review for students who have had courses in the subject and/or in human anatomy. This part has been expanded to include more fundamentals of joint motion analysis and myology. The authors believe that the cause and effect relationships between anatomic kinesiology and biomechanics, as these involve human motion, can best be learned if the student understands the anatomic language and concepts first. Consequently, this part has been expanded to include the basic terminology for planes and joint motions, muscle contraction, and segmental myology. Accurate motion description of moving body segments is essential to the overall analytic process.

The student should be fully cognizant of the basic concepts and language of anatomic kinesiology. For biomechanic analyses, skeletal landmarks, joint motion potentialities, and joint ranges of motion must be known. This knowledge is necessary to make accurate observations and to locate specific points on levers or within joints. Plotting of joint and body segment motion on film is dependent on this in order, as examples, to calculate such factors as angular and linear velocities of limbs, acceleration of the center of mass, and force interactions between the body as a sport object and a sport implement being moved by internal force and gravity. Analytic kinesiology involves articulate observations utilizing concepts from biomechanics *and* anatomic kinesiology. Quality teaching is dependent upon those observations. The *effect* of the outcome of the performance can be explained in the language of biomechanics, but the *cause* lies in the nature of the timing, degree, and sequence of movement in joints. Adjustments during the teaching process to improve performance must be made in the sequential motion pattern of the involved joints. Therefore, the student of physical education must have functional knowledge of anatomic kinesiology.

Part 3, "Physics of Sport," is written for the student who has not had college course work in mathematics and physics. This part serves as the theoretical foundation for comprehending biomechanic analysis. Elements of physics are presented. These are the fundamental physics required to describe motions observed in exercise, sport, and dance whether the performances are on land, in air, or in water. The last chapter of part 3, "Principles of Biomechanics," summarizes those concepts from physics in a nonmathematical framework.

For students who do not have backgrounds in physics and mathematics, chapter 11 can be used as an outline for the first three chapters of this part of the book. As a teaching progression, the principles can be presented and discussed in a nonmathematic context first. This can be followed by discussion of the principles, using the appropriate physics. Verbal familiarity with the concepts tends to lessen the numerical "symbol shock" for some students.

Part 3 of the book is designed primarily to assist the reader in gaining an understanding of biomechanic analyses through the intermediate cinematographic level. Three procedures are used to convey biomechanic principles to the reader. First, principles are explained in a fundamental manner to facilitate comprehension. Second, to reinforce the word description of principles, graphic illustrations are included where applicable. Third, where deemed necessary to facilitate comprehension and add precision, mathematical formulae and examples are included to further reinforce the ideas conveyed. Numerals as symbols have a distinct advantage over words as symbols. Words, especially in the English language, have some degree of ambiguity; mathematical symbols are more precise ways of expressing ideas. Some people tend to succumb to a syndrome known as "symbol shock" if they are asked to work extensively in the field of mathematics. This is especially true when students have had limited mathematical training at the secondary school and university levels. It is believed that the best method of presenting biomechanic concepts initially is in a nonmathematic context. After this has been accomplished, one may then proceed to the very essential equations necessary to describe motion adequately at the higher levels of the analysis hierarchy. In such a teaching progression the learner would at least have a frame of reference or general understanding of biomechanic principles on which to apply the necessary mathematics.

The appendices are designed to support Parts 3 and 4. They include conversion tables for mechanical units, a review of the basic mathematical relationships most used, and three examples of electronic calculator use in the analytic process.

Part 4, "Theory to Practice: Techniques of Sport Motion Analysis," leads the student to practical applications of biomechanics for teaching-coaching situations. In order to accomplish the practical objective of utilizing biomechanic subject matter in instruction situations, *a hierarchy for biomechanic analysis is introduced.* Each level is designed to increase the degree of precision in controlled observations of performers in classes, athletic situations, or as research subjects. The four levels of biomechanic analysis are (1) noncinematographic analysis, (2) basic cinematographic analysis, (3) intermediate cinematographic analysis, and (4) biomechanic research. This text provides the student with the potential for a functional understanding of the first three levels within the hierarchy of biomechanic analysis. The first two levels should be used in day-to-day teaching and coaching by physical educators in elementary, secondary, and undergraduate university positions. Intermediate cinematographic analysis can be applied in special teaching-coaching situations on a less frequent basis.

Part 4 includes a chapter (13) on "Extracting Data Manually From Film." The manual data extraction process is viewed by the authors as a logical learning progression from simple to more complex procedures involving computer techniques. The student should have a greater understanding and appreciation of more sophisticated data extraction theory, procedures, and hardware if he or she has learned the manual techniques initially. Also, the majority of physical educators working in secondary school and some college situations will not have the computer hardware for film data extraction. Therefore, they need to know manual techniques.

Conducting biomechanic research requires extensive multidisciplinary professional preparation at both undergraduate and graduate levels. This involves course work in mathematics through calculus, advanced physics, computer science, human engineering, and other related fields. As a consequence, the most advanced analytic step in the hierarchy, biomechanic research, lies beyond the scope of this book (except to inform the reader that this level does exist and logically follows after the student has attained an understanding of the first three levels of biomechanic analysis). Most professional physical educators will utilize the first three levels of biomechanic analysis within the hierarchy and be consumers as opposed to producers of biomechanic research.

Acknowledgment is made to the following people for their artistic contributions. David Andereck contributed the vector diagrams and line drawings in Part 3. Bill Armstrong provided the plane of motion drawings in Chapter 3. The anatomic drawings which appear in Part 2 were drawn by Philip J. Van Voorst and Gene A. Logan. Many illustrations in this text are from *Anatomic Kinesiology*, Third Edition by Gene A. Logan and Wayne C. McKinney. © 1970, 1977, 1982 Wm. C. Brown Company Publishers, Dubuque, Iowa and from *Adapted Physical Education* by Gene A. Logan. © 1972 Wm. C. Brown Company Publishers, Dubuque, Iowa. Reprinted by permission.

Acknowledgment for this edition is made to Don Anderson, USC Athletic News Service; Dr. Dick Perry, USC Athletic Director; Dick Bank and Phil Bath of Visual Track and Field Techniques; Art Evans, photographer; R. D. Freeborg, President of Instrumentation Marketing Corporation; Glen Bridges of Instrumentation Marketing Corporation; Robert E. Pritchard of Lafayette Instrument Company; and Dr. L. Dennis Humphrey. In addition, the authors acknowledge the exceptional athletes who appear in the photos throughout the text.

The invaluable assistance of Shirley A. Minger in preparing the manuscript is appreciated. The content and purposes of this book, together with possible errors, are the sole responsibility of the authors.

John W. Northrip, Ph.D.

Gene A. Logan, Ph.D.

Wayne C. McKinney, Ph.D.

1

Introduction to Kinesiology

1

Human Motion Analysis in Physical Education

The scientific study of human motion is known as *kinesiology*. Kinesiology has many dimensions; however, it is most commonly divided into two special study fields: (1) anatomic kinesiology and (2) mechanical kinesiology. During recent years, the term *mechanical kinesiology* has been replaced by *biomechanics*. The term *biomechanics* more accurately describes this field of study. The prefix *bio* is from the Greek, and it means "life." The suffix *mechanics* is an area of Newtonian physics designed to study the effect of forces on bodies and motion. The bodies studied by people concerned with physics per se are mostly inanimate. Physical educators concern themselves primarily with forces and motions of human beings. A body in motion may literally be a performer's body, or it may be a sport object projected into motion by a series of sequential joint and lever actions by the performer. *Studying and analyzing humans in motion, sport object motion, and forces acting upon these animate and inanimate bodies defines the field of biomechanics.*

It is desirable that a thorough understanding of anatomic kinesiology precede the study of biomechanics. An understanding of anatomic planes of motion, the motions possible at major joints of the body, ranges of motion possible at each body joint, and comprehension of the myologic internal force components for controlling the extent of motion with and against gravity at each major joint is of utmost importance. There are positive and reciprocal relationships between anatomic kinesiology and biomechanics. An understanding of anatomic kinesiology assists the learner in comprehending aspects of biomechanics, and vice versa. For the student who has not had course work in anatomic kinesiology or for the individual who needs a review, part 2 of this book covers some of the basic concepts of anatomic kinesiology. Accurate motion description of moving body segments is essential to the overall analytic process.

The utilization of biomechanic subject matter is both common and diversified in the area of sport. The timed, sequential joint motions of the performer executing a skill are of prime concern. The performer, through his or her joint motions, is one major source for generating and summating the forces necessary to meet the skill objectives. Therefore, detailed analyses of these joint motions must be made qualitatively and quantitatively in order to determine skill effectiveness. However, as the working definition of biomechanics implies, analyses of sport performances often include critical observations of both the performers and their sport objects.

There are many sport skills which require that sport objects, implements, equipment, and apparatus be utilized. These components and requisites of sports are of keen interest to physical educators and biomechanists. The physical educator in a teaching-coaching situation is primarily interested in how effectively the student-athlete uses such sport paraphanalia. Biomechanists employed in research, design, and development for major manufacturers of sporting goods are especially interested in quality, design, and function. Thus, there are mutual interests and concerns of members of these two professional groups, because there is a positive relationship between quality engineered and built sporting goods, and efficient learning and performance of neuromuscular skills. Therefore, the study of biomechanics by physical educators must include cause as well as effect relationships which exist between the sequential joint motions of the performer and the motion of the inanimate objects which he or she strikes, throws, kicks, holds, wears, rides, hangs on, falls on, runs on, swims through, leaps over, or manipulates.

A sport object is anything propelled into motion by a performer using a sport implement, body part, or total body motion at the point of impact, time of release, or take-off. The effects of the forces caused by the sequential, timed joint motions of the performer are summed and transferred to the sport object. Balls, pucks, javelins, arrows, shuttlecocks, Frisbees, hammers, bowling pins, shots, and the human body itself in some sport skills, are examples of sport objects.

The total analysis of a sport skill must include the cause and effect relationships between the human and the sport object being manipulated or propelled into motion. As an example, if a ball is thrown, the analysis of the sport skill must include the sequential joint motions of the thrower, plus such biomechanic considerations as angle of projection, trajectory, linear velocity, angular velocity and spin direction of the ball. All factors must be studied in terms of the skill objectives of the throw. If problems are noted in the performance of the skill, where did they originate? Within the performer? The sport object? Both? What precise changes must be made to attain the skill objectives? The answer to the last question leads directly to what is known as *quality teaching*. The directions for improvement given to the performer must be based on scientific and technical analyses of the total skill.

The quality of sport objects has a direct bearing on the timing and execution of skills. Therefore, sport objects must be selected with care to assure proper learning. An athlete can perfectly execute the proper sequential joint motions dictated by the stereotype of perfect mechanics for executing a skill, yet still be unsuccessful due to an inferior or worn sport object. As an example, an inelastic tennis ball put into play would greatly disrupt the timing and execution of the ground strokes. Simply stated, the ball would not have the expected timed reaction in rebounding from the racket face following impact. It would tend to remain on the surface of the racket a fraction of a second too long because of its inelasticity. Rotary joint motion within the shoulder of the tennis player while such a ball is in contact with the racket face during a forehand stroke would cause the ball to be driven into the net instead of clearing it by a few inches. When poor quality sport objects are used, time and space errors may be introduced despite correct joint motions by the performer. As a consequence, analyses of sport performances should effectively combine the motions of the performer with the quality and motion of the sports equipment used to execute the skills.

A sport implement is anything utilized by a performer to increase the speed or the force applied to a sport object. Examples are bats, golf clubs, field hockey sticks, bows, pool cues, kayak paddles, fishing rods, vaulting poles, diving boards, rackets, and jai-alai cestas. The way in which the performer manipulates the sport implement during the execution of the skill must be analyzed carefully, using anatomic and biomechanic parameters. It is equally important to evaluate initially the quality and suitability of the sport implement for the performer. Poor quality or improperly fitted sport implements have an adverse effect upon learning and performance.

The sport of target archery presents a classic example of the importance of the relationships which exist between sport objects, implements, and the performer. Archery tackle (equipment) must be perfectly matched in three respects if the archer is to have any success in the sport: (1) the sport implement (bow) must be designed for the strength and other anatomic features of the archer; (2) the sport objects (arrows) must be structurally compatible with the sport implement; and (3) the sport objects must be of specific design and length to meet the sport objectives, anthropometric measurements of the archer, and the specific shooting techniques of the archer. The physics of archery are complex, and they must be given considerable thought by the physical educator prior to teaching or coaching the sport. The same is true of most sports. Beginners in physical education classes as well as expert performers need to have appropriately fitted and quality sport implements in order to interact with sport objects in all areas of athletics. The responsibility for this lies directly with physical educators who teach/coach, and with biomechanists who research, design, and test sporting goods and apparatus.

Biomechanic Analysis of Performance

One of the major objectives of physical education often cited in the literature is to improve skill levels of students. In order to attain this objective, the physical educator must be thoroughly cognizant of a variety of analytic techniques. Analyzing performers in motion to improve the efficiency of their movements, whether it be with males or females in the context of sport, exercise, or dance, always precedes the communicative process between teacher and student known as *teaching*. The physical educator, therefore, must have a variety of professional experiences and course work at both undergraduate and graduate levels in order to analyze motion by using a variety of scientifically based procedures. One of these procedures is known as biomechanic analysis. Anatomic and biomechanic analyses of sport skills are ongoing processes and integral functions for a physical educator fulfilling professional responsibilities in instructional as well as athletic situations, with students of all ages, both sexes, and varying skill levels. Biomechanic analysis is a vital "working tool" for physical educators, and it should be utilized extensively as a part of all teaching, coaching, and motor learning situations.

Biomechanic analysis, the application of the principles of physics to human motion in sport, is essential because sport motion occurs in a precise framework of time and space. An understanding of temporal and spatial relationships provides the foundation for analysis of sport strategies and performances. The rules of sport, the sizes of playing areas, the design of equipment, and the optimum decisions for performance of skills, all are based on the abilities of human beings to move and make specific spatial changes in precise time frames.

The human being is geared to operate in time intervals of fractions of seconds. Reaction times, movement times, muscle contractions, and even the processing of a single visual image, occur in approximately one-tenth to three-tenths of a second. Therefore, complex sport motions often succeed or fail within this critical time frame. The winning of a sprint race, the successful steal of second base in baseball, the faking motions of a defensive player in field or court sports, usually depend on specific and critical timing within a fraction of one second.

Likewise, many successes in sport depend on a spatial precision involving a few millimeters or a fraction of an inch. As an example, this small space interval in striking a baseball or a golf ball determines much larger space trajectories in the flight of the object. A half inch variation in the hand or body position in wrestling or the martial arts is the difference between a successful or failing maneuver.

These small space and time factors are often overlooked due to the sheer size of a playing field or the longer periods of time involved in a complete contest or game. It should be remembered that winning or losing a complete game often depends on time and space precision in the execution of a single skill! Thus, the understanding of the relationship of time and space quantities is the essence of biomechanic analysis.

Biomechanic analyses should be used any time a physical educator is attempting to improve performances of students so they may attain their full potential as performers. Quality instruction in physical education always includes some form of kinesiologic analysis. One purpose of biomechanic analysis is to increase the precision of the instructor's observations for teaching-coaching. Some have maintained that biomechanic analysis procedures and techniques which do not meet scientific detail and precision based on biomechanic research criteria cannot be used in teaching-coaching situations. The contention is that data obtained would be limited to the point of negating the findings of the analysis. This is a point of view limited to the research situation. Significant biomechanic analyses can be made in instructional and athletic situations with and without cinematographic equipment. Furthermore, the findings of these analyses will greatly assist the physical educator in communicating to students, in language understandable to them, exact changes that should be made to improve their performances. In this context, the fact that research criteria were not met for biomechanic analysis is irrelevant. *Utilization of the subject matter from biomechanics is not restricted to the research laboratory.*

Techniques for performing analyses will be presented herein for use with and without photographic equipment. Procedures will be shown to implement analyses without high-speed research cameras, sophisticated projection equipment which needs to be connected to computers, and timing equipment accurate into the range of nanoseconds. Obviously, this type of equipment is not available in the physical education department of the average secondary school. Therefore, since analyses are so basic to the teaching process, adjustments must be made in analytic techniques to utilize the type of instrumentation available to physical educators in elementary and secondary schools as well as in colleges and universities.

The precision and control of the observations made by physical educators who are teaching and coaching differ considerably from analyses made by physical educators conducting research. Although differences do exist, these two analytic levels are compatible. The physical educator in the field should be a knowledgeable consumer of research in biomechanics and other fields related to human performance; and the professionally oriented teacher-coach should utilize scientific principles as the bases for decisions to change motion patterns in students and student-athletes. Any teaching suggestion, exercise, or drill utilized in any situation should be valid, logical, and appropriate to the students involved. The professional physical educator makes analyses of a student's performance, contemplates teaching suggestion alternatives, and ultimately makes adjustments in a student's performance, based on knowledge instead of guesswork.

During the professional preparation of a physical educator, a gradual change must occur from being a fan of sport to becoming a teacher-coach of students involved in sport. What are the basic differences between a fan and a physical educator responsible for teaching and coaching? One of the principal differences lies in the fact that the physical educator understands motion so thoroughly that a detailed analysis of any sport skill can be undertaken. The physical educator, unlike the fan, observes a skill in terms of its components as well as its subsequent outcomes. The average sport fan appears to be interested only in the outcome of a given performance. As an example, during a field goal attempt in American football, the average fan tends to watch the trajectory of the ball as it moves toward the goalpost. The fan's only concern is whether the ball passes between the uprights and over the crossbar. On the other hand, the physical educator, watching from the sidelines and subsequently viewing the same field goal attempt on game film, is fully aware that sequential movements by the kicker are of utmost importance. Once the ball has been contacted by the kicker's foot, the end result is very predictable. The physical educator, unlike the fan, is very much aware of the summation of motions generating forces to be applied to the ball. The relationships of biomechanic principles to these sequential motions are understood by the physical educator, and the implications of using them either effectively or ineffectively during the execution of a performance are known. Unlike the sport fan, the physical educator and the performer do not have to follow the flight of the football in the example to know whether or not the kick was successful. The kinetic chain of sequential movement by the performer indicates to the skilled observer, most of the time, whether the performance was effective or ineffective. The teacher must be concerned with both cause and effect related to a skill, while a fan is most concerned with effect.

From a technical or techniques perspective, there is a stereotype of perfect mechanics for each skill. *The stereotype of perfect mechanics for any sport skill, adjusted to allow for individual differences, is that procedure which will allow individual velocities to sum together most effectively to produce an optimal result.* The technical or techniques aspect of any skill plus subtle variations must be thoroughly understood by the physical educator prior to (1) undertaking a scientific analysis of the skill, and (2) correcting the performance faults of the student or student-athlete. *Understanding the stereotype of perfect mechanics for a skill serves as the starting point for scientific analyses and subsequent quality improvement in the teaching-coaching process.*

Kinesiologic knowledge on the part of physical educators is an absolute necessity when stereotypes of sport skills change as a result of pragmatic experiences by athletes or coaches. Efficiency evaluations of new performance techniques and equipment must be made biomechanically. Judgments based on scientific criteria should be made when, for example, high jumpers start "flopping" and "flipping" over the bar instead of "straddling." Which style is best? For whom? Will a baseball hitter hit the ball with greater velocity utilizing an aluminum or wooden bat? What type of ski wax is best for specific snow conditions? What type of shoe sole is best on artificial surfaces? How safe is a football helmet against impacts at all angles? Can a long jumper attain greater distance by doing a "flip" during the trajectory? Should a shot putter integrate angular motion, analogous to the discus thrower, with linear motion across the ring? Answers to such questions, which are directly related to coaching decisions, should be derived from kinesiologic analyses.

Physical educators will acquire professional preparation in a wide variety of sports. The actual involvement in performing and learning skills is a vital aspect of the analytic process. The physical educator who has not performed skills he or she attempts to analyze as a teacher or coach is at a disadvantage technically. In other words, personal involvement with the skills and procedures of the sports one is teaching or coaching is also a prerequisite of biomechanic analysis. Once knowledge of the techniques for each skill is attained, the physical educator must be able to utilize the appropriate level of biomechanic analysis for any given teaching-coaching or research situation.

Biomechanic analyses are divided into four levels based on frequency of use and sophistication: (1) noncinematographic analysis, (2) basic cinematographic analysis, (3) intermediate cinematographic analysis, and (4) biomechanic research. The physical educator performing routine responsibilities in elementary, secondary, and university teaching will utilize the first two levels primarily. The third level, intermediate cinematographic analysis, can also be used extensively in teaching-coaching situations to objectify analytic observations which tend to be subjective. The first three levels of biomechanic analyses are of prime concern within this book because they are applicable to the physical educator outside the research laboratory.

Hierarchy of Biomechanic Analyses

Noncinematographic analysis is the most common analytic technique used by physical educators. As the term implies, no film or videotape is utilized during this observation of performers. This means that the physical educator must have a disciplined approach to observing and analyzing motions at various body articulations. This type of approach is presented in chapter 12. Noncinematographic techniques are used most of the time during the teaching-coaching process. Therefore, knowing how to observe human motion without the assistance of film or videotape is of utmost importance to teaching and achieving the skill-improvement objective of physical education.

The implications of utilizing biomechanic principles in the noncinematographic technique are often overlooked. The reason for this lies in the fact that some persons involved in the professional preparation of physical educators believe that biomechanic ideas can only be transmitted by using intricate mathematical calculations of physics principles. From the standpoint of time, it is not always feasible in teaching-coaching situations at the high school level, for example, to undertake detailed mathematical analyses of performances being observed. Such an analysis would be desirable to greatly objectify the observation. However, if the underlying biomechanic principles are fully understood by the teacher-coach, communication of ideas for improving performance based on biomechanic principles can be conveyed to the performers in succinct, nonscientific terms. This applies, regardless of which biomechanic analysis level is being utilized. It is often quite difficult to communicate to a performer an idea regarding the implications of adhering to a biomechanic principle, without having the assistance of making calculations from film. Noncinematographic analysis and the subsequent teaching from the observations are challenging!

The physical educator should have a systematic approach for viewing human performances while utilizing his own field of vision. If not, teaching-coaching suggestions to the performer will result in considerable guesswork. Communication of violated biomechanic principles by the physical educator to the performer should be conveyed in a *positive manner*. In the context of the learning situation, it is not always necessary for the student to understand fully the physics principle being violated and adversely affecting the performance. The physical educator, however, will have an in-depth understanding of the biomechanic principles underlying the performances observed and their implications for the ultimate outcome of the performance. The communication with the student (teaching) should be a positive recommendation designed to eliminate the performance problem. *A good teacher never dwells on the negative aspects of performance.*

As an example, a baseball coach should have a functional grasp of the biomechanic principle that states, "Most efficient sport movements should be viewed in terms of a linear motion of the center of mass of the body combined with angular rotations of the limbs and body segments around the center of mass." This relates directly to the coaching of pitchers (fig. 1.1). The pitcher must move his center of mass linearly during the wind-up and delivery, and this must be performed according to a strict rule involving contact with the pitching rubber

Figure 1.1 Jim Barr, USC pitcher, starting a "wind-up" to effectively integrate linear and angular motions. *Courtesy USC Athletic News Service.*

with one foot at all times. The shifting of weight, linear "stride," and foot placement are very important to integrating the angular rotations of the pelvis, lumbar-thoracic spine, and upper limbs. If the foot placement is improper, these rotations will be impeded. The result would be a decreased summation of velocity within the moving body articulations, reduced ball velocity, and diminished accuracy. The coach can eliminate this problem by giving the athlete a simple drill to focus his attention on consistency of foot placement for a given pitch. This should be done in a *positive manner* without dwelling upon the performance problems. There are also psychological advantages to this approach to coaching. Good physical educators do not have to communicate everything they know about a skill to their students to provide an effective change in performance.

The analysis of the above example could be made noncinematographically during practices and games. Precision and objectivity could be greatly enhanced by including basic cinematographic observations discussed below.

The major limitation of noncinematographic analysis lies in the time frame in which critical joint motions occur. The very complex processing of visual perception takes approximately one-tenth of a second to occur within the central nervous system. Many sport motions are executed faster than that; therefore, the multiple facets of most ballistic skills cannot be objectively observed. The teacher-coach must have a disciplined analysis technique to attain precision and objectivity while watching performers with the eyes only. The physical educator *must* focus attention on one motion parameter, joint motion, or moving body segment, instead of trying to view total body motion. Sport skills actually occur in fast time frames. As an example, the entire sequence of motions by the sprinter in figure 12.2 occurred in approximately 0.5 second! Many coaches give very complex suggestions to performers based on what they thought they observed noncinematographically, as opposed to what they actually observed. Such coaches without a disciplined analysis technique are guessing about what they thought they saw. The students are the losers when receiving instruction based on this kind of subjective evidence, since the instruction lacks precision, tends to be ambiguous, and often frustrates students attempting to learn neuromuscular skills.

In summary, the noncinematographic analysis technique discussed in chapter 12 involving the use of biomechanic principles is a vital function of the physical educator. Without the aid of cinematographic equipment, the teacher-coach must have a *disciplined procedure* for watching humans moving, because the greater part of time spent in physical education classes and athletic assignments involves observing motion with the "unaided eye." The individual without a disciplined noncinematographic analysis technique may not be able to see the critical joint motions for the overall movement! This is analogous to not being able to see the trees due to the forest.

Basic Cinematographic Analysis

Basic cinematographic analysis involves utilization of film or videotape specifically for the purpose of improving the performances observed. This level of biomechanic analysis does not involve any mathematical calculations. The use of film tends to objectify, substantiate, or refute what has been observed by noncinematographic techniques. For practical purposes, the way performers are visualized through basic cinematographic and noncinematographic techniques is essentially the same.

Film has the distinct advantage of retaining permanent images of even the fastest limb actions; and projection techniques enable the physical educator to stop these images on a timed, frame-by-frame basis. This adds precision to the basic cinematographic analysis process, as compared with the noncinematographic technique. *Film allows the observer to see what has actually occurred as contrasted with what he or she thought took place within the moving joints of the performer.* The implications for increasing the quality of instruction are evident.

The fact that we are interested in motion components involving the human body, sport objects, and sport implements which occur in tenths of seconds dictates the filming speeds used for basic cinematographic analysis. The common filming rates of 24 and 32 frames per second allow such motions to appear on from two to five frames of the film. For more precise analysis of rapid motion changes as seen in highly ballistic skills, framing rates of 80 to 200 or more frames per second are readily available. Framing rates of this level are utilized for greater precision in intermediate analyses and biomechanic research situations.

To return to the example of the field goal kicker, the coach utilizing noncinematographic techniques on the practice field or during a game has severe limitations as to what he can observe during a once-only performance of the skill. At best, since place-kicking is a ballistic action of the lower limb, the coach would only observe the pendular action of the kicking leg and its linear force; whereas in viewing game film, the coach could make a much more detailed analysis of all major joint components utilized during the kicking process. In addition, various biomechanic principles should serve as a frame of reference for the development of questions as the coach watches film: (1) What are the force-counterforce components between upper and lower body segments of the kicker? (2) Was conservation of angular momentum utilized at the knee joint within the kicking limb? (3) Was an optimum position of the foot applied to the ball at the point of contact for force to be transferred from the kicking limb to the ball? (4) Was transformation of linear motion of the body and angular motion of the kicking limb adequate to execute the skill? (5) What were the angle of projection and subsequent trajectory and flight pattern of the ball? These questions could not be answered during a noncinematographic observation.

Basic cinematographic techniques are also very useful in any learning situation, whether it be in a physical education class at the elementary or secondary-school level or in competitive athletics. The physical educator who relies strictly on an *undisciplined,* noncinematographic technique has a tendency to do a considerable amount of guessing during analysis and subsequent teaching. Therefore, film and videotape should be utilized as much as possible during the teaching-coaching process. Most elementary- and secondary-school districts as well as colleges and universities have communications media or audiovisual departments. A wide variety of eight-millimeter and sixteen-millimeter cameras and projectors are obtainable in most schools. In addition, there is a trend toward having portable videotape cameras and monitors available for use by school personnel. Since this type of cinematographic equipment is available, basic cinematographic analyses should be made, as much as possible, during instructional and athletic situations.

Basic cinematographic analysis does not involve utilization of mathematical computations to objectify observations made on film. During this type of analysis, the physical educator must be capable of observing the performer who is effectively or ineffectively applying biomechanic principles. Where a student is obviously violating a biomechanic principle, or several principles, the physical educator must decide precisely what joint motions are causing violations of principles and which principles are being violated. The coach/instructor should also determine which changes or alterations should be made in the performer's style to bring about a more efficient application of biomechanic principles. *The quality of instruction in physical education classes and athletic situations is directly related to the quality of these motion analyses.*

The basic cinematographic analysis technique should be used extensively by coaches analyzing individual skills of athletes on game film. Obviously, such film will not meet the very rigid film criteria established for biomechanic research. However, recognizing the fact that game film does have limitations, very significant and critical biomechanic analyses can be made, even on a basic level, to help improve performances. As one example of how this can be accomplished, tennis coaches can use basic cinematographic analyses to modify or improve serving techniques (fig. 1.2). Film of the athlete's serves should be taken during ten-

Figure 1.2 The tennis serve can be analyzed thoroughly in teaching-coaching situations by utilizing basic cinematographic analysis. *Courtesy USC Athletic News Service.*

nis matches. The film of the "effective serves" is spliced together, and the same should be done with the "ineffective serves." This would allow the coach to compare and contrast the two levels of performance by the same athlete; and, a comparison should be made of both levels of performance with what the coach knows to be the "stereotype of perfect mechanics" for the tennis serve.

The analysis can be done by concurrently using two single-frame projectors of the type shown in figure 12.7. The joint motions of the tennis player during the phases of the serve can be analyzed frame by frame. The projectors would be synchronized to show comparable frames during "effective" and "ineffective" serves. Motion variations should be noted between the two performances, with respect to such factors as (1) major joint angles in the stance, (2) shoulder, upper limb, and racket variations during the "backswing" or preparation phase, (3) extent of the ranges of motion for pelvic and lumbar-thoracic rotations, (4) ball toss differences, (5) weight shift and direction, (6) shoulder joint motions and racket angles prior to and at impact, as well as (7) the extent and direction of the follow-through and recovery phases of the serves.

On the basis of this type of basic and thorough cinematographic analysis, the tennis coach will also have objectified some observations made noncinematographically. Positive decisions can be made in regard to teaching suggestions to adjust joint motions within the serve to help the student become more consistent as an effective server. The adjustments, in whichever form chosen by the coach, would be communicated to the student during the practice or class session following the analysis.

Having a permanent record on film of both levels of performance tends to eliminate guesswork. This is a distinct advantage of basic cinematographic analysis over even the best noncinematographic techniques. From practical and efficiency standpoints, these two analytic levels must be used together as much as possible to objectify observations and improve performance of students in physical education classes and athletic situations.

Intermediate cinematographic analysis involves some mathematical computation to enhance the precision of observations by the analyst. All procedures and knowledge utilized by the physical educator in performing noncinematographic and basic cinematographic analyses are employed during intermediate cinematographic analyses.

Intermediate
Cinematographic
Analysis

In analysis at the intermediate level, data are numerically extracted from the film by measurement either in real or scaled units. These numbers can then be used for comparison with results from the same or other performers, or they can be the basis for calculation of the physical parameters characterizing the performance. The extraction of this numerical data may range from a simple counting of film frames to precision stereographic measurement on an automated x-y plotter. In general, however, analysis at this level requires only fundamental measuring techniques. The extraction of data from film for intermediate cinematographic analysis purposes is the subject of chapter 13.

Intermediate cinematographic techniques can be applied to film taken in instructional situations or to game film taken during athletic contests. There are distinct advantages for the physical educator to make mathematical computations at times to assist in communicating ideas to performers. This is particularly true in those sports not requiring measurements in precise units. As an example, a diving coach may have some difficulty communicating the importance of an extreme "tuck position" to a diver who is attempting to improve a two and one-half somersault diving maneuver. From an intermediate cinematographic analysis of the dive, calculations could be made to indicate the diver's angular velocity in precise units. When the diver follows the coach's suggestions regarding the basic body movements to "tighten the tuck" and, therefore, increase the subsequent angular velocity by conserving angular momentum, film can again be made and subjected to intermediate analysis techniques. Following this, the change in velocity achieved by the performer can be communicated to him in precise units. This type of communication plus actual viewing of the film helps motivate some student-athletes. Intermediate cinematographic analysis techniques are discussed in chapter 14.

Biomechanic Research

Biomechanic research techniques for analytic purposes involve highly sophisticated instrumentation. Most of this equipment can be found in only a few specialized laboratories in universities throughout the United States. High-speed cameras for triaxial cinematographic analysis with associated velocity measurement devices, stroboscopic devices, electromyographic units, electrogoniometers, force plates, force transducers, and computers usually are not found in elementary and secondary schools. These and other types of equipment are absolutely essential to conducting biomechanic research.

Biomechanic research techniques bring a great amount of precision to the task of solving a motion problem. Also, research facilitates discovery of new information in the area of biomechanics. When research techniques are utilized, considerable attention is given to working with a delimited portion of the motion problem. This facilitates a deeper understanding of the factors underlying the motion observed. In order to conduct biomechanic research properly, a background of highly specialized professional preparation is needed. At the present time, this preparation usually involves earning a doctorate in biomechanics. Biomechanic research takes a great amount of time to solve rather minute motion problems. The latter fact alone makes it unrealistic to think that all people in physical education will engage in biomechanical research even during their professional preparation through the first graduate degree. All physical educators, however, must have formal course work in biomechanics and be exposed to research in order to gain an understanding of the subject matter and its relationship to improving performances in sport, exercise, and dance.

What would be involved in doing a "simple" research project to analyze the anatomic and biomechanic components of shot putting? It would be desirable to utilize some form of triaxial cinematographic technique. One way this is accomplished is to synchronize three cameras at precise angles relative to three critical planes of motion traversed by the shot-putter. These cameras would have to be of sufficient quality to include precision-timing on the film. A scaling device within the photographic background would have to be included to help determine limb velocities. A special shot-putting ring would be set up with a variety of force transducers to determine weight-shifting factors by the shot-putter as he traversed the ring. If ranges of motion were studied, electrogoniometers would be required. If the electromyographic parameters were to be evaluated concurrently, this equipment would have to be set up to determine precisely the internal forces generated by specific muscles or muscle groups during the execution of the skill. This, in turn, would necessitate a fourth camera being synchronized with the three cameras for triaxial analysis. The fourth camera would be set on the oscilloscope for reading muscle-action potentials elicited by the muscles during various points within the execution of the skill. It is rather obvious that it takes considerable time to work out the electronic details and actually set up for an experiment of this nature with the type of equipment indicated. An electronic technician is indispensable as a member of the biomechanic research team.

After the data are collected, the information is processed statistically according to the model and experimental design decided upon prior to undertaking the research. Computers are utilized to analyze data extracted from the film and elsewhere, as well as for the statistical treatment of the data. The findings of such research may or may not have implications for teaching and neuromuscular conditioning of shot putters.

The requirements of research do not lend themselves to being "daily working tools" to enhance the teaching-coaching processes conducted by physical educators at all educational levels. However, the professional physical educator in the field should be cognizant of the research literature and utilize research findings where applicable to improve performances. Biomechanic research will be discussed briefly in chapter 14. *The main emphasis in this book is on biomechanics subject matter and analytic techniques, and how they relate to improving the quality of teaching-learning in the assignments given to physical educators in elementary schools, secondary schools, and universities.*

Part 3 of the book is designed primarily to assist the reader in understanding biomechanic analyses through the intermediate cinematographic level. Three procedures are used to convey biomechanic principles to the reader: (1) Principles are explained in a fundamental manner to facilitate comprehension; (2) To reinforce the word description of principles, graphic illustrations are included where applicable; and (3) Where deemed necessary to facilitate comprehension and add precision, mathematical formulae and examples are included to further reinforce the ideas conveyed. Numerals as symbols have a distinct advantage over words as symbols. Words, especially in the English language, have some degree of ambiguity; mathematical symbols are more precise ways of expressing ideas. Some people tend to succumb to a syndrome known as "symbol shock" if they are asked to work extensively in the field of mathematics. As a consequence, it is believed the best way to initially convey ideas regarding biomechanics to the majority of potential users in teaching-coaching is to limit the mathematics to the essential equations necessary to describe motion adequately through the intermediate cinematographic analysis level. This procedure is followed in part 3. Part 2 provides the reader with the background to describe motion from the anatomic perspective.

Recommended Reading

Atwater, Anne E. "Kinesiology/Biomechanics: Perspectives and Trends," *Research Quarterly for Exercise and Sport,* 51 (March, 1980), 193–218.

Dillman, Charles J. and Sears, Ronald G. (Eds.). *Proceedings Kinesiology: A National Conference on Teaching.* Urbana-Champaign: University of Illinois, 1977.

2

Basic Concepts of Anatomic Kinesiology

2

Fundamentals of Joint Motion Analysis

The major purposes of this chapter are to provide the reader with some of the basic concepts needed to undertake joint motion analyses. These include: body positions for describing motions; definitions of reference terminology commonly used to describe motion; a glossary of the principal joint motion terms; an introduction to segmental analyses; and a delimited list of the most important external skeletal landmarks for approximating joint centers, objectifying motions of joints, and marking various body segments to quantify the extent of motion.

The human body is the ultimate *sport object*. Therein lies a compelling reason why the joint motion capabilities of the human organism must be thoroughly understood by physical educators, if meaningful skill analyses are to be conducted. The neuromuscular and skeletal systems are the chief producers of the internal forces needed to execute sport skills; therefore, basic concepts of the functions of these systems plus their interaction related to joint motion must be understood to facilitate quality skill analyses. The motion capabilities of each joint must be a part of the physical educator's professional knowledge, and the instructor should understand how each joint motion contributes force in the timed sequence for each skill to be taught. Such knowledge is essential before skill analyses can be made. Ultimately, quality skill analyses which preceed teaching require an intellectual synthesis of knowledge involving exercise sciences (biomechanics, kinesiology, exercise physiology, etc.) and technical information (stereotypes of perfect mechanics, etc.).

All joint motions are angular in nature, i.e., movement of a given joint will produce a rotary or angular motion. This means that joint motions are quantitatively described in units of angle. Every joint has a normal *flexibility* range through which it may turn. This *range of motion* for all joints is commonly expressed in degrees. For example, the normal range of motion for elbow flexion is designated as 150 degrees. For some individuals angular motion beyond that range could result in trauma to the elbow. If a throwing skill demands maximum elbow flexion during the motion phase of the performance, the coach analyzing the skill should be looking for 150 degrees of angular motion during elbow flexion. Any more or less motion would be contraindicated for proper skill execution.

Most skills integrate angular and linear motion components. The locomotor skills provide a classic example of such integration. For example, spectators usually focus on a sprinter moving linearly down the track from the starting blocks to the finish tape at the end of the 100 meter dash. They would describe the motion as purely linear, because the total body mass was moved in a straight line over the 100 meters from start to finish. This is true, but what caused this linear motion? Joint motion analyses of the sprinter would indicate that it was primarily a series of timed, sequential angular motions at the ankles, knees, hips, and shoulders which provided the internal force necessary to move the body mass of the sprinter linearly. The linear-angular motion relationships become more complex in the ballistic skills, so the physical educator must understand joint motion from the perspective of anatomic kinesiology.

Performers in sport and dance move their bodies in a complex sequence of motion on land, through the air, or in water. They rotate their bodies through various planes of motion and often dive through the air face down or inverted to the earth. The motion capabilities, combinations, and sequences for the human body are infinite. However complex the series of motions may be during a performance, the physical educator analyzing the performance must define the planes and joint motions by utilizing the anatomic and/or attention positions as a reference. All definitions of joint motions for skill analyses are based on these two analysis positions.

Analysis
Positions

The *anatomic position* as shown in figure 2.1 serves as one of the major reference positions for joint motion analyses. This position can be utilized as the

Figure 2.1 Anatomic Position—anterior view. Bipedal weight bearing.

reference for all joint motions, but it is most commonly used to describe the motions at the shoulder, elbow, radio-ulnar, wrist, and hand articulations. This position is borrowed from anatomists. If the human body (cadaver or skeleton) is hung from a point in the center of the superior aspect of the skull, it will hang with the shoulder joints laterally rotated and with the remaining joints in the upper limbs extended. The ankles would be plantar flexed. The latter is not the case in the weight bearing situation shown in figure 2.1.

The *attention position* is another reference for starting joint motion analyses (figure 3.9). It is usually employed as the reference position for all joint motions except those of the upper limb. The chief difference between the bipedal, weight bearing anatomic and attention positions appears in the upper limbs. In the attention position the shoulder joints are medially rotated to the point that the palmar surfaces of the hands are in contact with the lateral aspects of the thighs. All other joints are in the same positions as shown for the anatomic position in the weight bearing situation. The attention position has been used for centuries by the military, and it is utilized frequently as the starting stance by gymnasts when they begin their routines.

| Reference Terminology | There are six basic terms which need to be understood. These terms are used primarily to *describe relative position* of one anatomic structure to another segment, bone landmark, joint, or to the total body. These terms are used in a wide variety of ways in analyses of human motion. The general reference points for these terms are the traditional planes of motion described in Chapter Three: |

Medial. The cardinal anteroposterior plane shown in figure 3.1 is the midline or medial aspect of the body. Any structure positioned relatively close to the midline of the body is medial, e.g., the sternum (figure 6.5) is medial to the arm or upper limb. Relative positions of all body parts are described in this manner.

Lateral. This term is used in the same context as "medial," but it is used to describe the opposite position. To continue the above example, the upper limb is lateral to the sternum. Skeletal landmarks should also be understood in their relative positions to each other. For example, the trochanters of the femur as shown in figure 5.20 are located laterally and medially to each other. The greater trochanter is lateral in relation to the lesser trochanter and the center of the hip joint. It is necessary to know these kinds of relative spatial relationships between anatomic structures.

Anterior. The lateral plane of motion shown in figure 3.6 is the general reference line for differentiating between what is anteriorially or posteriorly positioned on the body. (*Ventral* is a synonym for anterior.) Anterior can be used in a wide variety of ways. As one example, palpation of the belly of the biceps brachii muscle and its tendon in the anatomic position would confirm the fact that the tendon has an anterior relationship to the center of the elbow joint. The location of muscle-tendons to joints has determining implications in terms of the function of muscles during contraction to joint motion.

Fundamentals of Joint Motion Analysis

Posterior. This term is used in the same context as "anterior" relative to the lateral plane, but in the opposite manner. (*Dorsal* is a synonym for posterior.) Posterior can be used extensively to describe relative positions of large and small body parts. As one example, the coranoid and olecranon processes are located on the ulna as shown in figure 7.29. Their relative position would be described as follows: The olecranon process is posterior to the coranoid process. Also, the olecranon process is posterior to the center of the elbow joint (figure 7.29). A muscle-tendon attached to the olecranon process would have a posterior spatial relationship to the elbow joint.

Superior. This term, like the others above, is used to describe relative position. It is not used in a qualitative context. In the anatomic position, the skull is superior to the feet. That denotes position only; it is not a statement regarding the relative value of skull-brain over feet. Superior can be used to describe relative position of any body structure. As an example, the olecranon process on the ulna is located superior to the ulna styloid process (figure 7.29).

Inferior. This term is utilized to describe anatomic parts as the opposite of superior, e.g., the styloid process of the ulna is inferior to the olecranon process. *The tendons of the extrinsic muscles of the feet are described as being either inferior or superior to the interphalangeal and metatarsophalangeal joints* (figure 5.7).

The reference terms are often used in conjunction with each other to precisely describe location of body structures. Examples are shown in figure 5.23 regarding the iliac spines on the iliac crests of the pelvic girdle. Using these reference terms in their proper context helps to objectify and define more precisely the anatomic structural relationships. Ultimately this leads to better motion descriptions of performances in sport.

Joint Motion Definitions

In order to communicate adequately, it is essential to have a thorough comprehension of the vocabulary commonly used by people working in a given field. All specialized areas of endeavor, whether they be scientific or nonscientific, have specialized vocabularies. This is just as true for anatomic kinesiology and biomechanics as it is for American football. A specialized vocabulary is necessary to enable individuals working in a common area adequately and precisely to communicate ideas to each other. Those same ideas conveyed in the very specialized vocabulary of a given area may not have meaning to someone untrained in the discipline.

Scientific terminology has the advantage of being more precise and meaningful when taken completely out of context of the discipline. As examples, let us take the terms *biomechanics* and *love* out of their respective areas of kinesiology and tennis. As one views these two terms out of context, *biomechanics* informs the reader who has not been trained in the area of biomechanics that it is an area having some relationship with life—*bio*—and a relationship between force and motion of this living matter—*mechanics*. Most scientific terms have meaningful prefixes and suffixes of Latin or Greek derivation that help clarify their meaning; terms or words from other disciplines do not have this communication advantage. For example, the term *love* taken out of a tennis context connotes to the reader many different ideas that may be completely unrelated to tennis! There is no hint in the term *love* that would lead an individual without a tennis background to deduce that it is defined as a zero score in tennis.

Many of the prefixes and suffixes utilized in "scientific language" are well known and used in our common language. This facilitates understanding if one stops to analyze an unfamiliar word or term and looks for familiar components. Some students tend to make the study of scientific subjects too difficult, because they do not look for the familiar.

A few examples will help clarify this point. The terms "bi" and "tri" are commonly used in everyday language, e.g., a bicycle is a two-wheeled vehicle. There is nothing mysterious or difficult about that term. Therefore, a scientific term such as *biceps femoris* from the area of anatomic kinesiology should not be difficult to understand either. "Bi" relates to two of something. In this case, it means two "ceps." Most educated individuals have used the term "cephalic," which pertains to the head of something. With limited knowledge of human anatomy, one can logically deduce that a "biceps" is a two-headed muscle. But, where is it located? Is it an upper or lower limb muscle? There is no mystery, because the term "femoris" indicates that it is located next to the *femur* on the lower limb. Consequently, the biceps femoris is a large, two-headed muscle located in the thigh next to the femur. A biceps brachii is a two-headed muscle located next to the brachium or humerus of the arm. Thus, many scientific terms relevant to motion may be new to the reader and appear to be somewhat confusing at first, but they have the advantages of a good language base, and at least some elements that may already be familiar to the student. It is therefore much easier to learn scientific terms than some terms from sport and dance such as "love," "red dog," "assemblé," "fast break," "monster man," and thousands of others. Their connotations are diverse, especially when used out of context.

One purpose of this chapter is to assist the student in the development of a motion-description vocabulary. This will enable the student to observe human motions and accurately describe what is seen. The use of appropriate and accurate motion terms is essential during the analyses that precede teaching; but the use of appropriate language is even more significant during actual instruction, when the communication of ideas must occur between the instructor and student. Therein lies the art within the field known as physical education.

Some students are utterly confused by the directions they receive during skill instruction. For example, a coach who tells an athlete that, "you need to bend your back more," is an expert in ambiguity! First, the performer has not been given a positive suggestion for improvement. A negative generalization has been passed along, based upon what the coach thought he or she saw during the analysis. That is not teaching or coaching! Second, such a statement lacks precision in terms of direction of motion. The spinal column can be flexed, extended, rotated, laterally flexed, hyperextended, and circumducted. Precisely what does the term "bend" mean in terms of the desired motion direction to meet the skill objective? Finally, the statement lacks the qualitative elements of desired degree and velocity of motion within the spinal column. "Bend your back more," as a teaching suggestion, means nothing in terms of these criteria.

To avoid confusing students, professional physical educators provide teaching-coaching suggestions that accurately tell the students *what* motion adjustments must be made for improving performance, *how* these can be accomplished, *what* ranges of motion are desired, and *how* slow or fast the motions must be executed to reach the skill objective. They leave the clichés and slang terms for use by the so-called "coaches" who are best defined as technicians instead of professionals.

This section contains definitions for all of the major joint motions, many of which are illustrated. The remainder of this part of the book will specifically relate these general definitions to motion capabilities within each joint. This is done from the standpoint of joint motions through planes of motion, and in terms of muscle function causing or controlling each major joint motion in the body. A working knowledge of joint motion definitions is the starting point for undertaking joint motion analyses.

Abduction is movement of a body part or limb away from the midline of the body. An example of abduction is movement of the upper limb away from the side of the body through the lateral plane of motion. Figure 2.2 shows shoulders in the abducted position.

Figure 2.2 Shoulder joint abduction in a stabilized condition by Don Elshire, USC gymnast. *Courtesy USC Athletic News Service.*

Adduction is movement of a body part toward the midline of the body. Adduction is the return of the abducted limb to the anatomic position through the lateral plane of motion.

Circumduction is movement of a limb or body part in a manner that describes a cone. Circumduction involves a combination of four basic movements: (1) flexion, (2) extension, (3) abduction, and (4) adduction. Also, circumduction is a combination of diagonal adduction and diagonal abduction (fig. 2.3).

Depression is downward movement of the shoulder girdle. Very little depression within the shoulder girdle can occur from the anatomic position. Depression is the return movement of the shoulder girdle from elevation to the anatomic position. Figure 2.4 illustrates depression of the shoulder girdle.

Figure 2.3 Circumduction of both shoulders occurs during the "fly stroke" as executed by Jim McConica, USC swimmer. *Courtesy USC Athletic News Service.*

Figure 2.4 Shoulder girdle depression during a ring maneuver by Gareth Burk, USC gymnast. *Courtesy USC Athletic News Service.*

Fundamentals of Joint Motion Analysis

Figure 2.5 Diagonal abduction of the left shoulder joint by Steve Cameron, USC swimmer, during the back crawl. *Courtesy USC Athletic News Service.*

Figure 2.6 Diagonal adduction of the right shoulder joint of Tom Seaver, former USC pitcher. Roy Smalley, former USC shortstop, has started moving by plantar-flexing both ankle joints. *Courtesy USC Athletic News Service.*

Diagonal abduction is movement by a limb through a diagonal plane across and away from the midline of the body. Figure 2.5 illustrates diagonal abduction of the left shoulder joint during the back-crawl stroke in swimming.

Diagonal adduction is movement by a limb through a diagonal plane toward and across the midline of the body. Figure 2.6 shows diagonal adduction of the right shoulder joint at the completion of the follow-through by an excellent baseball pitcher.

Figure 2.7
Dorsiflexion of the right
ankle adds to the
aesthetic quality of this
modern dancer's
position.

Dorsiflexion is movement at the ankle joint of the "top" of the foot toward the lower limb, i.e., flexion of the ankle or talocrural joint. The dancer's right ankle in figure 2.7 is in a dorsiflexed position.

Elevation is upward movement of the shoulder girdle. Elevation occurs within the shoulder girdle during the execution of the "fly" swimming stroke as shown in figure 2.3.

Eversion is movement of the sole of the foot outward or laterally. This movement takes place within the subtalar and transverse tarsal joints. It is not an ankle-joint movement.

Extension is any movement resulting in an increase of a joint angle. Most major joints are in extension while the individual is in the anatomic position. Complete extension of a body part approximates 180° of motion. Figure 2.8 shows extension in stabilized elbows, hips, and lumbar-thoracic spine.

Figure 2.8 Elbow, hip, and lumbar-thoracic spine extension by Mike McKinney.

Figure 2.9 Knee flexion-extension is a vital aspect of cycling, as shown by Jeff Spencer, member of the U.S.A. Olympic cycling team.

Flexion is any movement resulting in a decrease of a joint angle. For example, when the elbow is being flexed from the extended position, the number of degrees within the joint angle is decreased as the hand moves toward the shoulder. The cyclist in figure 2.9 is flexing the right knee and moving the left knee from a position of flexion to extension. These motions are through anteroposterior planes of motion that lie parallel to the cardinal anteroposterior plane.

Horizontal abduction is movement of an upper limb through the transverse plane at shoulder level away from the midline of the body.

Figure 2.10
Horizontal adduction of both shoulder joints by Mike McKinney, SMS wrestler, working against resistance of an ExerGenie Exerciser.

Figure 2.11
Hyperextension of both shoulder joints by Don Elshire, USC gymnast, during a ring maneuver. *Courtesy USC Athletic Service.*

Horizontal adduction is movement of an upper limb through the transverse plane at shoulder level toward the midline of the body. Figure 2.10 illustrates horizontal adduction of both shoulder joints by the performer working against resistance provided by an ExerGenie Exerciser.

Hyperextension is movement of any joint beyond the joint's normal position of extension. The gymnast in figure 2.11 is hyperextending at both shoulder joints.

Figure 2.12 Right lateral flexion of the lumbar-thoracic spine occurring during a defensive move by Mike Westra, USC center. *Courtesy USC Athletic News Service.*

Inversion is movement of the sole of the foot inward or medially. If both feet are inverted, the soles of the feet will be toward each other. Inversion occurs at the subtalar and transverse tarsal joints. Inversion and eversion are both inefficient movements for many performers during the process of running.

Lateral flexion is movement of the head and/or trunk laterally away from the midline of the body. Figure 2.12 shows lateral flexion of the lumbar-thoracic spine of the defensive basketball player. *Reduction* is the return motion to the anatomic position.

Opposition of the thumb is a diagonal movement of the thumb across the palmar surface of the hand. This is commonly observed in the grips utilized to handle sport implements or objects. Figure 2.13 shows a catcher's grip on a baseball; opposition of the thumb is utilized.

Pelvic anteroposterior rotation is movement of the pelvic girdle through the anteroposterior plane of motion. In *anterior pelvic rotation* the iliac crests move forward. Conversely, *posterior pelvic rotation* results in the iliac crests being moved posteriorly. Figure 2.14 shows posterior rotation of the pelvic girdle.

Pelvic lateral rotation is movement of the pelvic girdle through the lateral plane of motion. Lateral pelvic rotation takes place either right or left through the lateral plane, and the axis is anteroposterior. Lateral rotation occurs unilaterally when the weight is borne either by the hands or one leg. This is a common pelvic girdle motion during the pivoting actions observed in sport. Figure 2.15 shows left lateral pelvic rotation. Lateral pelvic rotations occur during each stride while walking or running.

Pelvic transverse rotation is movement of the pelvic girdle through the transverse plane of motion. Transverse pelvic rotation can occur either right or left through the transverse plane around the longitudinal axis of the body. Transverse rotation of the pelvic girdle is an absolute necessity for high ballistic performances such as javelin throwing and baseball pitching. For the right-handed athlete throwing the javelin, for example, left transverse pelvic rotation must be an integral aspect of the sequential body motions for optimal results. Figure 2.16 shows left pelvic transverse rotation.

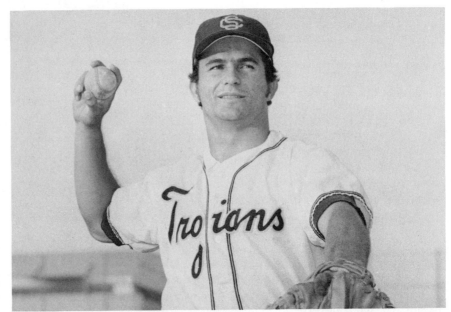

Figure 2.13
Opposition of the right thumb by Sam Ceci, USC catcher, while gripping the baseball. *Courtesy USC Athletic News Service.*

Figure 2.14 Left, posterior pelvic girdle rotation.

Figure 2.15 Center, left lateral pelvic girdle rotation.

Figure 2.16 Right, left transverse pelvic girdle rotation.

Fundamentals of Joint Motion Analysis

Figure 2.17 Plantar flexion of both ankle joints by Willie Deckard, USC sprinter, *Courtesy USC Athletic News Service.*

Plantar flexion is movement at the ankle joint of the sole of the foot downward. The term *plantar flexion* is an exception to the previous definition of flexion. In reality, plantar flexion is extension of the ankle. Most body movements from the vertical or erect body position begin with plantar flexion (fig. 2.6). The sprinter crossing the finish line in figure 2.17 has plantar-flexed both ankle joints.

Pronation is movement of the "back of the hand" forward. Pronation takes place at the radio-ulnar joint. The swimmer in figure 2.18 has pronated both radioulnar joints.

Prone position is the face-downward position by the entire body. The body does not have to be lying face downward on the ground or on some other supportive surface to be in a prone position. The prone position can be assumed, for example, in midair while rebound tumbling or diving.

Protraction is forward movement of the shoulder girdle. Scapular abduction occurs during protraction. This motion can be observed during the "up phase" of the "push-up" exercise.

Figure 2.18 Jim McConica, USC swimmer, has medially rotated both shoulder joints, radially flexed both wrists, and grasps the starting stand by flexing the interphalangeal joints of the toes prior to a racing start. *Courtesy USC Athletic News Service.*

Radial flexion is movement at the wrist on the thumb side of the hand toward the forearm. Figure 2.18 also illustrates radial flexion of both wrist joints.

Retraction is backward movement of the shoulder girdle. Scapular adduction occurs during retraction. The catcher in figure 2.13 has retracted his right shoulder girdle concurrent with right shoulder joint diagonal abduction, i.e., the right scapula has been adducted or moved toward the spinal column during the preparation phase of the "snap throw."

Rotation downward is rotary movement of the *scapula*, with the inferior angle of the scapula moving medially and downward. The glenoid fossa is moved downward to accommodate the head of the humerus. Downward rotation of the scapula always accompanies any downward movement of the upper limb. Figure 2.19 shows downward rotation of both scapulae during the downward pull on the latissimus bar.

Rotation laterally is movement around the longitudinal axis of a bone away from the midline of the body. Lateral rotation of the humerus, for example, must occur prior to executing the overhead serve in volleyball, serving a tennis ball, or throwing a baseball. This action helps place muscles on stretch and increases the range of motion. In arm wrestling, figure 2.20, lateral and medial rotations of the shoulder are very important.

Fundamentals of Joint Motion Analysis

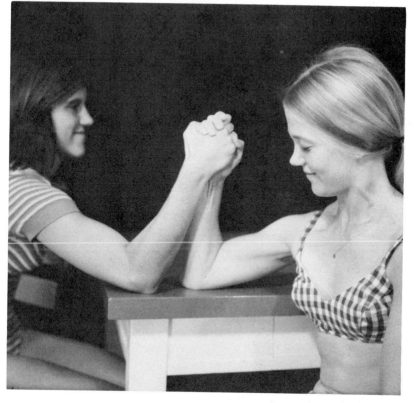

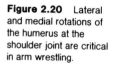

Figure 2.19
Downward rotation of
both scapulae occurs
concurrent with shoulder
joint adduction during a
latissimus pull by Mike
McKinney.

Figure 2.20   Lateral
and medial rotations of
the humerus at the
shoulder joint are critical
in arm wrestling.

Basic Concepts of Anatomic Kinesiology

Rotation medially is movement around a longitudinal axis of the bone toward the midline of the body. As an example, a well-timed throw through the high diagonal plane of motion involves medial rotation of the humerus immediately prior to the release of the thrown object. Medial rotation of the humerus is the most critical movement in the sport of arm wrestling as seen in figure 2.20. The swimmer in figure 2.18 has also medially rotated his shoulders.

Rotation upward is rotary movement of the *scapula*, with the inferior angle moving laterally and upward. The glenoid fossa is being moved upward to accommodate the head of the humerus. Upward rotation of the scapula always accompanies any upward movement of the upper limb.

Supination is the "palms forward" position of the hands in the anatomic position (fig. 2.1). Supination is the return movement from pronation, and it occurs at the radio-ulnar joint. A flat handball serve involves a supinated hand.

Supine position is lying with the body in a face-up position. The body does not have to be lying face upward on a supportive surface. It may be suspended in air in a supine position. This is observed in diving and rebound tumbling.

Ulnar flexion is movement of the little finger side of the hand toward the forearm. Ulnar flexion of the wrist is a vital movement for the baseball pitcher throwing a curve ball. The left wrist of the pole-vaulter in figure 2.21 is ulnar flexed through its complete range of motion.

Figure 2.22 lists motions possible at selected joints of the body. These should be studied with care, and the student should do the motions at each joint while learning the definitions.

The motions within the feet and hands are excluded from figure 2.22. There are similarities between the motions of the feet and hands. The toes and fingers are known as *phalanges*; therefore, the joints connecting the distally located bones

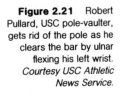

Figure 2.21 Robert Pullard, USC pole-vaulter, gets rid of the pole as he clears the bar by ulnar flexing his left wrist. *Courtesy USC Athletic News Service.*

Fundamentals of Joint Motion Analysis

Joint or Segment	Bones Involved	Motions Possible
Ankle	Tibia, fibula, talus	Dorsiflexion, plantar flexion
Knee	Tibia, femur, patella	Flexion, extension, rotation (when knee is flexed)
Hip	Femur, pelvis	Flexion, extension, adduction, abduction, diagonal adduction, diagonal abduction, medial and lateral rotation, circumduction
Pelvis	Ilium, ischium, pubis (pelvic rotations)	Anterior rotation, posterior rotation, lateral rotation left and right, transverse rotation left and right
Intervertebral (spine)	Vertebrae	Flexion, extension, rotation, lateral flexion, hyperextension, circumduction
Shoulder	Humerus, scapula	Flexion, extension, adduction, abduction, diagonal adduction, diagonal abduction, medial and lateral rotation, circumduction, hyperextension, horizontal adduction, horizontal abduction
Sternoclavicular (shoulder girdle)	Clavicle, sternum, scapula	Elevation, depression, protraction, retraction, rotation, circumduction
Sternoclavicular (scapula)	Scapula (scapular movements)	Elevation, depression, adduction, abduction, upward and downward rotation
Elbow	Humerus, radius, ulna	Flexion, extension
Radio-ulnar	Radius, ulna	Pronation, supination
Wrist	Radius, navicular, lunate, triangular	Flexion, extension, adduction, abduction, circumduction (adduction and abduction also called ulnar and radial flexion, respectively)
First carpometacarpal (thumb)	Multiangular first metacarpal	Flexion, extension, adduction, abduction, rotation, opposition

of each toe or finger are known as *interphalangeal joints*. Flexion and extension are the only two motions possible at those joints. Logically, the flexion motions of the interphalangeal joints are used for a wide variety of grips and grasping functions in sports.

The long bones of the foot are known as the metatarsals; therefore, the joints formed by the articulations between the phalanges and metatarsals are known as *metatarsophalangeal joints* (fig. 5.3). Similarly, the long bones of the hand are named metacarpals. The large knuckles formed by these bones and the phalanges of the hand are known as *metacarpophalangeal joints* (fig. 7.31). The motions possible in these large foot and hand "knuckle joints" are extension, flexion, abduction and adduction. The latter two motions are not as evident in the metatarsophalangeal joints, but the motion capabilities are there.

The foot contains several small bones which articulate with each other and allow motion. These articulations are known as the *transverse tarsal* and *subtalar joints*. The motions of inversion and eversion occur at these joints. As noted in figure 2.22, they do not occur within the ankle joint.

Figure 2.22
Selected major joints and their motions.

Anatomic Aspects of Segmental Analysis

The extent, direction, and velocity of each joint motion used to execute a skill is of prime importance in biomechanic analysis. The result of these joint motions is movement of body segments through planes of motion. How effectively these segments are moved as a result of muscular contractions at joints to reach the skill objective is one of the major purposes of *segmental analysis*. This technique is used when film is available. It is not applicable for noncinematographic analysis.

Most individuals who have not studied biomechanics think of the measurement of such a factor as human speed only in terms of motion of the total body. We are conditioned to think of the linear velocity of the sprinter in terms of the numbers of seconds it takes to cover the dash distance. Baseball players are timed by scouts in terms of seconds needed to move the total body from home to first base. Football coaches take great pride in relating forty yard times of their fastest athletes, although the relationship to football playing ability has not been proven scientifically! Nevertheless these measurements serve their purposes. For in-depth analyses, however, one must make critical observations of the body segments as they move angularly through their planes of motion to produce linear motion of the total body or the integrated linear-angular motion needed to produce skilled motion.

The body has 25 *major* moving segments. These are outlined in figure 2.23. The fingers and toes are not included as major body segments, although their motions must be analyzed as important parts of many sport skills.

The number and location of segments to be used in a segmental analysis of a sport skill is determined by the biomechanic parameters and anatomic motions to be studied. The film is projected onto a screen or writing surface. Surface anatomic landmarks and approximate joint centers are utilized as reference points to estimate the position of each body segment on every frame or at selected time

Figure 2.23 Major body segments used in segmental analysis.

Spinal Column Segments	Upper Limb Segments	Lower Limb Segments
Head and Cervical Spine	Left Humerus, Forearm and Hand	Left Femur, Lower Leg and Foot
Trunk (Lumbar-Thoracic Spine)	Left Humerus	Left Femur
Head and Trunk (No Limbs)	Left Radio-Ulnar Joint (Forearm)	Left Tibia-Fibula (Lower Leg)
Lumbar Spine and Pelvic Girdle	Left Forearm and Hand	Left Lower Leg and Foot
Pelvic Girdle	Left Hand	Left Foot
	Right Humerus, Forearm and Hand	Right Femur, Lower Leg and Foot
	Right Humerus	Right Femur
	Right Radio-Ulnar Joint	Right Tibia-Fibula
	Right Forearm and Hand	Right Lower Leg and Foot
	Right Hand	Right Foot

Fundamentals of Joint Motion Analysis

Segment	Anatomic Landmark	Figures	Pages
Head	Midnose, maxilla, mandible, occipital protuberance, mastoid process, ear lobule	6.11 6.12	124 124
Spinal Column	Spinous processes	6.4	114
Rib Cage	Sternum, xiphoid process	6.5	115
Pelvic Girdle	Iliac crest, pubis, anterior superior iliac spine	5.23	99
Shoulder Girdle	Acromion process, clavicle	7.14	137
Humerus	Greater tubercle, medial epicondyle, lateral epicondyle, deltoid tuberosity	7.15	138
Radio-Ulnar Joint	Olecranon process	7.29	150
Femur	Great trochanter, lateral condyle, medial condyle, patella	5.20 5.13	97 93
Tibia-Fibula	Fibular head, medial condyle, lateral condyle, fibular malleolus, tibial malleolus	5.14	93
Foot	Calcaneus, great toe, fifth lesser toe	5.3	85

intervals on the film. (Figure 2.24 lists the surface anatomic landmarks which should be used to determine segment positions.) Once the landmarks have been located and joint centers estimated as accurately as possible, point and line drawings should be made by drawing straight lines between the appropriate reference points.

Two examples are presented to demonstrate how the line drawings are made from film based on the available anatomic information. More detailed examples of segmental biomechanic analyses are presented in Chapters 12 and 14. Figures 2.25 through 2.28 are *contour drawings* of a world class fast pitch softball hitter, Bonus Frost. These were made from film taken at 32 frames per second. It took this athlete one second to complete the skill. The body segments are shown in figure 2.29 depicting the motion in time intervals of 0.25 seconds. (The sport implement is included as a moving segment in figure 2.29.) From these rather

Figure 2.24
Selected surface anatomic landmarks to be used as positions of reference to approximate joint centers and determine body segment lines for segmental analysis.

Figure 2.28 Follow-through.

Figure 2.27 Movement—point of impact.

Figure 2.26 Preparatory.

Figure 2.25 Stance (left-handed hitter).

Figure 2.29 Point and line illustration—(numbers indicate film frames—film taken at 32 frames/second), left-handed hitter.

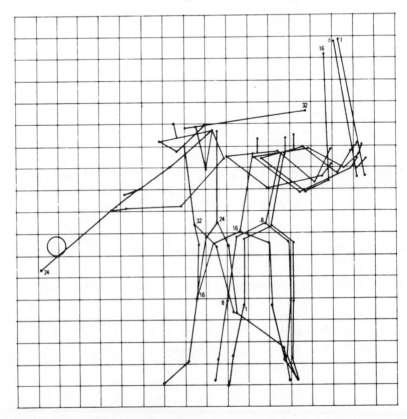

Fundamentals of Joint Motion Analysis

simple line drawings depicting the totality of skill, the analyzer can accurately extract information related to such factors as the extent of motion at joints, direction of moving segments, and velocity of the upper limb and sport implement segments. An analogous procedure is also derived by computers from film.

Lower limb segment motions are critical for the hurdler (fig. 2.30). Body segment information from figure 2.30 was extracted and placed in a more objective form on figure 2.31. Data such as the angular positions and velocities of the femur and tibia-fibular segments of the right lower limb can be extracted from segmental drawings such as these. These procedures will be detailed in Chapters 13 and 14.

Figure 2.30 Hurdling trajectory.

Fundamentals of Joint Motion Analysis

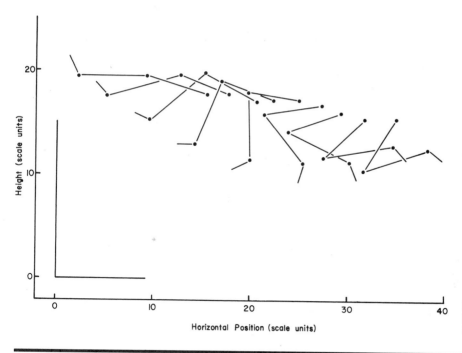

Figure 2.31 Lower limb segments of hurdler in figure 2.30.

Adrian, M. J. "An Introduction to Electrogoniometry." In *Kinesiology Review* 1968. Washington, D.C.: American Association of Health, Physical Education, and Recreation, 1968.

Dempster, W. T. *Space Requirements of the Seated Operator,* WADC Technical Report 55–159, Wright-Patterson Air Force Base, Ohio, 1955.

Hirt, Susanne P. "Joint Measurement." *American Journal of Occupational Therapy* 1 (1974): 209–14.

Holland, George J. "The Physiology of Flexibility: A Review of the Literature." In *Kinesiology Review* 1968. Washington, D.C.: American Association of Health, Physical Education and Recreation, 1968.

Logan, Gene A. and McKinney, Wayne C. *Anatomic Kinesiology,* (Third Edition). Dubuque: William C. Brown Company, Publishers, 1982.

Salter, N. "Methods of Measurement of Muscle Function." *Journal of Bone and Joint Surgery* 37–B (1955): 474–91.

Recommended Reading

3

Anatomic Planes of Motion

One purpose of this chapter is to introduce the reader to the traditional planes of motion which have appeared in the medical literature for centuries. In addition, the diagonal planes of motion are presented. These were introduced into the literature by Logan and McKinney in 1970. The joint motions discussed in Chapter 2 are presented in tabular form with their corresponding planes of motion. Finally, due to the fact that joint motions are described in the literature in several ways, the reader is provided with a table of common motion synonyms for many of the motion terms presented in this text.

As discussed in Chapter 2, all joint motions are angular in nature. The motion of a joint causes a body segment to turn through a range of motion quantitatively expressed in degrees or radians. The body segment moves through a plane of motion. *Therefore, a plane of motion is defined as an imaginary circular two-dimensional surface through which a body segment is moved.* In order to conceptualize and comprehend the relationship between a plane of motion and the moving body segment, the plane should be visualized as a circular and transparent windowpane which disects the center of a joint or area of the body. This is depicted artistically in figure 3.1 and other figures in this chapter.

The important fact to remember regarding an anatomic plane is the direction in which it lies relative to the body, as opposed to the point at which it intersects the body. Since joint motion is angular in nature, the moving body segment is rotated around an imaginary line perpendicular to the plane. That line is known as the *axis of motion.* In this context, there is always a 90° relationship between the plane of motion and its axis.

To conceptualize a theoretical plane and axis at a joint, one must first make an accurate estimate of the location of the joint center. The plane and axis both pass through the joint center. The exact joint center cannot be determined with precision unless radiologic filming techniques are utilized; except in rare research situations this is not feasible for numerous reasons.

Joint centers can be estimated with relative precision on film and videotape, by utilizing one's knowledge of joint anatomy including bone and muscle relationships as well as the motion capabilities of the joint. As an example, the ankle joint is formed by the articulation between the talus and tibia (figure 5.10). This joint has two motion capabilities: (1) dorsiflexion and (2) plantar flexion. (Joint motions are defined in Chapter 2.) These motions are through an anteroposterior plane of motion around an axis of motion perpendicular to that plane. The axis of rotation can be conceptualized by drawing an imaginary line through the medial malleolus to the lateral malleolus. The foot turns through an anteroposterior plane of motion as it dorsiflexes and plantar flexes. The theoretical plane would be anterior and posterior through the center of the ankle joint formed by the talus and tibia.

A simpler method of actually visualizing a plane with its axis of rotation is to tie a low weight object such as a ring on a piece of string six to eight inches in length. The other end of the string is then tied to the center of a pencil. The weighted string should be moved to allow it to turn or rotate through 360 degrees around the center of the shaft of the pencil. The pencil simulates the axis of rotation, and the weighted string allows one to visualize the plane of motion. Flexion and extension of the elbow, as a joint motion example, constitutes the same axis-plane relationship, except that the elbow motion (hand and forearm segments) is limited to approximately 150 degrees through its plane due to joint structure and surrounding tissue mass. (figure 3.3).

Traditional Planes of Motion

Anatomists and kinesiologists have for centuries recognized, in the literature, the existence of three anatomic reference planes. These traditional planes of motion are: (1) anteroposterior or saggital plane, (2) lateral or frontal plane, and (3) transverse or horizontal plane. The term *cardinal* is used at times in connection with these planes; this term means that the discussion is centered on only the two vertical planes (anteroposterior and lateral) as they are shown in figures 3.1 and 3.6. These two planes meet at a theoretical point within the body known as the *center of gravity* when the anatomic position is assumed (refer to figure 2.1). The center of gravity of the total body shifts continuously as one performs in sport and dance. In some sport activities such as high jumping, diving, and pole vaulting, the theoretical point for the center of gravity may lie outside the body (refer to figure 8.20).

The center of gravity of the total body is where the weight of the body is concentrated. *It is a mathematic construct rather than a reality.* The center of gravity may be calculated for each body segment as well as for the total body in order to study such aspects as segment displacement patterns and equilibrium. However, these calculations are not made during kinesiologic analyses at the noncinematographic and basic cinematographic levels. With these analytic techniques, which are used on a daily basis, it is not necessary or feasible to mathematically calculate the location of the center of gravity. However, scientific study of centers of gravity of the human body and its segments during motion is of prime concern to many research investigators in anatomic kinesiology and biomechanics.

The general position of the center of gravity for the total body in the anatomic position was described above. For adult males this point is located at approximately 56 to 57 percent of their total height. Usually, this is near the level of the umbilicus at a central point of the body. Due to greater distribution of subcutaneous adipose tissue in adult females around the anterior, lateral, and posterior pelvic-hip area, the center of gravity lies at approximately 55 percent of their total height.

The terms *horizontal* and *vertical* tend to be somewhat confusing when used in connection with planes of motion. Since these terms are usually used to discuss a position relative to the surface of the earth, they can be misleading in regard to sport movements. As an example, the transverse plane is a plane horizontal to the earth when the individual is in an upright position. The axis of motion is vertical. But performers do not always assume upright positions, and this is where confusion can exist. When the performer is supine and suspended in air, as during a dive, the transverse plane becomes a vertical plane, and the axis of motion becomes horizontal. To avoid this type of communication problem, the terms *horizontal* and *vertical* are not used here to describe planes of motion of performers in relation to the earth. Instead, the reference for planes is the body itself.

Anteroposterior or Sagittal Plane of Motion

The cardinal anteroposterior plane divides the body into equal, bilateral segments (figure 3.1). This theoretical plane bisects the body through the center of the skull, spinal column, rib cage, pelvis, and lower limbs. Motions through the cardinal anteroposterior plane involve the cervical, thoracic, and lumbar regions of the spinal column. Motions through this plane occur, for example, during a sit-up exercise (figure 3.2).

Figure 3.1 Cardinal anteroposterior plane of motion. This is also known as the midsagittal plane.

Figure 3.2 The "hook-lying sit-up." This exercise involves lumbar-thoracic flexion. The trunk-head segment moves through the cardinal anteroposterior plane of motion (Ruth Miller).

Figure 3.3 The "biceps curl" exercise. Elbow flexion and extension moves the forearm-hand segment through an anteroposterior plane of motion which lies parallel to the cardinal anteroposterior plane of motion (Ardie McCoy).

Figure 3.4 Anteroposterior plane motions at the left hip and knee are important to long jump performance as demonstrated by Sandy Crabtree. *Courtesy USC Athletic News Service—Michael R. Harriel.*

Anteroposterior plane motions are not restricted to movements involving the spinal column. When the term cardinal is not used with this or any other traditional plane, it means that the motion described causes the joint or limb to move in a plane *parallel* to the cardinal plane of motion. An example of this would be the biceps curl weight-training exercises shown in figure 3.3. The weight plus the hand and forearm segments move through an anteroposterior plane as the elbow is flexed and extended. In this illustration, the anteroposterior plane being traversed is parallel to the cardinal anteroposterior plane of motion.

The long-jumper shown in figure 3.4 is flexing her left hip and knee joints through anteroposterior planes parallel to the cardinal anteroposterior plane. These joint motions, with others, are important for the long-jumper at take-off, because they help determine the critical angle of take-off and the subsequent trajectory for the flight of the body. Maximum distance in flight is determined, in part, by the efficiency of these anteroposterior plane motions. For example, extreme hip flexion tends to lift the body too much vertically and increase the take-off angle. That will decrease horizontal speed and adversly affect the distance of the jump. In all sport skills, there are optimum joint motions through appropriate planes at correct velocities which will produce the desired results.

The balance beam in women's gymnastics is a very demanding piece of apparatus, because it is only 10.16 centimeters wide. This requires that several anteroposterior plane motions must be executed precisely by an elite performer

Figure 3.5 Gale Wyckoff, USC gymnast, executing a balance beam skill which requires several joint motions through both the cardinal anteroposterior plane of motion and anteroposterior planes parallel to the cardinal plane. *Courtesy USC Athletic News Service.*

during a routine, or she will lose her equilibrium. The gymnast in figure 3.5 is shown moving several major joints and segments through anteroposterior planes: (1) left hip joint—extension, (2) right hip joint—flexion, (3) lumbar-thoracic spine—hyperextension, (4) cervical spine—hyperextension, and (5) both shoulder joints—flexion. The spinal segments move through the cardinal anteroposterior plane, and the other segments noted move through anteroposterior planes parallel to the cardinal plane.

Lateral or Frontal Plane of Motion

This cardinal plane divides the body into anterior and posterior halves. It bisects the body laterally through the ear, shoulder, spinal column, pelvis, hips, knees, and ankles (figure 3.6).

Motion capabilities for movement through the lateral plane are somewhat limited, as compared to the other planes of motion. Body areas involved are the upper and lower limbs, cervical, and lumbar-thoracic spinal regions. The jumping-jack calisthenic exercise shown in figure 3.7 is a good example of upper and lower limb movements through the lateral plane of motion.

Figure 3.6 Cardinal lateral plane of motion.

Figure 3.7 The "jumping jack" exercise moves the upper and lower limb segments through the cardinal lateral plane of motion by abducting and adducting the shoulder and hip joints (Ruth Miller).

Numerous calisthenic exercises utilize the lateral plane motions at the shoulder and hip joints as well as the spinal column. The "thrust phase" of the elementary back stroke in swimming utilizes powerful shoulder and hip joint lateral plane motions (adductions) to propel the body forward in the water. Motions of the upper and lower limbs through the lateral plane are often used to enhance body stability or equilibrium both in moving and nonmoving situations. The gymnast in figure 3.8 has moved her upper limbs through the lateral plane during a floor exercise routine. This helps her maintain equilibrium on a relatively small unilateral weight-bearing base of support (the toes of her left foot).

The cardinal transverse plane of motion theoretically divides the body into superior and inferior halves when the individual is in the anatomic position. This plane is shown in figure 3.9. The position of the cardinal transverse plane differs slightly in the average adult male and female. As noted, the transverse plane for the adult female lies at a point slightly below the umbilicus. The reason for this is that adipose tissue distribution following sexual maturation of women is usually below the iliac crests posteriorly and laterally. Adipose tissue is also distributed anteriorily and inferior to the umbilicus. This means that the center of gravity in the anatomic position is slightly lower in the average adult female than in the

Transverse or Horizontal Plane of Motion

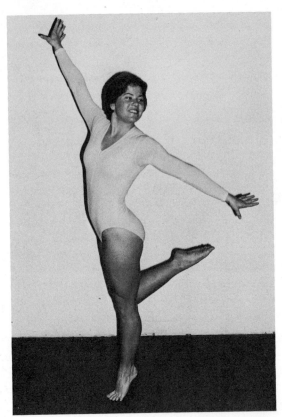

Figure 3.8 Holley Donaldson, USC gymnast, abducts both shoulder joints which move both upper limb segments through the lateral plane of motion. This assists her in maintaining equilibrium. *Courtesy USC Athletic News Service.*

average adult male. Men have a higher center of gravity and cardinal transverse plane due to the fact that adipose tissue, if any excess is found on the trunk, usually is immediately superior to the iliac crests. This causes the cardinal transverse plane to separate the superior and inferior halves of the "ideal" male body at the umbilical level.

The "twisting motions" of the spinal column (rotations) are performed through the cardinal transverse plane of motion (figure 3.10). There are numerous calisthenic type exercises for flexibility development, which depend upon spinal column rotations. In addition, there are many ballistic sport skills, gymnastic routines, and diving maneuvers in which spinal column rotations through the cardinal transverse plane are essential to acceptable performance levels (figure 3.8).

Rotations through a transverse plane are also possible within joints other than the spinal column. As examples, the medial and lateral rotations at the hip and shoulder joints while in the anatomic position, move through transverse planes. The bench press is another example of an exercise performed through a transverse plane with the axis longitudinal to the body (figure 3.11).

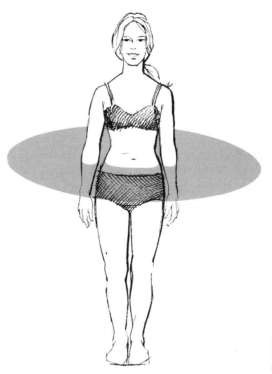

Figure 3.9 Cardinal transverse plane of motion.

Figure 3.10 Left lumbar-thoracic rotation; a spinal column motion which moves the lumbar-thoracic segment through the cardinal transverse plane of motion (Ruth Miller).

Figure 3.11 The "bench press" exercise—an exercise through a transverse plane parallel to the cardinal transverse plane of motion. The critical shoulder joint motions are horizontal adduction and horizontal abduction.

| Diagonal Planes of Motion | The description of motions during analyses of sport skills is very limited when restricted to the movements strictly defined by the three traditional planes of motion. Therefore, to add to motion terminology and enhance the accuracy of motion description in the area of analytic kinesiology, Logan and McKinney introduced the diagonal planes of motion in 1970. These planes and the motions through the diagonal planes, were established to be utilized with, not to replace, the three traditional planes of motion. |

The diagonal planes are: (1) high diagonal for upper limbs at the shoulder joints, (2) low diagonal for upper limbs at the shoulder joints, and (3) low diagonal for lower limbs at the hip joints. In regard to diagonal plane motion at the hip joint, the muscular and ligamentous structures do not allow motion through a high diagonal plane which would be analogous to shoulder joint motion. A high diagonal plane of motion for an upper limb is shown in figure 3.12. The corresponding low diagonal plane of motion of the opposite upper limb is shown in the same illustration. A low diagonal plane of motion for the left lower limb is shown in figure 3.13. Figure 3.14 is a composite of the three diagonal planes at the hip and shoulder joints.

It is useful to note that the sports which tend to utilize the traditional planes of motion the most are the so-called "judged sports." Many times these events, such as gymnastics, diving, and portions of weight lifting, are scored on the basis of aesthetic factors. Perhaps there is a relationship between this type of activity and formal programs of physical education. Through 1927 the formal programs seen extensively in the past throughout Germany, Sweden, and in America had as their bases movement patterns associated with traditional planes of motion which were described by and borrowed from anatomists.

Many sport motions do not conform precisely to movement through the anteroposterior, lateral, and transverse planes of motion. On the other hand, activities such as marching, wands, and calisthenics include numerous motions through the traditional planes. Physical education programs utilizing these formal activities in the latter part of this century in America would not be too popular. Instruction in a wide variety of team and lifetime sports is the present trend.

The longitudinal axis of the body is the reference point for diagonal limb motions, i.e., a limb moving through a diagonal plane moves in an arc oblique or diagonal to the cardinal anteroposterior plane. It is a commonly observed fact that there are numerous occasions in sport when the upper and lower limbs move diagonally through planes toward or away from the longitudinal axis of the body. Most throwing, striking, and kicking techniques employ these motions. The act of "crossing the legs" while seated or standing requires diagonal motion of one lower limb involving the hip joint. Since these planes and motions are so common, it was deemed necessary to identify them formally. *This was done to enhance skill analysis, motion description, and the communication process related to the teaching and learning of neuromuscular skills.*

Structural observations of the shoulder and hip joints provide some of the rationale for diagonal planes of motion. These joints are classified as enarthrodial

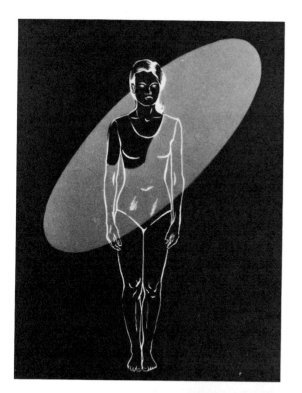

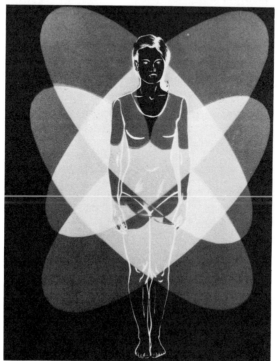

Figure 3.12 High diagonal plane of motion for the left shoulder joint.

Figure 3.13 Diagonal plane of motion for the left hip joint.

Figure 3.14 Composite of diagonal planes at both hip and shoulder joints.

(ball-and-socket) joints. This means that they have motion capabilities of a multiaxial nature; motion is not restricted to the three traditional planes only. The hip joint arrangement of the head of the femur (ball) being received into the acetabulum of the pelvis (socket) allows the lower limbs to be moved diagonally. This is analogous to an automobile-trailer hitch arrangement; the reader is familiar with the versatility of motion of a trailer when one tries to back it into a parking space. For a closer analogy to the versatility of motion within the hip and shoulder joints, imagine that the trailer is hitched to an airplane flying through the air. Depending upon the external forces, that trailer can move in any direction; it is not restricted to up-and-down (anteroposterior), back-and-forth (lateral), or limited rotation around its longitudinal axis (transverse) motions. The same can be said of the hip and shoulder joints; they can be moved in any direction through space. The shoulder joint structure involving the head of the humerus (ball) articulating with the glenoid fossa of the scapula (socket) allows the same motion versatility as the hip joint. Therefore, the bone structure of these joints allows the limbs to be moved diagonally as well as through the three traditional planes of motion.

Muscular charcteristics of the shoulder and hip joints provide additional rationale for diagonal planes of motion. Muscle contraction provides the internal force to move the body, its limbs, or other segments. The muscles are attached proximally and distally to bones, and the tendons of muscles cross the joints at a point either anterior, posterior, medial, or lateral to the joint center. The line-of-pull and the spatial relationship of muscles and muscle groups to joints are important considerations affecting motion capabilities of any joint.

The arrangement of the muscles at the hip joint indicates that several powerful muscles have diagonal lines-of-pull. The gluteus maximus (lower fibers), three large adductor muscles, and other muscles of the hip, by virtue of lines-of-pull and spatial relationship to the center of the hip joint, have the capability of moving the lower limb diagonally.

The pectoralis major, latissiumus dorsi, teres major, deltoid, and other muscles have diagonal lines-of-pull at the shoulder joint; therefore, they are optimally placed to exert their contractile forces, if desired, on the upper limbs and move them through diagonal planes.

The multiaxial nature of the shoulder and hip joints allows the limb levers to be pulled diagonally during muscle contraction when the proximal attachments of the muscle group are momentarily stabilized, allowing the force to be exerted distally. This diagonal bone-muscle arrangement enables the performer to apply more force to a sport object if desired. In addition, this arrangement facilitates more functional dissipation of force during the follow-through phases of ballistic motions such as those involving throwing, striking, and kicking skills (figure 3.17).

Another reason that diagonal motions are involved in most ballistic motions involving the limb segments lies in the fact that pendular levers (the upper and lower limbs) revolving on multiaxial articulations involve angular momentum. As an attempt is made to move a limb through a range of motion in one plane, there is a tendency for the limb/lever to describe a circular motion. This causes the limb segment to move in an arc diagonal to the longitudinal axis of the body. It should be remembered that the involved joint is also literally moving through space. This complexity of motion is possible by virtue of the multiaxial bone configuration and spatial relationships of the muscles to the shoulder and hip joints.

There also appears to be some neurologic basis for diagonal movement patterns in humans. One reflex adds some credibility to the diagonal movement pattern. This is known as the crossed-extensor reflex. Simply stated, this reflex is a combination of a flexion reflex in one limb and an extensor reflex occurring almost simultaneously in the opposite limb. This crossed-extensor reflex is responsible for an "automatic relationship" in diagonal-type movements. There is a constant reciprocal relationship between flexion and extension of alternating limbs as observed in most locomotor activities. Figure 3.15 shows this type of relationship in upper and lower limbs of a good football player.

What problems would be encountered by a performer if the shoulder and hip joints were only structured to execute motions through the three traditional planes? Right-handed baseball or softball hitting will serve as an example of some of the difficulties one would encounter. (The reader should attempt to perform the skill while using the following motions.) First, the bat could not be

Figure 3.15
Crossed-extensor reflex limb action—Mike Garrett, USC Heisman Trophy winner. *Courtesy USC Athletic News Service.*

gripped initially with both hands. The stance is assumed, and the bat is held in the left hand. The left shoulder joint and limb is moved through the anteroposterior plane (flexed) to the level of the shoulder where motions are possible through a transverse plane involving the shoulder joint. The bat is then moved through the transverse plane (horizontal adduction) to a position of readiness to await the pitch. The bat can now be gripped with the right hand. The spinal column is also rotated through the cardinal transverse plane during the stance. The pitch is delivered. There is a weight shift involving lateral plane motion of the left hip joint, and the bat and upper limbs are moved through a transverse plane at the shoulder to the point of impact with the ball. It is rather obvious that such motion limitations would greatly inhibit effective performance. The necessary force, angular velocities, and other components would not be generated and the "traditional plane hitter" would always have to make contact with a ball out of the strike zone! Obviously, humans use motions through traditional as well as the diagonal planes to perform skills. Therefore, it is recommended that physical educators analyze and describe all skills using traditional and diagonal plane motion nomenclature. This will greatly enhance the objectivity of motion description.

High Diagonal Plane of Motion (Shoulder)

Figure 3.16 shows a baseball pitcher utilizing high diagonal plane motions involving the left shoulder joint. The left arm would follow the plane during release, and the follow-through would be diagonally across the pitcher's longitudinal axis.

The highly skilled pitcher in figure 3.17 shows the typical diagonal plane motion of the throwing limb during the follow-through phase of the skill. For almost all the ballistic throwing and striking skills involving shoulder joints, variations of this motion pattern through the diagonal plane are dependent upon the direction of spin applied to the sport object. If "back spin" has been applied to

Figure 3.16 High diagonal plane of motion—left shoulder joint.

Anatomic Planes of Motion

Figure 3.17 Tom Seaver, former USC pitcher, moves his total right upper limb segment through a high diagonal plane of motion. The limb motion is seen as diagonal adduction in the follow-through phase of the pitch. *Courtesy USC Athletic News Service.*

the baseball (sport object), the throwing limb motion through the diagonal plane, as seen in figure 3.17, is observed. Spin application to sport objects is critical in most sports, and it can ultimately determine the skill level the performer will attain in his or her sport specialty.

High diagonal plane motions occur at any degree above the transverse plane at the shoulder joint, i.e., the plane is determined by the angle traversed by the upper limb during the motion. A person throwing a sport object such as a ball or javelin "over the top" would be using a very high diagonal plane. The term many coaches give for the diagonal plane being used by the pitcher in figure 3.16 is three-quarters overhand. "Overhand," regardless of the angle or degree of motion traversed by the shoulder joint and limb, is considered to be a high diagonal plane. A true "sidearm" motion moves through a transverse plane at the shoulder joint; this is not an anatomically or biomechanically sound way to perform throwing or striking motions. Throwing by using shoulder joint motions which bring the upper limb through the transverse plane can cause considerable stress due to centrifugal force, and this can lead to injuries to the "rotator cuff" of the shoulder, as well as to other structures and articulations in the throwing limb.

Low Diagonal Plane of Motion (Shoulder)

The upper limb moves through this plane of motion when performers execute some "underhand" skills. Usually, but not in all cases, the low diagonal plane is traversed when considerable force or velocity is desired by the performer. The low diagonal planes are below the transverse plane at the shoulder joint; figure 3.18 shows a discus thrower utilizing a low diagonal plane of motion using the right shoulder and upper limb.

Simple anteroposterior, pendular motions of the shoulder joint and upper limb are commonly observed when force and velocity are not prime objectives.

Figure 3.18 Low
diagonal plane of
motion—right shoulder
joint.

Figure 3.19 Denise
Strebig, USC golfer,
demonstrates the
utilization of low diagonal
plane motions. Both
upper limb segments plus
the golf club move
through low diagonal
planes during the golf
shot. *Courtesy USC
Athletic News Service.*

However, a discus thrower, for example, throws through a low diagonal plane to maximize velocity and to give the discus the proper type of aerodynamic release and trajectory. It would be very inefficient for the athlete to release a discus underhand while flexing the shoulder and moving the limb through an anteroposterior plane of motion. A slow pitch softball pitcher, on the other hand, might throw underhand using anteroposterior plane motions. However, good fast pitch softball pitchers utilize low diagonal plane motions.

Golf is a game which demands extensive use of low diagonal plane motions involving the shoulder joints. As seen in figure 3.19, the golfer is employing diagonal abduction of the left upper limb and diagonal adduction of the right upper limb. These motions through low diagonal planes are necessary to transfer the summated forces from the contraction of muscles to the sport implement (golf club) and subsequently apply those forces to the sport object (golf ball).

Low Diagonal Plane of Motion (Hip)

Figure 3.20 shows a soccer player using the diagonal plane of motion for a kick involving the right hip joint and lower limb. This type of movement pattern is observed rather frequently in soccer kickers and punters. It is being applied more often among American football kickers and punters who are interested in maximizing force application to the ball to attain greater heights and distances. It should be noted, however, that low diagonal plane motions involving the hip joint need not be ballistic in nature. The crossing of one's legs while seated, or the scissors kick of the side stroke in swimming, involve motions through the low diagonal plane at hip joints, performed at a low velocity.

Figure 3.20 Diagonal plane of motion—right hip joint.

As indicated, the hip joint is multiaxial in nature. Therefore, it has a very extensive potential range of motion. This potential is often reduced owing to the lack of flexibility within the noncontractile tissue surrounding the hip joints. Dance directors, gymnastic coaches, and diving coaches have long known the value of hip flexibility to range of motion and performance. It now appears that some coaches in team sports are also concerned to increase flexibility in their student-athletes to enhance motion at the hip joint through anteroposterior, transverse, lateral, and diagonal planes.

There are a relatively few individuals who, by virtue of special flexibility training and, perhaps, genetic or familial predisposition for flexibility, have the capability of moving the lower limb through a diagonal plane at the hip in a position analogous to the high diagonal plane at the shoulder. This plane of motion is difficult to attain at the hip joint, and it is not often observed. Therefore, it is not included in this text for discussion purposes; one should be aware, however, that a high diagonal plane potential at the hip joint may exist.

Joint Motions and Planes

Figure 3.21 is provided to summarize and review the joint motions. In conjunction with that review, the plane of motion for each joint motion is listed.

One major motion which is not listed in figure 3.21 is circumduction. If a joint is structurally capable of flexion, extension, abduction and adduction, it can also be circumducted. Or another criterion to determine whether circumduction is possible involves diagonal motion; diagonal abduction constitutes 180° of circumduction and diagonal adduction contributes the remaining 180° of motion. Circumduction is 360° of motion of a body segment in a conical manner.

Figure 3.21 Joint motion summary with planes of motion.

Segment/Joints	Motions Possible	Plane of Motion
I. Foot		
A. Interphalangeal Joints	Flexion	Anteroposterior
	Extension	Anteroposterior
B. Metatarsophalangeal Joints	Flexion	Anteroposterior
	Extension	Anteroposterior
	Abduction	Lateral
	Adduction	Lateral
C. Transverse tarsal and subtalar Joints	Inversion	Transverse
	Eversion	Transverse
II. Talocrural Joint	Dorsiflexion	Anteroposterior
	Plantar flexion	Anteroposterior
III. Knee Joint	Flexion	Anteroposterior
	Extension	Anteroposterior
	Lateral rotation	Transverse
	Medial rotation	Transverse

Figure 3.21—*Continued*

Segment/Joints	Motions Possible	Plane of Motion
IV. Hip Joint	Flexion	Anteroposterior
	Extension	Anteroposterior
	Abduction	Lateral
	Adduction	Lateral
	Lateral rotation	Transverse
	Medial rotation	Transverse
	Diagonal abduction	Diagonal
	Diagonal adduction	Diagonal
V. Pelvic Girdle	Anterior rotation	Anteroposterior
	Posterior rotation	Anteroposterior
	Lateral rotation left & right	Lateral
	Transverse rotation left & right	Transverse
VI. Lumbar-Thoracic Spine	Flexion	Anteroposterior
	Extension	Anteroposterior
	Hyperextension	Anteroposterior
	Rotation left and right	Transverse
	Lateral flexion and reduction left and right	Lateral
VII. Cervical Spine	Flexion	Anteroposterior
	Extension	Anteroposterior
	Hyperextension	Anteroposterior
	Rotation left & right	Transverse
	Lateral flexion left & right	Lateral
VIII. Shoulder Girdle	Elevation	
	Depression	
	Upward rotation	
	Downward rotation	
	Abduction	
	Adduction	
IX. Shoulder Joint	Flexion	Anteroposterior
	Extension	Anteroposterior
	Hyperextension	Anteroposterior
	Abduction	Lateral
	Adduction	Lateral
	Lateral rotation	Transverse
	Medial rotation	Transverse
	Horizontal abduction	Transverse
	Horizontal adduction	Transverse
	Diagonal abduction	Diagonal—Low & High
	Diagonal adduction	Diagonal—Low & High
X. Elbow Joint	Flexion	Anteroposterior
	Extension	Anteroposterior
XI. Radio-Ulnar Joint	Pronation	Transverse
	Supination	Transverse

Figure 3.21—*Continued*

Segment/Joints	Motions Possible	Plane of Motion
XII. Wrist Joint	Flexion	Anteroposterior
	Extension	Anteroposterior
	Hyperextension	Anteroposterior
	Radial flexion	Lateral
	Ulnar flexion	Lateral
XIII. Hand		
A. Interphalangeal Joints (Thumb & Fingers)	Flexion	Anteroposterior
	Extension	Anteroposterior
B. Metacarpophalangeal Joints (Four fingers)*	Flexion	Anteroposterior
	Extension	Anteroposterior
	Hyperextension	Anteroposterior
	Abduction	Lateral
	Adduction	Lateral
C. Carpometacarpal Joint (Thumb)	Flexion	Anteroposterior
	Extension	Anteroposterior
	Abduction	Lateral
	Adduction	Lateral
	Opposition	Diagonal
	Rotation	Transverse

*The middle finger lateral plane motions are called radial and ulnar flexions instead of abduction and adduction.

Motion Synonyms The motion terminology presented in this chapter should become an integral part of the professional physical educator's vocabulary. As indicated, this is necessary to communicate and describe skills involving joint motions. It is also essential to know these terms in order to understand the literature in biomechanics and anatomic kinesiology. The motion terms defined in this chapter are used extensively, *but not exclusively,* by authors of kinesiology literature to describe movement. Consequently, there is a need to know the definitions of other terms and synonyms for terms which also appear in the literature. Some of the more common terms are included in figure 3.22.

Figure 3.22
Definitions and synonyms
of motion terms.

Motion Terms	Definitions and/or Synonyms
I. Subtalar and Transverse Tarsal Joints	
A. Pronation of the foot	A. Eversion, talonavicular abduction plus eversion
B. Supination of the foot	B. Inversion, talonavicular adduction plus inversion
II. Talocrural Joint	
A. Dorsal flexion	A. Dorsiflexion, foot flexion
B. Extension	B. Plantar flexion, foot extension

Figure 3.22—*Continued*

Motion Terms	**Definitions and/or Synonyms**
III. Hip Joint	
A. Hyperadduction	A. Diagonal adduction
IV. Pelvic Girdle	
A. Decreased inclination	A. Posterior pelvic rotation, backward tilt, decreased tilt, a backward rotation
B. Increased inclination	B. Anterior pelvic rotation, forward tilt, increased tilt, forward rotation
C. Lateral tilt	C. Lateral pelvic rotation
D. Lateral twist	D. Transverse pelvic rotation, rotation
V. Spinal Column	
A. Abduction	A. Lateral flexion
B. Adduction	B. Return to the anatomic position from lateral flexion or abduction, reduction
VI. Shoulder Girdle	
A. Lateral tilt	A. Slight rotation of the scapula around its vertical axis
B. Reduction of upward tilt	B. Return to the anatomic position from upward or forward tilt, backward tilt
C. Scapular abduction	C. Protraction
D. Scapular adduction	D. Retraction
E. Upward tilt	E. Backward protruding of the inferior angle of the scapula which occurs with shoulder joint hyperextension in some individuals, forward tilt
VII. Shoulder Joint	
A. Horizontal extension	A. Horizontal abduction
B. Horizontal flexion	B. Horizontal adduction
C. Hyperflexion	C. Flexion beyond 180 degrees
D. Inward rotation	D. Medial rotation
E. Outward rotation	E. Lateral rotation
VIII. Radio-ulnar Joints	
A. Lower arm lateral rotation	A. Supination
B. Lower arm medial rotation	B. Pronation
IX. Wrist Joint	
A. Wrist abduction	A. Radial flexion, radial deviation
B. Wrist adduction	B. Ulnar flexion, ulnar deviation
X. Thumb-Carpometacarpal Joint	
A. Hyperadduction	A. Posterior motion at right angles to the hand
B. Hyperflexion	B. Medial motion from a position of slight abduction
C. Reposition	C. Return to the anatomic position following opposition

Basic Concepts of Anatomic Kinesiology

Recommended Reading

American Academy of Orthopaedic Surgeons. *Joint Motion Method of Measuring and Recording.* Chicago: American Academy of Orthopaedic Surgeons, 1965.

Cochran, Alastair and Stobbs, John. *The Search for the Perfect Swing.* New York: J. B. Lippincott Company, 1968.

Logan, Gene A. and McKinney, Wayne C. *Kinesiology.* Dubuque, Iowa: Wm. C. Brown Company Publishers, 1970.

Fundamentals of Muscle Functions

The major purpose of this chapter is to assist the reader in gaining insight regarding the study of muscle functions. The spatial relationship concept is presented to help the student understand the interrelationships which exist between and among the relative position of the muscle-tendon in relation to its joint center, leverage, and joint motion. Muscle action terminology with synonyms is presented to help communicate muscle function and continue to develop the student's vocabulary in the area of anatomic kinesiology. Biomechanic and kinesiology vocabularies are both needed if motions are to be effectively analyzed, described, and articulated. Finally, fundamental concepts of muscle contraction are presented. The emphasis is placed on describing the basic differences between concentric and eccentric contractions and how they function in causing and/or controlling sequential joint motion.

For centuries the functioning of muscles has intrigued scholars. Some of the techniques for studying muscle action have withstood the test of time, and they are still functional today. Fortunately, advances in electronics have greatly improved the "scientific hardware" to study muscle function. This has led to new knowledge and helped to validate or refute observations made in the past.

The study of muscle function through dissection techniques has been practiced for only about six hundred years. In the earlier stages the joints of the cadavers were moved through their ranges of motion in order to determine muscle actions. This method of determining muscle function has several limitations, because cadaver joint motions bear little resemblance to the motions observed in the joints of the living human being.

Another technique for studying muscle function is to study muscle action by omission. In certain pathologic conditions, such as poliomyelitis, some muscles become paralyzed while others remain normal. Observation of the loss of function at a given joint is an indicator of what the muscle does when its neurologic mechanisms are intact in the normal individual. The major limitation in studying muscle action by omission is that normal musculature in the damaged area or near the damged joint will have a tendency to supplement or replace the loss of function.

Determining the action of a muscle or muscle group by using the sense of touch is known as palpation. The major limitation of this technique is that the only muscles which can be palpated are the superficial muscles located immediately under the surface of the skin. Some of the deeper muscles can be palpated on extremely lean individuals, but most deep muscles, even on people of lean body mass, are too difficult to palpate. Even with this limitation, palpation is recommended as a study technique for the student. This will help the individual conceptualize spatial relationships and to become more aware of muscle contractions. Since each student has a pair of all the muscles under study, each person is a potential motion laboratory for palpation, i.e., muscles can be palpated on one's self. However, using the palpation technique with a consenting adult has the infinite advantage of making the study of muscle function more meaningful, lively, and sensual!

The most accurate method of studying muscle function is by using electromyographic (EMG) techniques. The electromyograph provides a direct electronic readout (by means of graph or oscilloscope) of the amount and degree of muscle-action potentials occurring during the contraction of a muscle. These muscle-action potentials are perceived by electrodes either placed on the skin or implanted with needles in the muscle fibers. Both of these electrode techniques have limitations and tend to be impractical at times, especially when one is trying to study an actual ballistic performance.

Skin electrodes have a tendency periodically to record many movement artifacts. These are unwanted recordings of electrical activity from the surrounding area, instead of being limited to the muscle under study; consequently, it can be difficult in a complex movement analysis to be certain that the electromyographic data are accurate and valid. It should be noted, however, that there is very little problem with this technique when analyzing simple joint motions in a clinic or laboratory situation.

Needle electrodes can be more accurate, but they introduce trauma and psychologic factors which tend to inhibit the performer or human subject. For example, there are not many performers, who would serve tennis balls as hard as possible with needle electrodes implanted in their pectoralis major muscles! This simple problem of performing while wired to a machine is also a limitation. However, electromyographic analysis employed in a laboratory situation to study motion problems is the most scientific method for determining the functions of muscles.

Spatial
Relationship
Concept

The comprehension of muscle function can be enhanced by thoroughly understanding the spatial relationship concept. The essence of this concept is knowing the literal location or spatial relationship of each muscle-tendon to the center of the joint(s) it crosses. If this is known for each muscle, it becomes much easier to classify and identify muscles according to their common, aggregate motion function at each joint. *The spatial relationship concept is the most important set of ideas to help the student gain an initial understanding of muscle function.*

The spatial relationship concept is based upon a synthesis of factual background material derived from six related areas. These basic facts, when synthesized, will help the student understand this concept. As a consequence, the student will be able to understand the leverage of muscles and to deduce the functions of most of them, either independently or in their common aggregate action to produce joint motion.

Fundamentals of Muscle Functions

First, the material presented in Chapters 2, 5, 6 and 7 relating to skeletal landmarks and articulations must be known thoroughly. The exact locations of bone landmarks and their relationships to surrounding anatomic parts are essential to the subsequent understanding of the position or route a muscle-tendon may take as it attaches to the skeletal landmark or passes over the articulation. For example, the tibial tuberosity is an important anatomic landmark shown in figures 5.13 and 5.14. It is located on the anterior, superior aspect of the tibia inferior to the center of the knee joint. It is an important landmark for the quadricep femoris tendon, because that is where concentric contraction force is applied to extend the knee against resistance. (Muscle contraction is discussed later in this chapter.) If one does not know the exact location of the tibial tuberosity and its relationship to the knee, it will be impossible to comprehend the functions of the muscles which attach there. The same is true in regard to all of the other important anatomic structures described in Chapters 2, 5, 6 and 7.

Second, it is important to have a functional grasp of the planes of motion. The axes of rotation and planes at each joint must be thoroughly understood. This knowledge, and knowing the locations of specific bony landmarks, help one determine the centers of joints as accurately as possible; *the body segments move through these planes.*

Third, the motions possible through the planes at each joint must be known. Combined with this knowledge there should be awareness of the structure of each joint. As an example, one would not expect lateral plane abductions and adductions of the elbow joint solely because of its bone structure (figure 7.24); *joint motions cause body segments to move through the planes.*

Fourth, the line of pull for each muscle or muscle group should be known. This line of pull is from attachment to attachment. (This material is covered in Chapters 5, 6 and 7.) The *exact* positions of the muscle's tendon into its proximal and distal attachment points must be known in order to understand the leverage needed for muscle functions. As an example, if one thought the ischial tuberosities (figure 5.21) were anterior and superior on the pelvic girdle, it would be impossible to understand "hamstring" muscle group function, because those muscles are all attached there proximally. (The ischial tuberosities are posterior and inferior on the pelvic girdle.) The line of pull for leverage and force application must be known to deduce the function of a contracting muscle.

Fifth, along with the line of pull of the muscle, one needs to identify the exact position of the tendon in relation to the center of the joint(s) it crosses. *This is the most important factor within the spatial relationship concept in terms of determining muscle function.* With the exception of the joints within the feet, tendons can be described as passing the center of the joint anteriorly, posteriorly, laterally or medially. For the interphalangeal and metatarsophalangeal joints of the feet, the terms superior and inferior are more appropriate than anterior and posterior. One would describe the biceps femoris tendon in figure 5.18 as having a *posterior/lateral spatial relationship* to the center of the knee joint. The quadriceps femoris tendon has an *anterior spatial relationship* to the knee joint's axis of rotation. These relationships are highly significant in terms of the motion functions of these muscles.

Finally, to fully comprehend the spatial relationship concept and motion at joints, the reader must understand internal force applications at the muscle-tendon attachments. *Internal force is a result of isotonic and isometric (static) contractions of muscles.* The differences between concentric and eccentric (isotonic) contractions are presented later in this chapter.

Figure 4.1 is presented as an example, to show the spatial relationships of all the muscle-tendons which cross the knee joint. The following are some basic considerations related to the spatial relationship concept as applied to this joint:

1. *Important skeletal landmarks:* Tibia, fibula, tibial tuberosity, medial and lateral condyles of the tibia, fibula head

2. *Planes of motion:* Anteroposterior and transverse

3. *Motions possible:* Flexion and extension; medial rotation and lateral rotation when the knee is flexed

4. *Lines of pull:* The quadriceps femoris group is from the anterior, inferior iliac spine (rectus femoris) and superior femur (vastii) proximally (figure 5.18) to the tibial tuberosity distally. The "hamstrings" (semitendinosus, semimembranosus, and biceps femoris) arise on the ischial tuberosities proximally (figure 5.18). The biceps femoris attaches distally on the head of the fibula. The semimembranosus attaches distally on the posterior and superior aspect of the medial condyle of the tibia, and the semitendinosus attaches distally on the medial aspect of the shaft of the tibia inferior to the medial condyle. The sartorius has a line of pull from the anterior, superior iliac spine (figure 5.27) to the same location distally as cited for the semitendinosus. The gracilis has the same distal attachment area as the sartorius and semitendinosus, but its proximal attachment is inferior on the symphysis pubis (figure 5.28). The small popliteus extends diagonally from the lateral condyle of the femur to the posterior and medial aspect of the tibia. The gastrocnemius attaches proximally to the posterior, inferior femur condyles, and extends to the posterior calceneus distally by way of the Achilles tendon (figure 5.18).

Figure 4.1 Spatial relationships of the right knee joint musculature— cross sectional superior view.

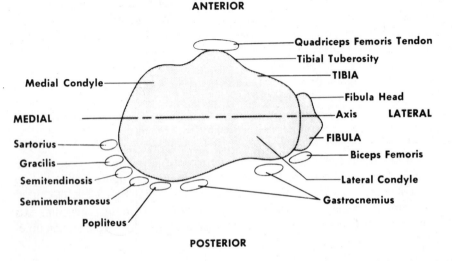

ANTERIOR

Quadriceps Femoris Tendon

Tibial Tuberosity

TIBIA

Medial Condyle

Fibula Head

MEDIAL

Axis   LATERAL

Sartorius

FIBULA

Gracilis

Biceps Femoris

Semitendinosis

Semimembranosus

Lateral Condyle

Popliteus

Gastrocnemius

POSTERIOR

Fundamentals of Muscle Functions

5. *Spatial relationship and function:* For purposes of this example, assume that each of these muscle groups undergoes *concentric contraction.* With reference to figure 4.1, the muscles located *anterior* to the joint would cause the knee to extend against resistance. The muscles with a *posterior* spatial relationship to the axis of rotation would flex the knee against resistance. With the knee flexed, the *lateral* muscle (biceps femoris) would be capable of laterally rotating the knee joint. Conversely, the five muscles with *medial* spatial relationships would be the muscles most involved for medial rotation of the knee joint.

Several of the large knee muscles and their tendons can be palpated. It is recommended that the reader while seated palpate the concentrically contracting quadriceps femoris muscle group with its *anterior spatial relationship* to the joint as the knees are extended. Stand with the weight on the left foot, palpate the medial and lateral tendons of the muscles posterior to the right knee as it is flexed. Their functions related to the *posterior spatial relationship* will become obvious. While seated with the knee flexed, laterally rotate the knee while holding the biceps femoris tendon near its distal attachment lateral on the fibula head. The importance of the lateral spatial relationship should be noted. The large medial and posterior tendons of the semimembranosus and semitendinosus muscles should be palpated during medial rotation of the knee. Their *medial spatial relationships* to the knee joint provide them with the leverage position to rotate the knee medially when it is flexed. The muscles and muscle groups palpated in these motions are undergoing concentric contractions to elicit the joint motions observed.

Muscle contraction will be presented later in this chapter, but the differences between concentric and eccentric contraction need to be described in this spatial relationship example. This will be done by using the motions of knee flexion and extension only. The reader should refer to figure 4.1:

1. All muscles with an anterior spatial relationship will extend the knee against resistance when they contract concentrically.

2. All muscles with an anterior spatial relationship will control the extent and velocity of knee flexion with gravity when they contract eccentrically.

3. All muscles with a posterior spatial relationship will flex the knee against resistance when they contract concentrically.

4. All muscles with a posterior spatial relationship will control the extent and velocity of extension with gravity when they contract eccentrically.
The *reversal of muscle function concept* is discussed in detail in this chapter following the presentation of muscle contraction.

It should be noted that a drawing of a joint such as figure 4.1 can serve only as a *guide* for study purposes. The exact location or spatial relationship of a tendon as it passes a joint is dependent upon the precise point where the transverse drawing or section is made. As an example, the tendons (or part of the muscle) of the semimembranosus and semitendinosus change their relationship to each other as they go to their distal attachments. This cannot be shown in an illustration; therefore, a drawing such as figure 4.1 is merely a guide indicating general spatial relationships; it is not a precise anatomic diagram.

Ultimately the reader must study the bones, skeletal landmarks, proximal and distal muscle attachment points, spatial relationships, and contracting muscles on both a fully articulated skeleton and a consenting adult in order to comprehend joint motions. Rote memory is essential during the early stages of learning anatomic terminology and structures. However, if kinesiology is to become a functional "working tool" for the physical educator, the subject matter must be learned, utilized, and understood to the point of thorough comprehension if it is to be retained. *Comprehension goes beyond memorization!*

Relying upon the study of muscle function by using the spatial relationship concept places the emphasis on logical deduction and conceptualization rather than rote memory. If the reader uses this approach to the study of muscle function, better retention of the subject matter and subsequent application of kinesiologic theory in practice are more likely. Obviously, this has positive implications regarding the improvement of performance, conditioning of athletes, and development of health related fitness.

Muscle Action Terminology

Several terms to better describe muscle actions are introduced here. *The objective in using these terms is to communicate important concepts of anatomic kinesiology.* Some of the traditional muscle action terms tend to be misleading and confusing. The terms are presented with their synonyms from traditional scientific terminology, because the reader must know both.

Muscles MOST Involved

Muscles which cause or control joint motion through a specified plane of motion are referred to in this book as *muscles most involved* (MMI) (Wallis and Logan 1964). The muscles most involved have a spatial relationship to each joint they cross. As a consequence, their lines of pull lie within or parallel to the plane of motion through which the body part will ultimately travel when contraction occurs within the MMI.

The term *muscles most involved* describes muscle function, and it can easily be integrated into the physical educator's regular vocabulary. As an example, when explaining the "biceps curl" exercise to a secondary school physical education class, the instructor may state, "The muscles most involved in this exercise are located in the front of your upper arm, and they are the biceps brachii, brachialis, and brachioradialis." The term fits easily into a common usage context, and that type of anatomic information can be passed on to high school students without any difficulty *if* the physical educator presents the ideas in a logical and progressive sequence.

As noted above, an emphasis is placed on the word *most* within the term *muscles most involved*. The reason for this is that this term *(MMI)* is delimited to include only those muscles involved in causing or controlling joint motion which have the best strength, size, and/or leverage potential. Several muscles may collaborate in producing any joint motion, but not all muscles are *most* involved. For example, by virtue of their size, strength, and leverage advantages, the MMI for planter flexion at the ankle joint are the gastrocnemius and soleus. There are five other muscles with posterior spatial relationships to the ankle joint, which contribute force for plantar flexion, but they are not defined as MMI. They do not meet the aforementioned criteria for MMI. The MMI for causing or controlling joint motions, as well as their assistant muscles, are all discussed in Chapters 5, 6 and 7.

The expression MMI is used at any time when a muscle group causes or controls a joint motion. It was noted that the MMI for elbow flexion are the biceps brachii, brachialis, and brachioradialis during a "biceps curl" exercise. This would denote a concentric contraction, because the elbow is being *flexed against gravity* plus the resistances of the weight of the arm and barbell. If the elbow is extended *slowly* back to the anatomic position from extreme flexion, the MMI for this elbow *extension with gravity* are the eccentrically contracting biceps brachii, brachialis, and brachioradialis. These are the muscles most involved for controlling the speed and extent of motion within the extending elbow.

The reader is encouraged to consult other kinesiology textbooks and to keep abreast of kinesiology literature. Therefore, it is important to know that the most common synonyms for "muscles most involved" are the terms *agonists* and *prime movers*. These are functional terms, but they are not as descriptive as MMI. Nor are they as easy to integrate into common usage to describe muscle functions. One of the principal objectives of teaching is to communicate ideas to students in an understandable context. Physical educators know that to teach a given skill to a class it may have to be verbalized and demonstrated in several different ways before learning or true communication occurs. Thus, it is a good idea to learn how to phrase and rephrase any concept which an individual has learned or is learning. This facilitates comprehension.

The term *contralateral* as used in this book means *opposite side. Contralateral muscles are the muscles located on the opposite side of the joint from the muscles most involved for causing or controlling joint motion.* While the muscles most involved are contracting, the contralateral muscles are being reciprocally innervated and relieved of some tension to allow the MMI to move the body segment through the desired range of motion. The extent of the "tension release" depends upon the joint velocity and tension desired for skill execution.

Contralateral Muscles

That type of cooperative interaction between the muscle groups results in functional, sequential muscle action and skilled movement. There must be *teamwork between* the muscles most involved and their contralateral muscles. This connotation of teamwork is lacking in the term *antagonists,* which is the synonym for *contralateral muscles.* Antagonistic muscle action connotes that those muscles are *working against* the agonists or muscles most involved for a joint motion. The opposite, of course, is true, because the contralateral muscle group *cooperates with* the muscles most involved to produce the desired motion. In the "biceps curl" example, the muscles most involved for elbow flexion against resistance have an anterior spatial relationship to the elbow joint. Therefore, the contralateral muscle has a posterior spatial relationship to the same joint (triceps brachii). *The term contralateral indicates location, once the MMI have been determined.* It also denotes cooperative function with the MMI.

The functions of the contralateral muscles often vary, depending upon the skills being performed. Two major functions are: (1) control of the speed of motion within the muscles most involved and (2) protection of the joint.

Guiding Muscles

The main function of guiding muscles is to rule out undesired motions. Guiding muscles is a descriptive term which connotes the function of these muscles, i.e., they guide a body segment through the desired plane of motion providing the necessary force and counterforce to eliminate extraneous motion. Synonyms for *guiding muscles* are *synergists, helping synergists,* and *true synergists.* These are excellent terms, but they do not fully indicate the function of these muscles.

Guiding muscles are located parallel to the plane of motion through which the body segment will move. Most joints have several different motion capabilities. When a desired motion through a specific plane of motion is performed, the guiding muscles provide a balance of pull on either side of the plane of motion. Therefore, the muscles most involved are allowed to perform as directed volitionally by the central nervous system. This does not rule out the possibility that guiding muscles may also be directly involved in a portion of the volitional movement. As an example, several motions can be performed at the hip joint through four planes of motion, but it is possible to move through one plane of motion at a time. This requires teamwork among various muscle groups. When hip flexion through the anteroposterior plane of motion is desired, this exact movement would be impossible if there were not muscles other than the muscles most involved for hip flexion to guide the total action. The abductors and adductors of the hip work as a team with the hip flexors to rule out undesired rotary actions which might occur at the hip joint during flexion.

The hip flexors (MMI) against resistance have an anterior spatial relationship to the hip joint; therefore, the contralateral muscles are located posteriorly. The guiding muscles, in this example, have medial and lateral spatial relationships to the hip joint, i.e., they lie parallel to the anteroposterior plane of motion. They provide a force on one side of the plane and an equal counterforce on the opposite side to rule out transverse, diagonal and lateral plane motions.

Guiding muscle force is not a factor when body segments are stable during motions or exercise. As an example, no guiding muscle force is needed in the shoulder joints while doing a bench press (figure 3.11). The hands are stable while grasping the bar. On the other hand, a tennis player hitting a forehand stroke (velocity and accuracy factors) would need considerable guiding muscle force-counterforce to execute the skill.

Stabilizing Muscles

Stabilizing muscles surrounding a joint or body part contract to fixate the area in order to enable another limb or body segment to exert force and move. The terms *stabilizers* and *fixators* are synonyms; they are used commonly in anatomic kinesiology literature.

Stabilizing muscles undergo varying degrees of static contraction when they fixate a joint or body segment. The status of the stabilizing muscles undergoing static contraction can range from mild tension to extreme contractions with ac-

Fundamentals of Muscle Functions

Figure 4.2 Isometric abdominal strengthening exercise. Stabilization of the lumbar-thoracic spine (Jan Stevenson).

companying tremors. To illustrate this, the reader can stabilize the wrist and radio-ulnar joints with the elbow flexed. By gripping the hand with minimum to maximum force, the changes in static contraction and stability can be noted in the extrinsic hand musculature and affected joints respectively. There should be noticeable tremor in the forearm musculature if the interphalangeal and meta-carpophalangeal flexions are maximal.

Figure 4.2 shows an isometric (static) abdominal strengthening exercise which serves as another example of static contraction and stabilizing muscles. The cervical spine is flexed concentrically, moving the head-neck segment through the anteroposterior plane of motion as seen in figure 4.2. In order for the muscles most involved for cervical flexion against resistance to function, there must be a stabilized segment at which to exert force. The muscles surrounding the lumbar-thoracic spine segment are contracting statically or isometrically to keep this area of the spinal column in a nonmoving or stabilized state. Flexion of the cervical spine tends to add to the resistance. The person exercising can volitionally add to the extent of the static contraction in the abdominal musculature as the head-neck is flexed. For purposes of the strength exercise, the static contraction is held for 10–15 seconds and repeated several times. This static contraction and stabilizing effect of the abdominals can be palpated.

Stabilized body segments are essential for sequential joint motion. As an example, in some upper and lower limb motions, the lumbar-thoracic spine must be stabilized. This is seen during the "jumping jack" exercise (figure 3.7), where the shoulder and hip joints are abducted and adducted through the lateral plane of motion. The muscles on the anterior, posterior, medial, and lateral aspects of the lumbar-thoracic spine are statically contracting to hold the segment stable so the limbs will have a point at which to exert force. This type of force-counterforce situation is an ongoing process in the human body during motion.

When muscles undergo isotonic contraction, they exert equal force on the proximal and distal attachments, i.e., there is force to pull the muscle attachments toward each other. This is particularly evident in a joint such as the elbow. The part that moves is entirely dependent upon the relative stability of the two attachments and the objective of the motion (volitional control). Usually there will be momentary static contraction by stabilizers at the proximal end of a limb muscle. This allows the force of the muscle contraction to be exerted on the bony landmark (lever) at the distal attachment. This type of stability occurs in statically contracting muscles literally within nanoseconds, whereas stabilized body segments, such as shown in figure 4.2, can be maintained in a static or nonmoving state for relatively long periods of time. Finally, the degree of stabilizing force during muscle contraction can range from mild to extreme. As noted above, one does not have to maximally contract all muscles surrounding a joint to stabilize it.

In connection with joint stability, the effects of gravity and weight should also be considered. A joint can be placed in a stabilized status in a weight bearing situation with minimal or no static contraction by the muscles surrounding the joint. The reader should try standing in a balanced state with the body weight resting on the left leg only. Flex the right hip joint forty-five degrees, and keep the right knee joint extended. The left knee is extended and stable due to the weight bearing situation, primarily if equilibrium is maintained. The right knee is stabilized in the extended position by virtue of the static contraction of muscles surrounding the joint. In this example, the muscles with an anterior spatial relationship to the knee are exerting more force than their posterior counterparts. This is due to the gravitational situation and the motion capabilities within the knee joint through the anteroposterior plane.

| Dynamic Stabilization | As described above, stabilization is regarded in a nonmoving context. This type of stabilization is observed in many performances such as calisthenic exercises, weight training, and weight lifting. However, this type of nonmoving stabilization is the exception rather than the rule in most sport or movement activities. Stabilization must be regarded as a relative concept, because to a greater or lesser degree movement is always taking place within joints considered to be stabilized. Owing to this factor, a different conceptual approach to stabilization was introduced by Logan and McKinney in 1970. This concept is known as *dynamic stability*. This appears paradoxical since the terms *dynamic* and *stability* have opposite meanings. This is one of many paradoxes seen in muscle functions. This one is called the *Lomac Paradox*. |

A consideration of dynamic stabilization is essential to the understanding of the timing aspect of sequential movements. Dynamic stabilization occurs within stabilizing muscle groups during a sport skill in instantaneous and fleeting bursts of muscle contractions. There are three major functions of dynamic stabilization, which may occur independently of each other or interrelated during the performance of sport skills. These functions are: (1) to maintain the position of the body against gravity, other external forces, and internal force generated by muscle contractions; (2) to prevent trauma within joints due to elongation or compression of the joint structure; or (3) to provide a base from which the original force needed to perform a sport skill can be initiated and transferred from one body segment to another body part in the desired direction or sequence of the skill.

One way to gain an understanding of the concept of dynamic stabilization is to compare it to nonmoving stabilization. For example, the overhead press is a performance skill required in the sport of weight lifting. The athlete is required to bring the weight to the shoulders and then press the barbell over the head by flexing the shoulder joints and extending the elbows. In order for these forces to be exerted vertically, there must be nonmoving stabilization within the musculature surrounding the lumbar-thoracic spine and pelvic girdle as well as the hip, knee, and ankle joints. Assuming a maximum weight load being moved vertically, the level of static contraction in the stabilizing muscles would be great. If it were not, the upper limbs would not have a point at which to exert enough force to complete a good lift. There would be a strong combination of weight bearing and static contraction stability in this sport example.

What would happen to a distance runner in a 10,000 meter run, as an example, if he or she tried to simulate the same type of lumbar-thoracic spine stability used in the execution of an overhead press by a weight lifter? Myologically and physiologically, the results would be disastrous! The runner would not be able to perform, because that extent of stabilization in the lumbar-thoracic spine is not needed during the locomotor performance. Nevertheless, there is need for some degree of stabilization within the pelvic girdle and lumbar-thoracic spine of the runner, because of the forces and counterforces between the runner and the earth. Also, there is a stabilization relationship between the moving upper limbs, shoulder girdle, and trunk. The three body segments (pelvic girdle, lumbar-thoracic spine, and shoulder girdle) are moving under tension. They are providing critical concurrent motions with the hip and shoulder joint motions so the runner can perform in a skilled manner. The *moving tension* within these body segments means that they are providing *dynamic stability* so the upper and lower limbs will have points at which to exert force.

Dynamic stability is provided by the contraction tension within both the muscles most involved and the contralateral muscles, for movements within segments such as the pelvic girdle, lumbar spine, and shoulder girdle. These segments usually are moving slowly under tension, even during ballistic skills. To continue with the runner as an example, running involves concurrent lateral and transverse pelvic girdle rotations, with hip joint flexions and extensions. In the skilled runner those four motions must be intricately synchronized from several aspects. Compared to the hip joint motions, the pelvic segment has a low velocity. Therefore, the lateral and transverse pelvic girdle motions are of the slow tension variety, even for a sprinter. This means that the muscles most involved for these pelvic rotations, plus their contralateral muscles, are all under considerable contractile tension. *As a result, both sets of muscles are the dynamic stabilizers for the pelvic girdle.*

Finally, one should also remember that some stabilizing force is being contributed to the pelvic girdle in the process of running, by virtue of the momentary stability to the proximal attachments of the hip flexors and extensors. They are located anteriorly and posteriorly on the pelvic girdle.

For most skilled performers, dynamic stability is essential within the pelvic girdle, lumbar-thoracic spine, and shoulder girdle. This allows these segments to be optimally stable without impeding motion elsewhere or disrupting physiologic functions. Motions under tension concurrently within those segments are necessary, and complement the movements within upper and lower limbs.

Muscle Contraction	*The specific function of a muscle is to contract and develop tension or internal force.* Movement occurs due to changes in tension within muscle groups acting on the bony levers and articulations of the skeletal system.

How does a muscle group contract? *A muscle tends to pull from its proximal and distal attchments toward its center.* For movement to occur within a joint at either end of the muscle group contracting, one end or attachment of the muscle group must be stabilized momentarily. The range through which a muscle can change its length is known as its *amplitude.*

There are two general classifications of muscle contraction: (1) *isometric* and (2) *isotonic.* Isometric contraction is static contraction, that is, tension is developed within the muscles, but no change in length occurs. As a result, no motion occurs within the joints or body segment involved. Muscle groups surrounding a body segment often undergo static contraction to stabilize the area; therefore, isotonically contracting muscles will have a place at which to exert force and cause effective movement.

Isotonic contraction involves both shortening and lengthening of muscle fibers under varying degrees of tension. *Concentric contraction* takes place when the muscle fibers are *shortened,* and *eccentric contraction* occurs when the muscle tissue undergoes *lengthening* under tension. It is necessary to understand the differences between these two types of isotonic contractions. *The determination of which muscles are actually most involved in activating and controlling joint motion at all times during a performance of a skill is dependent upon knowing the differences between concentric and eccentric contractions.*

Concentric Contraction	When innervation to the muscle causes the fibers to shorten, this phenomenon is known as concentric contraction. The effect of this type of contraction is to *cause motion* at a joint *against gravity* and other resistive external forces.

Figures 4.3 and 4.4 provide a familiar example of concentric contraction within a muscle group (the abdominals). The critical joint motion in figure 4.4 is lumbar flexion through the cardinal anteroposterior plane of motion. This motion at the lumbar spine has been performed against gravity. Therefore, the muscles most involved in causing this motion have been contracted concentrically. Knowing the motions possible within the lumbar spine, it can be deduced that the muscles most involved for lumbar flexion against gravity have an anterior spatial relationship to the joints within the lumbar spine. This can be verified by palpation. The pelvic girdle in this example has been dynamically stabilized by the tension developed during contraction of both the muscles most involved and the contralateral muscles for anterior pelvic girdle rotation. The pelvis must rotate anteriorily concurrent with lumbar flexion.

This dynamic stabilization is necessary, because the concentrically contracting abdominals need a stable point at which to exert force so the motion can be completed by bringing the thoracic-rib cage forward through the anteroposterior plane of motion.

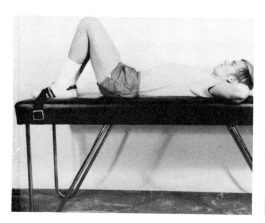

Figure 4.3 Flexed-knee sit-up—start.

Figure 4.4 Flexed-knee sit-up—finish. Concentric contraction of the abdominals resulting in lumbar flexion.

Eccentric contraction of muscle fibers maintains tension during the lengthening process; this is necessary for equilibrium and for skilled performance. In addition, this contraction protects joint structure and other body tissues. The most common effect of eccentric contraction is to *control motion* at a joint moving *with gravity*.

Eccentric Contraction

The reader should compare and contrast figures 4.5 and 4.6 with figures 4.3 and 4.4. In both sets of figures an exercise has been performed which involves lumbar flexion. Although lumbar flexion is a major joint motion in both exercises, the muscles most involved for causing or controlling lumbar flexion in these two examples are entirely different. The lumbar flexion shown in figure 4.6 has been performed through approximately 90 degrees *with gravity*. This means that the muscles most involved for controlling the extent and velocity of this lumbar flexion have been contracted eccentrically. Knowing the motions possible, and the structure of the lumbar spine, it is not difficult to deduce that the muscles most involved for lumbar flexion with gravity have a posterior spatial relationship to the lumbar spine. It is the erector spinae muscle group located posteriorly, not the abdominals, which are the muscles most involved in controlling the extent and velocity of lumbar flexion which occurs between figures 4.5 and 4.6.

It should be noted that the exercise shown in figures 4.5 and 4.6 is not the safest way to stretch the musculature posterior to the knee and hip joints. This lumbar flexion and extension from a standing position is relatively unstable, and the moving body segments can be moved with enough velocity to cause tissue trauma at the attachment points of the muscles on the pelvic girdle and near the knee. Exercises of this type should be done by placing the exerciser in a relatively stable position; this avoids equilibrium problems. The exercise shown in figures 4.7 and 4.8 is much safer than the one shown in figures 4.5 and 4.6, because the exerciser is in a more stable position.

Figure 4.5 Hamstring stretch—start (Billig).

Figure 4.6 Hamstring stretch—finish. Eccentric contraction of the erector spinae controlling lumbar flexion.

Figure 4.7 Hamstring stretch, sitting—start (Jan Stevenson).

Figure 4.8 Hamstring stretch, sitting—finish.

Fundamentals of Muscle Functions

One of the most essential functions of eccentric contraction is to help maintain equilibrium. If it is desired to control a moving body segment(s) through a range of motion within the gravitational field, then eccentrically induced tension must be maintained in the muscles most involved for the motion(s). If this were not the case, gravitational force would take over and move the body or body segment toward the earth and out of control. Gravity is a powerful external force, and skilled performers use gravity to their advantage. Persons unskilled in neuromuscular skills tend to allow gravity to control them. Eccentric contraction (internal force) is utilized extensively by athletes to control the extent and velocity of their motions with gravity. Evidently, this is a critical performance factor.

Eccentric muscle control of body segments is an absolute necessity in an event such as balance beam competition in women's gymnastics. The beam is only 10.16 centimeters wide; consequently, the margin of error related to control or loss of equilibrium is minute. The skilled gymnast must execute all skills aesthetically and with high-risk factors, an accomplishment which is not readily combined in a beam routine. The Olympic gymnast in figure 4.9 is executing a one-arm, front walkover. She has excellent flexibility and extension of several body segments; these are qualities the judges look for during a routine. As she brings the right foot down to the beam, the lumbar-thoracic spine will be hyperextended with gravity. This motion must be controlled by the eccentrically contracting abdominals to help the gymnast maintain equilibrium and the aes-

Figure 4.9 Kolleen Casey, Olympic gymnast for the U.S.A. and Southwest Missouri State University, executes a one-arm, front walkover on the beam. Eccentric muscle control for movement with gravity is of utmost importance in this event for equilibrium and aesthetic purposes. *Courtesy SMSU Public Information Office.*

thetic aspect of the performance. The abdominals are the muscles most involved in controlling the extent and velocity of the hyperextending lumbar-thoracic spine with gravity in this example; and the coach should make the athlete aware of this. Figure 6.10 in this text, as well as other anatomy and kinesiology books, indicate that the erector spinae are the muscles most involved for lumbar-thoracic extension-hyperextension. This myologic situation is only true when the motion is against gravity and the erector spinae are required to contract concentrically. One must observe joint motions carefully to determine the nature of the causing or controlling internal forces.

Reversal of Muscle Function Concept	One should never assume that a joint motion is always caused or controlled by the same muscle group. To make this assumption would be to overlook the differences between concentric and eccentric contractions, the effect of gravity, and body positioning, either in terms of bases of support or nonsupported positions in air. Determination of muscle function must take cognizance of the situation prevailing at each moment during the analysis of any performance.

Usually, anatomy and kinesiology textbooks describe muscle functions in terms of their concentric contraction capabilities only. Such descriptions are myologically correct only half of the time. They are correct when the motion is against gravity, but incorrect when the body segment is moved with assistance from gravity. A muscle group which is ordinarily described in anatomy and kinesiology textbooks as performing a given function can also contract to initiate control of the exactly opposite motion. This is the essence of the reversal of muscle function. As an example of this, the muscles with a posterior spatial relationship to the ankle joint cause plantar flexion when they contract concentrically and move the ankle joint against gravity. If dorsiflexion of the ankle is to be controlled slowly through the anteroposterior plane of motion with gravity, this motion is the result of eccentric contraction by the same group of muscles with a posterior spatial relationship to the ankle joint.

Most anatomy and kinesiology texts indicate that the prime movers for knee flexion are the semitendinosus, semimembranosus, and biceps femoris (hamstrings). These muscles are assisted by the gracilis, popliteus, gastrocnemius, and sartorius. This is true if resistance must be overcome, but there are situations where these muscles would not be the muscles most involved in controlling the extent of knee flexion. Such a case occurs when performing 90-degree knee flexion from a standing position. The effect of the force of gravity must be considered, because the mass of the body is being lowered primarily by gravitational pull. Consequently, the muscles most involved in controlling the velocity and extent or degree of knee flexion would be the quadriceps femoris muscle group on the anterior aspect of the thigh. By contracting eccentrically, the quadriceps femoris muscle group is actually controlling the degree of the knee flexion as well as the velocity. This is not the familiar function listed for the quadriceps femoris muscle group. (Anatomy books simply list this group as knee extensors.) They are receiving cooperative action by the "hamstrings" on the opposite side of the thigh. Therefore, in this example, there is a reversal of traditional muscle function by the two major muscle groups surrounding the knee joint. This is just

Fundamentals of Muscle Functions

another way to visualize the roles of concentric and eccentric contractions in a movement or performance. The reader should perform these movements and palpate the muscles.

Another example will help illustrate the effect of gravity upon the muscles most involved for a given joint action. In the knee joint example (above) it was noted that the body was being lowered in the direction of gravitational pull during knee flexion. What would happen to the muscle groups involved if the knee were flexed against gravity? With the individual standing on the right foot only, he or she is asked to flex the left knee through its full range of motion. Which muscles would be most involved to produce internal force for knee flexion? The answer is the "hamstring" muscle group, because they are moving the knee through a range of flexion against the resistance caused by the pull of gravity and weight of the lower leg. The muscles on the opposite or anterior side of the thigh, the quadriceps femoris muscle group, are being reciprocally innervated to allow the concentric contraction of the "hamstrings" to move the knee joint through the desired amount of flexion. In this example, the "hamstrings" are performing their familiar function of flexing the knee against resistance. Obviously, they are also capable, of controlling the extent and velocity of knee extension while contracting eccentrically.

Figure 4.10 shows two highly skilled American football players in action. The tailback (42) is following his fullback or blocking back (36) in this situation. Both knees of the blocking back (36) are being flexed, but there are distinct myologic differences between the muscles most involved for controlling and causing

Figure 4.10 Ricky Bell (42), USC tailback, follows his blocking back, Mosi Tatupu (36). Tatupu's left knee is flexed *against* gravity, and his right knee is being flexed *with* gravity. *Courtesy USC Athletic News Service.*

those knee flexions. His right knee is being flexed *with gravity*, and the muscles most involved are the eccentrically contracting quadriceps femoris muscle group on the anterior aspect of the knee. The left knee is being flexed *against gravity*, and the muscles most involved are the concentrically contracting "hamstring" muscle group with a posterior spatial relationship to the knee joint.

Muscle Length-Force Ratio

The force exerted by a a muscle is proportional to its length. This principle is reciprocally related to flexibility of joints and eccentric contraction. One can potentially exert a greater force during concentric contraction of a muscle group if the ranges of motion possible at the joints is greater than average. Eccentric contraction controlling joint motion with the gravitational field places the muscles in a stretched state under tension.

The muscle length-force ratio is also related to the force-time relationship principle of biomechanics. That principle states that *the total effect of force on motion of a body is the product of the magnitude of force and time over which it operates.* An increased range of motion in a joint can lead to the potential of greater internal force being generated by the muscle groups most involved for producing a motion, and it also allows for that force to be exerted over a greater time span. The interrelationships between these principles of anatomic kinesiology and biomechanics are important for most skilled performances, especially those of a ballistic nature.

The blocking back (36) in figure 4.10 must exert extreme force during the act of blocking against an opponent in order for the tailback (42) to gain yardage. He is in the process of *preparing* to do this (in figure 4.10) by lowering his center of mass through a series of flexions at the knees, hips, and lumbar-thoracic spine. These motions are with gravity; and this stretches the muscle groups eccentrically controlling these flexions: they are lengthened and placed under considerable tension.

The ultimate force will be applied to the opponent by the blocking back through a sequence of *timed extensions,* primarily at the ankles, knees, hips, and lumbar-thoracic spine. Those extensions would result from the concentric contractions of the muscle groups involved. The total summation of force in this example is greatly enhanced, due to the fact that these same muscle groups were placed on *stretch* by being flexed eccentrically immediately prior to volitional action to contract them concentrically. It is customary in sport skills to place critical force-producing muscle groups on stretch during the *preparation phase* of a performance as shown in figure 4.10. This preparatory stretching of muscles has critical quantitative and qualitative implications for the important *motion phase* of the performance which follows.

Fundamentals of Muscle Functions

Basmajian, J. V. *Muscles Alive: Their Functions Revealed by Electromyography.* Baltimore: Williams and Wilkins Company, 1978.

Evans, F. G. (Ed.). *Biomechanical Studies of the Musculoskeletal System.* Springfield, IL: Charles C Thomas Publisher, 1961.

Frankel, V. H. and Burstein, A. H. *Orthopaedic Biomechanics: The Application of Engineering to the Musculoskeletal System.* Philadelphia: Lea and Febiger, 1970.

Hawley, G. *An Anatomical Analysis of Sports.* New York: A. S. Barnes and Company, Inc., 1940.

Higgins, J. R. *Human Movement: An Integrated Approach.* St. Louis: C. V. Mosby Company, 1977.

Hill, A. V. *Muscular Movement in Man.* New York: McGraw-Hill Book Company, 1927.

Hubbard, Alfred W. "Homokinetics: Muscular Function in Human Movement." *Science and Medicine of Exercise and Sports,* edited by Warren R. Johnson. New York: Harper & Row, Publishers, 1960.

Logan, Gene A. and McKinney, Wayne C. *Anatomic Kinesiology.* Dubuque, Iowa: William C. Brown Company Publishers, 1982.

Steindler, A. *Kinesiology of the Human Body Under Normal and Pathological Conditions.* Springfield, IL: Charles C Thomas, 1955.

Zebas, C. J. "Kinesiologic Review of Muscle Function for Physical Education Majors." *Journal of Health, Physical Education, Recreation,* May, 1977, 60–61.

Recommended Reading

5

Lower Limb Segment Myology

Introduction

Myology is the area of anatomy concerned with the study of muscles. Chapters 5, 6 and 7 are delimited to include only the most basic musculoskeletal fundamentals. These chapters, as well as the previous chapters in Part 2, provide a brief review of anatomic kinesiology for the reader who may have had coursework in that field. For the person without professional preparation in anatomy, kinesiology, or functional myology, this part of the text will provide an introduction to some of the basic elements of anatomic kinesiology. This introduction will: (1) serve as a transition to more thorough coursework in anatomic kinesiology, and (2) provide the anatomic background necessary for integration with biomechanic principles in order to describe human motion accurately and to define motion problems during analyses.

The process of biomechanic analysis must ultimately identify and evaluate all elements and forces related to the motions involved. This goes beyond observations of the effects of external forces such as gravity. The trained observer must be able to make quantitative and qualitative evaluations of the internal forces (muscle contractions) which cause the joint motions. Furthermore, evaluations of total skilled movements as seen in sport must be made to determine the efficiency of interactions between internal and external forces. This is one reason why extensive backgrounds in both anatomic kinesiology and biomechanics are required. This is why subject matter from both academic areas is presented in this introductory textbook in analysis of sport skills. Accurate descriptions of human motion comprise a synthesis of facts and concepts from biomechanics and anatomic kinesiology.

Chapters 5, 6 and 7 present the bones, skeletal landmarks, major joints, and muscle functions which must be observed during biomechanic analysis. From the standpoint of cinematographic analysis, for example, it is imperative that these skeletal landmarks and joint functions be known, in order to locate levers, estimate joint centers, and mark moving body segments. This is necessary in order to make calculations of linear and angular velocities of limbs and body segments, determine acceleration rates, describe force interactions that occur in sport, and communicate this information derived from biomechanics and anatomic kinesiology to professional peers and students.

The myologic discussions in chapters 5, 6 and 7 do not include the important roles contributed by stabilizers, guiding muscles (synergists), and dynamic stabilizers during motion. These are beyond the scope of this book, which places a secondary emphasis on anatomic kinesiology, while its primary focus is on biomechanics. Several excellent anatomic kinesiology textbooks are listed in the bibliography.

As discussed in Chapter 4, there must be a thorough understanding of the differences between concentric and eccentric contractions to describe muscle function of a moving joint. The definitions of isotonic contractions should be reviewed as we start this chapter. Concentric contraction will *cause* joint motion *against gravity*. Eccentric contraction *controls* joint motion *with gravity*. *The figures in Chapters 5, 6 and 7 that list muscle function at major joints are limited to concentric function only.* One must be very observant during analyses to determine whether the body segments are moving with or against gravity. This is indispensable, if accurate myologic or internal force determinations are to be made. Exercise prescriptions are based on the specific myologic demands experienced by the athlete in his or her sport.

The motions possible at the interphalangeal joints of the toes are flexion and extension through anteroposterior planes of motion. These motions are caused by muscles located both entirely within the foot (intrinsic) and where the muscle bellies lie outside the foot on the tibia and fibula (extrinsic).

The extrinsic musculature is of prime concern in this introduction to foot myology; the intrinsic foot musculature is shown in figures 5.1 and 5.2. As il-

Interphalangeal and Metatarsophalangeal Joints

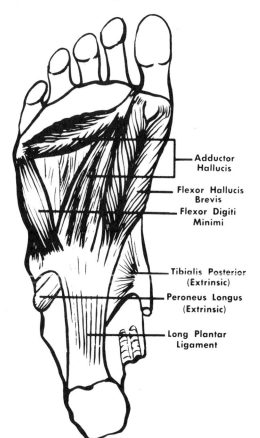

Figure 5.1 Intrinsic muscles of foot (deep)—inferior view.

- Adductor Hallucis
- Flexor Hallucis Brevis
- Flexor Digiti Minimi
- Tibialis Posterior (Extrinsic)
- Peroneus Longus (Extrinsic)
- Long Plantar Ligament

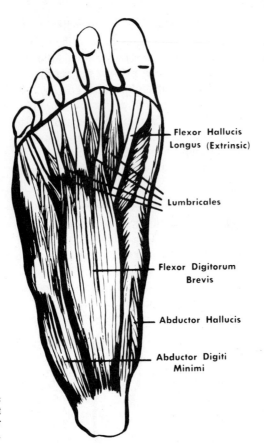

Flexor Hallucis
Longus (Extrinsic)

Lumbricales

Flexor Digitorum
Brevis

Abductor Hallucis

Abductor Digiti
Minimi

Figure 5.2 Intrinsic muscles of foot (superficial)—inferior view.

lustrated there, the muscle bellies and tendons are both contained within the foot. Most of these muscles have names which reflect their motion functions when they contract concentrically. One example, the adductor hallucis, is shown in figure 5.1. The term *hallucis* refers to the great toe; consequently, the adductor hallucis muscle adducts or moves the great toe toward the second toe. This motion, as well as movements caused by the other intrinsic muscles, occur at the metatarsophalangeal joint(s). The reader can identify concentric function in many of the intrinsic foot muscles on the basis of their names.

The major skeletal landmarks and bones of the foot are shown in figures 5.3, 5.5 and 5.6. The interphalangeal and metatarsophalangeal joints are shown in figure 5.3. The metatarsal bones are numbered one through five, from medial to lateral.

The motions of the metatarsophalangeal joints are flexion, extension, abduction, and adduction. At joints where these motions occur, circumduction is also a possibility; it is a minimal movement in these joints. Abduction and adduction are caused by intrinsic muscles of the foot. The extrinsic muscles that cause flexion and extension are listed in figure 5.4 and shown in figures 5.7 and 5.8.

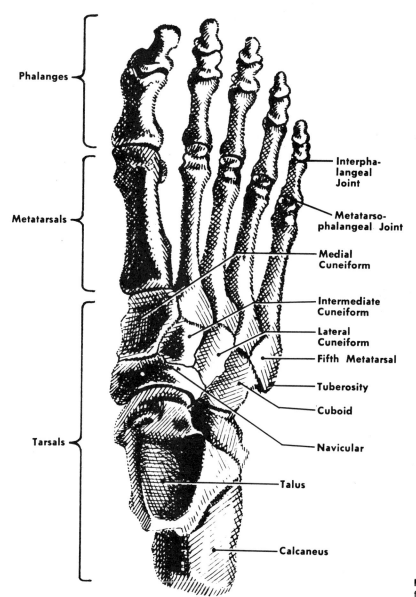

Phalanges

Metatarsals

Tarsals

Interpha-
langeal
Joint

Metatarso-
phalangeal Joint

Medial
Cuneiform

Intermediate
Cuneiform

Lateral
Cuneiform

Fifth Metatarsal

Tuberosity

Cuboid

Navicular

Talus

Calcaneus

Figure 5.3 Right foot—superior view.

Muscles	Proximal Attachment	Distal Attachment

1. Interphalangeal Joint Musculature (extrinsic)

A. Flexors

1. FLEXOR HALLUCUS LONGUS	Lower two-thirds of the posterior aspect of the fibula	Distal phalanx of the great toe (plantar surface)
2. FLEXOR DIGITORIUM LONGUS	Posterior surface of the tibia; tendon passes posterior to the medial malleolus and divides into four tendons	Bases of the distal phalanges of the four lesser toes

B. Extensors

1. EXTENSOR HALLUCIS LONGUS	Mid one-half of the anterior aspect of the fibula	Distal phalanx of the great toe (dorsal surface)
2. EXTENSOR DIGITORUM LONGUS	Lateral condyle of the tibia; upper three-fourths of the anterior aspect of fibula and divides into four tendons	Dorsal aspect of the four lesser toes

II. Metatarsophalangeal Joint Musculature (extrinsic)

A. Flexors

1. FLEXOR HALLUCIS LONGUS	Lower two-thirds of the posterior aspect of the fibula	Distal phalanx of the great toe (plantar surface)
2. FLEXOR DIGITORUM LONGUS	Posterior surface of the tibia; tendon passes posterior to the medial malleolus and divides into four tendons	Bases of the distal phalanges of the four lesser toes

B. Extensors

1. EXTENSOR HALLUCIS LONGUS	Mid one-half of the anterior aspect of the fibula	Distal phalanx of the great toe (dorsal surface)
2. EXTENSOR DIGITORUM LONGUS	Lateral condyle of the tibia; upper three-fourths of the anterior aspect of fibula and divides into four tendons	Dorsal aspect of the four lesser toes

C. Abductors (intrinsic)   See figures 5.1 and 5.2
D. Adductors (intrinsic)

Figure 5.4 Extrinsic muscles for interphalangeal and metatarsophalangeal joint motions. Muscles most involved are in uppercase letters and denote their function for concentric contraction.

The two primary bones involved in these articulations are the talus and calcaneus. These are major weight-bearing structures, i.e., the body weight while standing is transferred to the foot via the tibia through the talus. The subtalar joint lies inferior to the talus, and the transverse tarsal joint involves articulations between the cuneiforms, calcaneus, and talus, as shown in figures 5.5 and 5.6.

The motions possible at the transverse tarsal and subtalar joints are inversion and eversion. These medial and lateral rotary motions of the foot are problem movements for many people during locomotor activities such as walking, jogging, and running; tendons and ligaments within the foot are traumatized when there is too much inversion and eversion. The ankle is not designed to perform these motions. The inverters are located medially to the joints, while the everters have a lateral spatial relationship to the transverse tarsal and subtalar joints (figures 5.7 and 5.8). The muscles which cause inversion and eversion aginst resistance are listed in figure 5.9.

Transverse Tarsal and Subtalar Joints

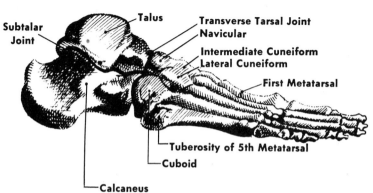

Figure 5.5 Right foot—lateral view— showing the transverse tarsal and subtalar joints.

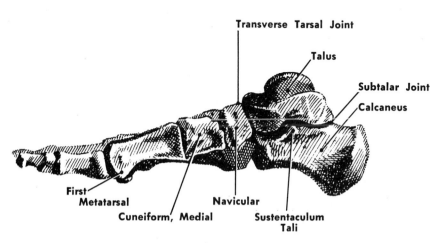

Figure 5.6 Right foot—medial view.

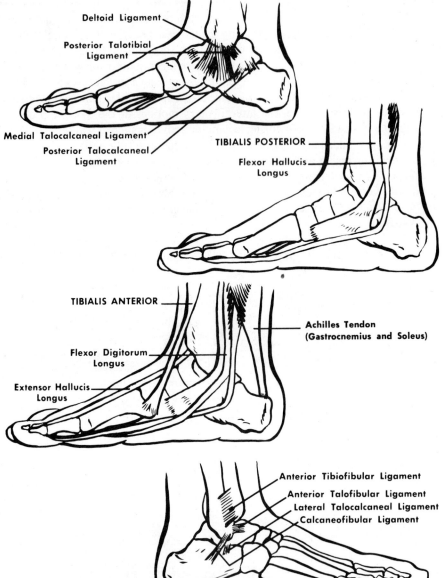

Deltoid Ligament

Posterior Talotibial
Ligament

Medial Talocalcaneal Ligament
Posterior Talocalcaneal
Ligament

TIBIALIS POSTERIOR

Flexor Hallucis
Longus

TIBIALIS ANTERIOR

Achilles Tendon
(Gastrocnemius and Soleus)

Flexor Digitorum
Longus

Extensor Hallucis
Longus

Figure 5.7 Medial
views of the ankle and
foot showing the medial
spatial relationships of
the inverters. The
muscles most involved
appear in uppercase
letters.

Figure 5.8 Lateral
views of the ankle and
foot showing the lateral
spatial relationships of
the everters. The
muscles most involved
appear in uppercase
letters.

Anterior Tibiofibular Ligament
Anterior Talofibular Ligament
Lateral Talocalcaneal Ligament
Calcaneofibular Ligament

PERONEUS LONGUS

PERONEUS BREVIS
PERONEUS TERTIUS

Achilles
Tendon

EXTENSOR DIGITORUM LONGUS

Muscles	Proximal Attachment	Distal Attachment
I. Transverse Tarsal and Subtalar Joint Musculature		
A. Inverters		
1. TIBIALIS ANTERIOR	Lateral condyle of the tibia and upper two-thirds of the lateral aspect of the tibia	Medial aspect of the medial cuneiform and base of the first metatarsal
2. TIBIALIS POSTERIOR	Lateral portion of the posterior aspect of the tibia and from the upper two-thirds of the medial aspect of the fibula	Navicular tuberosity with fiber connections to the three cuneiforms, cuboid, and bases of Metatarsals II, III, and IV
3. Extensor hallucis longus	Mid one-half of the anterior aspect of the fibula	Distal phalanx of the great toe (dorsal surface)
4. Flexor hallucis longus	Lower two-thirds of the posterior aspect of the fibula	Distal phalanx of the great toe (plantar surface)
5. Flexor digitorum longus	Posterior surface of the tibia; tendon passes posterior to the medial malleolus and divides into four tendons	Bases of the distal phalanges of the four lesser toes
B. Everters		
1. EXTENSOR DIGITORUM LONGUS	Lateral condyle of the tibia; upper three-fourths of the anterior aspect of fibula and divides into four tendons	Dorsal aspect of the four lesser toes
2. PERONEUS TERTIUS	Lower one-third of the anterior fibula	Base of the fifth metatarsal (dorsal surface)
3. PERONEUS LONGUS	Superior, lateral fibula and lateral condyle of tibia	Base of Metatarsal I and lateral aspect of medial cuneiform
4. PERONEUS BREVIS	Lower two-thirds lateral fibula	Tuberosity of Metatarsal V

Figure 5.9
Transverse tarsal and subtalar joint musculature. The muscles most involved for supplying force for the motion during concentric contraction are in upper case letters. Muscles supplying supplemental force are noted in lowercase letters.

Ankle (Talocrural Joint)

The most prominant skeletal landmarks in the ankle area are the lateral malleolus on the fibula and the medial malleolus of the tibia; they are shown in figure 5.10. These superficially prominent landmarks should be used to estimate the center of the ankle joint and to help make calculations of the extent of anteroposterior plane motions of the foot segment.

The only motions possible at the ankle joint are dorsiflexion and plantar flexion. These motions move the foot segment through an anteroposterior plane of motion. Figure 5.11 shows the spatial relationships of the extrinsic tendons of the foot and ankle. The tendons with anterior spatial relationships to the center of the ankle joint *cause dorsiflexion* when their muscles contract concentrically. These same muscles *control plantar flexion* (extent of motion and velocity) when they contract eccentrically. The tendons with posterior spatial relationships to the center of the ankle joint *cause plantar flexion* when their muscles contract concentrically. These same muscles *control dorsiflexion* when they contract eccentrically.

An example of this type of eccentric control for dorsiflexion is seen during stair running. The body weight is usually distributed over the metatarsophal-

Figure 5.10 Right talocrural joint—anterior view.

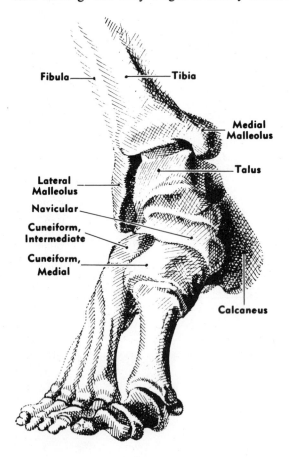

Fibula

Tibia

Medial Malleolus

Talus

Lateral Malleolus

Navicular

Cuneiform, Intermediate

Cuneiform, Medial

Calcaneus

Lower Limb Segment Myology

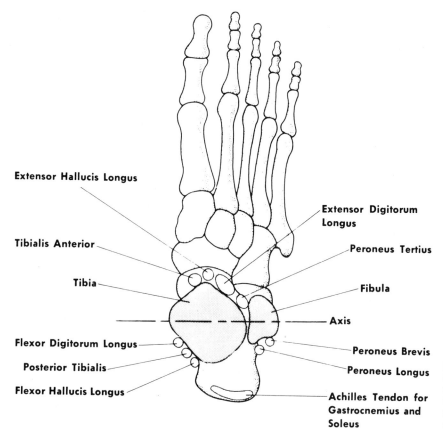

Extensor Hallucis Longus

Extensor Digitorum Longus

Tibialis Anterior

Peroneus Tertius

Tibia

Fibula

Axis

Flexor Digitorum Longus

Peroneus Brevis

Posterior Tibialis

Peroneus Longus

Flexor Hallucis Longus

Achilles Tendon for Gastrocnemius and Soleus

Figure 5.11 Spatial relationships of the extrinsic tendons of the ankle and foot—cross sectional, superior view of the right ankle.

angeal joint area on each stride. The calcaneus (heel) does not strike the stair, but the dorsum of the foot moves toward the tibia-fibula segement (dorsiflexes) under controlled tension. The eccentrically contracting muscles most involved for that motion are the gastrocnemius and soleus, with help from the remaining five muscles which lie posterior to the ankle joint (figure 5.11).

The reader should study figure 5.11 in conjunction with figures 5.7 and 5.8. The latter show the same tendons and in most cases their distal attachment points. It should be remembered that tendon attachment points are the focuses for force application by the contracting muscles. These tendons have spatial relationships with several joints; therefore, they have multiple motion functions. *If a muscle-tendon crosses a joint, it contributes force for moving that joint.* This should be kept in mind while studying muscle function. As one example of a multiarticular muscle, the extensor digitorum longus can extend the interphalangeal joints of the four lesser toes, dorsiflex the ankle, and contribute some force for eversion of the subtalar and transverse tarsal joints. This is very efficient from the standpoint of energy expenditure.

The ankle joint musculature for causing and controlling dorsiflexion and plantar flexion are listed in figure 5.12.

Basic Concepts of Anatomic Kinesiology   91

Muscles	Proximal Attachment	Distal Attachment
I. Ankle (Talocrural) Joint Musculature		
A. Dorsiflexors		
1. TIBIALIS ANTERIOR	Lateral condyle of the tibia and upper two-thirds of the lateral aspect of the tibia	Medial aspect of the medial cuneiform and base of the first metatarsal
2. EXTENSOR DIGITORUM LONGUS	Lateral condyle of the tibia; upper three-fourths of the anterior aspect of fibula and divides into four tendons	Dorsal aspect of the four lesser toes
3. PERONEUS TERTIUS	Lower one-third of the anterior fibula	Base of the fifth metatarsal (dorsal surface)
4. Extensor hallucis longus	Mid one-half of the anterior aspect of the fibula	Distal phalanx of the great toe (dorsal surface)
B. Plantar Flexors		
1. GASTROC-NEMIUS	Two heads attached to the medial and lateral condyles of the femur posteriorly	Achilles tendon to the calcaneus
2. SOLEUS	Upper one-third of the posterior tibia and fibula	Achilles tendon to the calcaneus
3. Peroneus longus	Superior, lateral fibula and lateral condyle of tibia	Base of Metatarsal I and lateral aspect of medial cuneiform
4. Peroneus brevis	Lower two-thirds lateral fibula	Tuberosity of Metatarsal V
5. Flexor digitorum longus	Posterior surface of the tibia; tendon passes posterior to the medial malleolus and divides into four tendons	Bases of the distal phalanges of the four lesser toes
6. Flexor hallucis longus	Lower two-thirds of the posterior aspect of the fibula	Distal phalanx of the great toe (plantar surface)
7. Tibialis posterior	Lateral portion of the posterior aspect of the tibia and from the upper two-thirds of the medial aspect of the fibula	Navicular tuberosity with fiber connections to the three cuneiforms, cuboid, and bases of Metatarsals II, III, and IV

Figure 5.12 Ankle joint musculature. The muscles most involved for the motions against resistance are in uppercase letters. Muscles supplying supplemental force during concentric contraction are in lowercase letters.

Knee Joint

The knee joint involves the tibia, femur, and patella. The tibia and femur are weight-bearing bones, but the laterally positioned fibula is not involved directly in weight bearing. The most important skeletal landmarks are shown in figures 5.13 and 5.14.

The knee joint moves through anteroposterior and transverse planes of motion. The most prominent motions are flexion and extension through an anteroposterior plane. When the knee is in a flexed state, the collateral ligaments of

Lower Limb Segment Myology

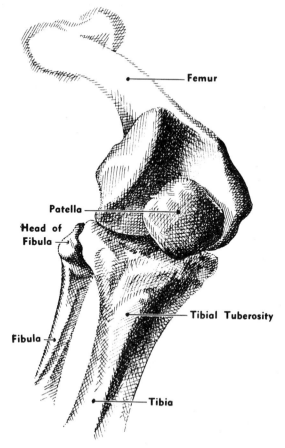

Femur

Patella

Head of
Fibula

Fibula

Tibial Tuberosity

Tibia

Figure 5.13 Flexed right knee joint—anterior view.

Figure 5.14 Skeletal landmarks of the tibia and fibula.

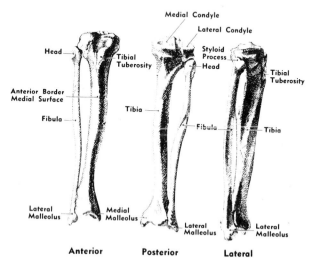

Medial Condyle

Lateral Condyle

Head

Tibial
Tuberosity

Styloid
Process

Head

Tibial
Tuberosity

Anterior Border
Medial Surface

Tibia

Fibula

Fibula

Tibia

Lateral
Malleolus

Medial
Malleolus

Lateral
Malleolus

Lateral
Malleolus

Anterior

Posterior

Lateral

the knee (figure 5.15) are relieved of some of their tension. For this reason, the knee (tibia-fibula segment) may be medially and laterally rotated through a transverse plane of motion. The knee joint musculature does not cause transverse plane motions when the knee is fully extended.

Normal flexion of the knee, which involves motion of the tibia-fibula and foot segment, is 130 degrees. Extension is the return from flexion. Figure 5.16 shows how these segments may be readily determined in order to calculate the extent of motion.

Figure 5.17 shows the spatial relationships of the muscle-tendons which surround the knee joint. The muscles with anterior spatial relationships extend the knee against resistance. The posterior muscles flex the knee when they contract concentrically. The lateral muscle rotates the knee against resistance, and its medial counterparts, when they contract concentrically, medially rotate the knee. These muscle groups reverse their functions as described here and in figure 5.19 when they contract eccentrically.

It should be remembered that a diagram of spatial relationships of tendons to joints as shown in figure 5.17 does not purport to show lines of pull for muscles. The lines of pull for some of the knee musculature are shown in figure 5.18, and

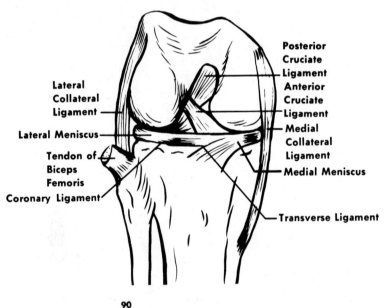

Figure 5.15 Right knee flexed—anterior view showing the cartilage and ligaments of the knee.

Posterior Cruciate Ligament
Anterior Cruciate Ligament
Medial Collateral Ligament
Medial Meniscus
Transverse Ligament

Lateral Collateral Ligament
Lateral Meniscus
Tendon of Biceps Femoris
Coronary Ligament

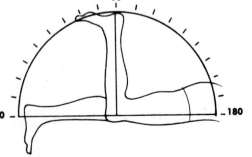

Figure 5.16 Knee flexion—extension range of motion determination. The knee has been flexed to 90 degrees. Segment relationships must be determined to calculate the extent of any motion at a joint.

Lower Limb Segment Myology

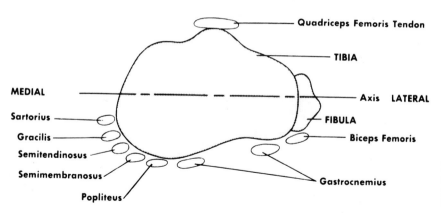

ANTERIOR

Quadriceps Femoris Tendon

TIBIA

MEDIAL

Axis   LATERAL

Sartorius

FIBULA

Gracilis

Biceps Femoris

Semitendinosus

Semimembranosus

Gastrocnemius

Popliteus

Figure 5.17 Spatial relationships of knee joint muscles—superior view, right knee.

Figure 5.18 Diagrammatic representation of some of the knee stabilizing musculature.

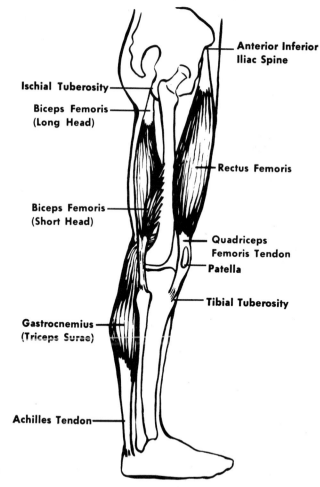

Anterior Inferior Iliac Spine

Ischial Tuberosity

Biceps Femoris (Long Head)

Rectus Femoris

Biceps Femoris (Short Head)

Quadriceps Femoris Tendon

Patella

Tibial Tuberosity

Gastrocnemius (Triceps Surae)

Achilles Tendon

Muscles	Proximal Attachment	Distal Attachment
I. Knee Joint Musculature		
A. Extensors		
1. RECTUS FEMORIS	Anterior, inferior iliac spine and superior to acetabulum	Tibial tuberosity via quadriceps femoris tendon
2. VASTUS LATERALIS	Superior aspect of lateral femur and upper portion of the linea aspera	Tibial tuberosity via quadriceps femoris tendon
3. VASTUS MEDIALIS	Medial lip of the linea aspera and medial supracondylar line	Tibial tuberosity via quadriceps femoris tendon
4. VASTUS INTERMEDIUS	Anterior and lateral aspects of the upper two-thirds of the femur	Tibial tuberosity via quadriceps femoris tendon
B. FLEXORS		
1. SEMI-TENDINOSUS	Ischial tuberosity	Medial aspect of the medial tibial condyle
2. SEMIMEM-BRANOSUS	Ischial tuberosity	Posterior aspect of the medial tibial condyle
3. BICEPS FEMORIS	*Long Head:* Ischial tuberosity *Short Head:* Lateral lip of the linea aspera	Head of the fibula and lateral condyle of the tibia
4. Sartorius	Anterior superior iliac spine	Medial tuberosity of the tibia
5. Gracilis	Inferior half of the symphysis pubis	Inferior to the medial condyle of the tibia
6. Popliteus*	Lateral condyle of femur; fibular head; lateral meniscus	Posterior surface of the superior and medial one-third of the tibia
7. Gastrocnemius	Two heads attached to the medial and lateral condyles of the femor posteriorly	Achilles tendon to the calcaneus
C. Medial Rotators		
1. SEMI-TENDINOSUS	Ischial tuberosity	Medial aspect of the medial tibial condyle
2. SEMIMEM-BRANOSUS	Ischial tuberosity	Posterior aspect of the medial tibial condyle
3. POPLITEUS	Lateral condyle of femur; fibular head; lateral meniscus	Posterior surface of the superior and medial one-third of the tibia
4. Sartorius	Anterior superior iliac spine	Medial tuberosity of the tibia
5. Gracilis	Inferior half of the symphysis pubis	Inferior to the medial condyle of the tibia
D. Lateral Rotator		
1. BICEPS FEMORIS	*Long Head:* Ischial tuberosity *Short Head:* Lateral lip of the linea aspera	Head of the fibula and lateral condyle of the tibia

*Initiates motion so knee flexion can occur

Figure 5.19 Knee joint musculature. The muscles most involved for the motions are in uppercase letters. Muscles supplying supplemental force during concentric contraction are in lowercase letters.

Lower Limb Segment Myology

they are all outlined in figure 5.19. The quadriceps femoris group, noted in figure 5.17, has an anterior spatial relationship to the knee and includes the four extensors listed in figure 5.19. The term "hamstring muscles" is used commonly; this refers to the three muscles most involved for knee flexion listed in figure 5.19.

The hip joint is formed by the articulation between the femur and acetabulum of the pelvis. This is a major weight-bearing joint, and the skeletal structure for this purpose is excellent. The femoral head fits into the acetabulum. Scientifically, this ball and socket type of arrangement is known as an *enarthrodial joint*. The structure of enarthrodial joints makes them multiaxial in terms of motion capabilities, and therefore there is great versatility for movement.

Hip Joint

The major skeletal landmarks of the femur and pelvis are shown in figures 5.20, 5.21, and 5.22. If the performer is executing a skill such as diving, the iliac crests, greater trochanters, and femoral condyles can be seen; they should be marked when feasible to facilitate subsequent noncinematographic and film analyses of femur and pelvic segment motions.

The hip joint, as shown in figure 5.23, is capable of motion through four planes.

Planes	*Motions*
1. Anteroposterior:	Flexion and extension
2. Lateral:	Abduction and adduction
3. Transverse:	Medial and lateral rotations
4. Diagonal:	Diagonal abduction and diagonal adduction

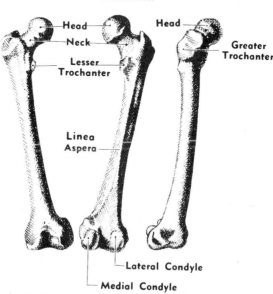

Figure 5.20 Skeletal landmarks of the right femur.

Head — Neck — Lesser Trochanter — Linea Aspera — Head — Greater Trochanter — Lateral Condyle — Medial Condyle

Anterior   Posterior   Lateral

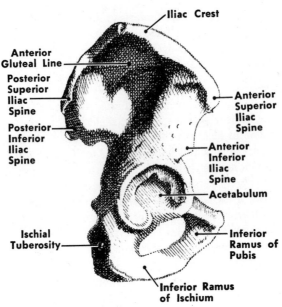

Figure 5.21   Skeletal landmarks of the right half of the pelvic girdle— lateral view.

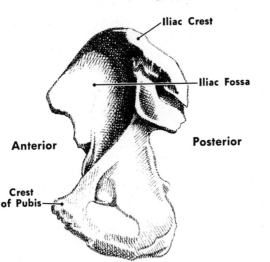

Figure 5.22   Skeletal landmarks of the right half of the pelvic girdle— medial view.

The joint can also be circumducted, but it cannot be hyperextended. What appears to be hyperextension of the hip is due to the compensatory motions of lumbar spine hyperextension and anterior pelvic rotation during hip extension.

Figures 5.24 and 5.25 provide a simple procedure for visualizing the extent and direction of motion through the anteroposterior and lateral planes of motion, respectively. The average extent of hip flexion with the knee flexed as shown is 125 degrees. (90 degrees of flexion is shown in figure 5.24). If the knee is extended, the average amount of hip flexion capability will be reduced by approximately 35 degrees. This is the result of the mechanical limitations and spatial arrangements of muscles on the posterior aspects of the hip and knee joints. For

Lower Limb Segment Myology

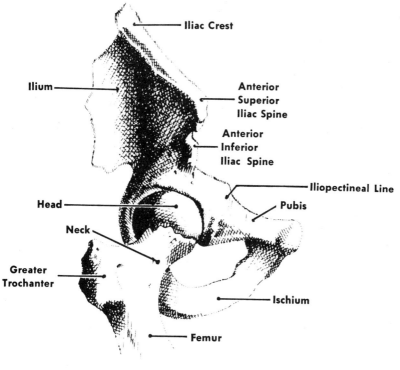

Iliac Crest

Ilium

Anterior
Superior
Iliac Spine

Anterior
Inferior
Iliac Spine

Iliopectineal Line

Head

Pubis

Neck

Greater
Trochanter

Ischium

Femur

Figure 5.23 Right hip joint—lateral view.

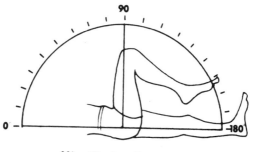

Hip Flexion-Extension

Figure 5.24 Conceptualizing the extent of hip flexion and extension by visualizing the femur segment and its anteroposterior plane relationships to the pelvic segment.

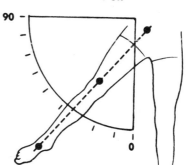

Hip Abduction-Adduction

Figure 5.25 Hip abduction—adduction of the total lower limb segment can move through approximately 45 degrees of lateral plane motion.

Basic Concepts of Anatomic Kinesiology

the average individual, hip abduction can extend through 45 degrees. The total lower limb segment is marked and being abducted in figure 5.25. It is very essential to mark or observe the moving segments properly. In order to approximate joint centers consistently and with as much accuracy as possible, (recognizing the limitations of this procedure) one must use external skeletal landmarks when they are observable. These simple motion-plotting techniques using segment lines and protractors can quite accurately provide answers to the extent of motion questions during basic and some intermediate cinematographic analyses. Contemporary computer hardware designed for motion analyses now make such manual techniques antiquated. But how many secondary school physical educators and coaches have sophisticated motion analysis computer systems available to them as a tool to study individiual performances on game film? The answer is virtually none. Therefore, manual techniques must be learned and used to enhance the objectivity of motion description at all joints.

As is the case with almost all joints, the major structural support for the hip joint comes from the muscles with spatial relationships to the articulation. Thus, the hip joint is structurally sound in terms of skeletal, muscle, and connective tissue support.

Due to the multiaxial complexities of the hip joint, a simple, two-dimensional drawing of this joint's muscles and their spatial relationships is not adequate, as it was at the ankle and knee joints. However, the spatial relationships of most of the major muscles can be observed in the figures which follow:

Figure 5.26 The anterior spatial relationships of the rectus femoris to the knee and hip joints make it a muscle most involved for hip flexion and knee extension during concentric contraction.

Figure 5.27 Anterior spatial relationships of the sartorius and tensor fasciae latae muscles mean that they contribute some force for hip flexion when they contract concentrically.

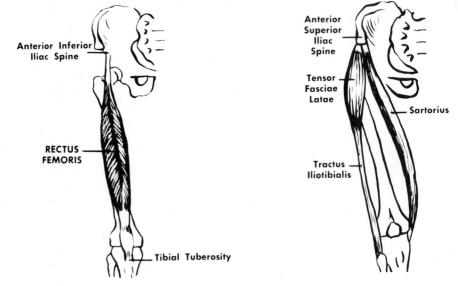

Anterior Inferior Iliac Spine

RECTUS FEMORIS

Tibial Tuberosity

Anterior Superior Iliac Spine

Tensor Fasciae Latae

Sartorius

Tractus Iliotibialis

Lower Limb Segment Myology

Spatial Relationship	Major Muscles	Figure
Anterior	Psoas Major, Iliacus, Sartorius, Tensor Fasciae Latae, Pectineus, Rectus Femoris	5.26, 5.27 5.28
Medial	Adductors, brevis, longus and magnus, Gracilis	5.28
Lateral	Gluteus medius	5.29
Posterior	Gluteus maximus, "hamstrings"	5.29

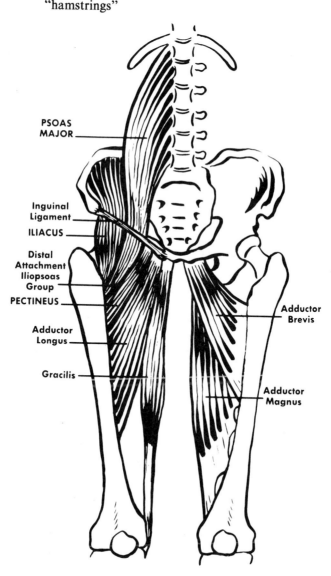

Figure 5.28 Muscles with anterior spatial relationships (uppercase letters) and medial spatial relationships (lowercase letters) to the hip joint.

PSOAS MAJOR

Inguinal Ligament
ILIACUS
Distal Attachment Iliopsoas Group
PECTINEUS
Adductor Longus
Gracilis

Adductor Brevis

Adductor Magnus

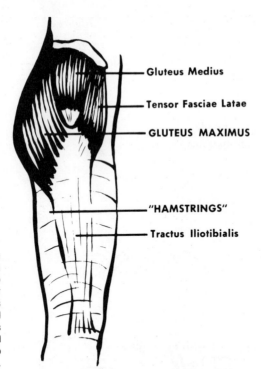

Gluteus Medius

Tensor Fasciae Latae

GLUTEUS MAXIMUS

"HAMSTRINGS"

Tractus Iliotibialis

Figure 5.29 The gluteus medius has a lateral spatial relationship to the hip joint. The lower fibers of the gluteus maximus and "hamstring" muscles have posterior spatial relationships to the hip joint.

The anterior muscles flex the hip joint when they contract concentrically. The muscles with medial spatial relationships adduct against resistance, and the lateral muscles abduct when they contract concentrically. The posterior muscles serve as the extensor mechanism against gravity. These muscles reverse such motion functions when they are needed to control hip movement with gravity by contracting eccentrically. For example, the posterior gluteus maximus eccentrically controls the extent and velocity of hip joint flexion with gravity. This is seen in a wide variety of sports when the athlete moves into a stance while standing with weight distributed through both lower limbs. The gluteus maximus is stretched in in this process; this involves the muscle length-force principle discussed in Chapter 4. This is an advantage, because forceful hip extension is usually the first hip motion required to overcome inertia and move out of the stance. The stretched gluteus maximus is the muscle most involved for that hip extension against resistance (concentric contraction). Thus, eccentric control of one motion by a muscle or muscle group can contribute to increasing concentric contraction force. The reader can probably think of numerous motions and skills where the muscle length-force principle is utilized at the hip joint.

Figure 5.30 provides a complete listing of all muscle functions for the hip joint motions.

Muscles	Proximal Attachment	Distal Attachment
I. Hip Joint Musculature		
A. Flexors		
1. PSOAS MAJOR	Bodies of the last thoracic and all lumbar vertebrae; transverse processes of the lumbar vertebrae	Lesser trochanter of the femur
2. ILIACUS	Upper two-thirds of the iliac fossa and internal lip of the iliac crest	Lesser trochanter of the femur
3. PECTINEUS	Between the tubercle of the pubis and the iliopectineal eminence	Pectineal line from the lesser trochanter to the linea aspera
4. RECTUS FEMORIS	Anterior, inferior iliac spine and superior to acetabulum	Tibial tuberosity via quadriceps femoris tendon
5. Tensor fasciae latae	Anterior aspect of the external lip of the iliac crest and the anterior superior iliac spine	Laterally into the tractus iliotibialis superficial to the greater trochanter of the femur
6. Sartorius	Anterior, superior iliac spine	Medial, inferior to tibial condyle
7. Gluteus medius (anterior fibers)	Lateral crest of the ilium to the gluteal line posterior and above and the gluteal line anterior and below	Greater trochanter of the femur
8. Gluteus minimus (anterior fibers)	Lower, lateral ilium	Anterior aspect of the greater trochanter of the femur
9. Gracilis	Inferior half of the symphysis pubis	Inferior to the medial condyle of the tibia
10. Adductor longus	Inferior pubic crest	Linea aspera of the femur between the vastus medialis and adductor magnus
11. Adductor brevis	Inferior ramus of the pubis	Upper aspect of the linea aspera
B. Extensors		
1. GLUTEUS MAXIMUS	Gluteal line posterior of the ilium and a portion of the posterior iliac crest	Upper fibers into the tractus iliotibialis and lower fibers into posterior femur
2. BICEPS FEMORIS (LONG HEAD)	*Long Head:* Ischial tuberosity	Head of the fibula and lateral condyle of the tibia
3. SEMITENDINOSUS	Ischial tuberosity	Medial aspect of the medial tibial condyle
4. SEMIMEMBRANOSUS	Ischial tuberosity	Posterior aspect of the medial tibial condyle

Figure 5.30 Hip joint musculature. The muscles most involved for the motions against resistance are shown in uppercase letters. Muscles supplying supplemental force during concentric contraction are in lowercase letters.

Basic Concepts of Anatomic Kinesiology

Figure 5.30—*Continued*

Muscles	Proximal Attachment	Distal Attachment
C. Abductors		
1. GLUTEUS MEDIUS	Lateral crest of the ilium to the gluteal line posterior and above and the gluteal line anterior and below	Greater trochanter of the femur
2. Gluteus maximus (upper fibers)	Gluteal line posterior of the ilium and a portion of the posterior iliac crest	Upper fibers into the tractus iliotibialis and lower fibers into posterior femur
3. Gluteus minimus (posterior fibers)	Lower, lateral ilium	Anterior aspect of the greater trochanter of the femur
4. Tensor fasciae latae	Anterior aspect of the external lip of the iliac crest and the anterior superior iliac spine	Laterally into the tractus iliotibialis superficial to the greater trochanter of the femur
D. Adductors		
1. ADDUCTOR BREVIS	Inferior ramus of the pubis	Upper aspect of the linea aspera
2. ADDUCTOR LONGUS	Inferior pubic crest	Linea aspera of the femur between the vastus medialis and adductor magnus
3. ADDUCTOR MAGNUS	Inferior ramus, anterior pubis and inferior ischial tuberosity	Entire linea aspera and the adductor tubercle on the medial condyle of the femur
4. GRACILIS	Inferior half of the symphysis pubis	Inferior to the medial condyle of the tibia
5. Gluteus maximus (lower fibers)	Gluteal line posterior of the ilium and a portion of the posterior iliac crest	Upper fibers into the tractus iliotibialis and lower fibers into posterior femur
E. Diagonal Abductors		
1. GLUTEUS MEDIUS	Lateral crest of the ilium to the gluteal line posterior and above and the gluteal line anterior and below	Greater trochanter of the femur
2. GLUTEUS MAXIMUS	Gluteal line posterior of the ilium and a portion of the posterior iliac crest	Upper fibers into the tractus iliotibialis and lower fibers into posterior femur
3. SEMITENDINOSUS	Ischial tuberosity	Medial aspect of the medial tibial condyle
4. SEMIMEMBRANOSUS	Ischial tuberosity	Posterior aspect of the medial tibial condyle

Figure 5.30—*Continued*

Muscles	Proximal Attachment	Distal Attachment
5. BICEPS FEMORIS (LONG HEAD)	*Long Head:* Ischial tuberosity	Head of the fibula and lateral condyle of the tibia
6. PIRIFORMIS	Anterior aspect of pelvis	Greater trochanter
7. OBTURATOR INTERNUS	Surrounds obturator foramen	Greater trochanter
8. GEMELLUS SUPERIOR	Outer aspect of the ischial spine	Greater trochanter
9. GEMELLUS INFERIOR	Upper aspect of the ischial tuberosity	Greater trochanter
10. QUADRATUS FEMORIS	External border of the ischial tuberosity	Intertrochanter line of the femur
11. OBTURATOR EXTERNUS	Medial aspect of the obturator foramen, inferior ramus of the pubis and ischial ramus	Trochanter fossa of the femur
F. Diagonal Adductors		
1. ILIOPSOAS	Lumbar vertebrae and iliac fossa	Lesser trochanter of the femur
2. RECTUS FEMORIS	Anterior, inferior iliac spine and superior to acetabulum	Tibial tuberosity via quadriceps femoris tendon
3. PECTINEUS	Between the tubercle of the pubis and the iliopectineal eminence	Pectineal line from the lesser trochanter to the linea aspera
4. ADDUCTOR BREVIS	Inferior ramus of the pubis	Upper aspect of the linea aspera
5. ADDUCTOR LONGUS	Inferior pubic crest	Linea aspera of the femur between the vastus medialis and adductor magnus
6. ADDUCTOR MAGNUS	Inferior ramus, anterior pubis and inferior ischial tuberosity	Entire linea aspera and the adducter tubercle on the medial condyle of the femur
G. Lateral Rotators		
1. GLUTEUS MAXIMUS	Gluteal line posterior of the ilium and a portion of the posterior iliac crest	Upper fibers into the tractus iliotibialis and lower fibers into posterior femur
2. SIX LATERAL ROTATORS		
a. PIRIFORMIS	Anterior aspect of pelvis	Greater trochanter
b. OBTURATOR INTERNUS	Surrounds obturator foramen	Greater trochanter

Figure 5.30—*Continued*

Muscles	Proximal Attachment	Distal Attachment
c. GEMEL-LUS SU-PERIOR	Outer aspect of the ischial spine	Greater trochanter
d. GEMEL-LUS IN-FERIOR	Upper aspect of the ischial tuberosity	Greater trochanter
e. QUAD-RATUS FEMORIS	External border of the ischial tuberosity	Intertrochanter line of the femur
f. OBTU-RATOR EXTER-NUS	Medial aspect of the obturator foramen, inferior ramus of the pubis and ischial ramus	Trochanter fossa of the femur
H. Medial Rotators		
1. GLUTEUS MINIMUS (ANTERIOR FIBERS)	Lower, lateral ilium	Anterior aspect of the greater trochanter of the femur
2. Gluteus medius (anterior fibers)	Lateral crest of the ilium to the gluteal line posterior and above and the gluteal line anterior and below	Greater trochanter of the femur
3. Tensor fasciae latae	Anterior aspect of the external lip of the iliac crest and the anterior superior iliac spine	Laterally into the tractus iliotibialis superficial to the greater trochanter of the femur

Recommended Reading

Andres, T. L., et al. "Involvement of Selected Quadriceps Muscles During a Knee Extension Exercise." *American Corrective Therapy Journal* 33 (July–August, 1979):111–114.

Basmajian, John V., and Stecko, George. "The Role of Muscles in Arch Support of the Foot." *Journal of Bone and Joint Surgery* 45-A (September, 1963):1184–90.

Bierman, William, and Ralston, H. J. "Electromyographic Study During Passive and Active Flexion and Extension of the Knee of the Normal Human Subject." *Archives of Physical Medicine and Rehabilitation* 46 (January, 1965):L71–75.

Bosco, C., et al. "Mechanical Characteristics and Fiber Composition of Human Leg Extensor Muscles." *European Journal of Applied Physiology* 41 (August, 1979):275–284.

Brewerton, D. A. "The Function of the Vastus Medialis Muscle." *Annals of Physical Medicine* 2 (1955):164–68.

Cavagna, G. A., and Margari, R. "Mechanics of Walking." *Journal of Applied Physiology* 21 (January, 1966):271–78.

Damholt, V., et al. "Asymmetry of Plantar Flexion Strength in the Foot." *ACTA Orthopaedica Scandinavica* 49 (April, 1978):215–219.

Dempster, W. T. "The Range of Motion of Cadaver Joints: The Lower Limb." *University of Michigan Medical Bulletin* 22 (1956):364–79.

Deutsch, H., et al. "Quadriceps Kinesiology (EMG) with Varying Hip Joint Flexion and Resistance." *Archives of Physical Medicine and Rehabilitation* 59 (May, 1978):231–236.

Elftman, H. "Forces and Energy Changes in the Leg During Walking." *American Journal of Physiology* 125 (1959):339–56.

Fugal-Meyer, A. R., et al. "Human Plantar Flexion Strength and Structure." *ACTA Physiologica Scandinavica* 107 (September, 1979):47–56.

Gollnick, Philip D. "Electrogoniometric Study of Walking on High Heels." *Research Quarterly* 35 (October, 1964):370–78.

Grieve, D. W., and Gear, Ruth J. "The Relationships Between Length and Stride, Step Frequency, Time of Swing and Speed for Walking for Children and Adults." *Ergonomics* 9 (September, 1966):379–99.

Hall, W. L., and Klein, K. K. "The Man, the Knee and the Ligaments." *Medicina Dello Sport* 1 (October, 1961):500–11.

Hallen, L. G., and Lindahl, O. "The Lateral Stability of the Knee Joint." *ACTA Orthopaedica Scandinavica* 36 (1965):179–91.

———. "The 'Screw-Home' Movement in the Knee Joint." *ACTA Orthopaedica Scandinavica* 37 (Fasc. 1, 1966):97–106.

Hooper, A. C. "The Role of the Iliopsoas Muscle in Femoral Rotation." *Irish Medical Journal* 146 (April, 1977):108–112.

Houtz, S. J., and Walsh, Frank P. "Electromyographic Analysis of the Function of the Muscles Acting on the Ankle During Weight Bearing with Special Reference to the Triceps Surae." *Journal of Bone and Joint Surgery* 41-A (1959):1469–81.

Inman, Verne T. "Functional Aspects of the Abductor Muscles of the Hip." *Journal of Bone and Joint Surgery* 29 (1947):607–19.

Jonsson, Bengt, and Steen, Bertil. "Function of the Gracilis Muscle." *ACTA Morphologica Neerlando-Scandinavica* 6 (1966):325–41.

Kaplan, Emanual B. "The Iliotibial Tract." *Journal of Bone and Joint Surgery* 40-A (1958):817–32.

Karpovich, Peter V., and Manfredi, Thomas G. "Mechanisms of Rising on the Toes." *Research Quarterly* 42 (1971):395–404.

Karpovich, Peter V., and Wilklow, Leighton B. "Goniometric Study of the Human Foot in Standing and Walking." *Industrial Medicine and Surgery* 29 (July, 1960):338–47.

Klein, Karl K. "The Deep Squat Exercise as Utilized in Weight Training for Athletics and Its Effect on the Ligaments of the Knee." *Journal of the Association for Physical and Mental Rehabilitation* 15 (January–February, 1961):6–11.

————. "The Knee and the Ligaments." *Journal of Bone and Joint Surgery* 44-A (September, 1962):1191–93.

Knight, K. L., et al. "EMG Comparison of Quadriceps Femoris Activity During Knee Extension and Straight Leg Raises." *American Journal of Physical Medicine* 58 (April, 1979):57–67.

Kroll, Walt, et al. "Muscle Fiber Type Composition and Knee Extension Isometric Strength Fatigue Patterns in Power and Endurance Trained Males." *Research Quarterly for Exercise and Sport* 51 (May, 1980):323–333.

Lawrence, Mary S.; Meyer, Harriet R.; and Matthews, Nancy L. "Comparative Increase in Muscle Strength in the Quadriceps Femoris by Isometric and Isotonic Exercises and Effects on the Contralateral Muscle." *Journal of the American Physical Therapy Association* 42 (January, 1962):15–20.

Mann, R. A., et al. "The Function of the Toes in Walking, Jogging, and Running." *Clinical Orthopaedics and Related Research* 142 (July–August, 1979):24–29.

Mann, Ralph and Sprague, Paul. "A Kinetic Analysis of the Ground Leg During Sprint Running." *Research Quarterly for Exercise and Sport* 51 (May, 1980):334–348.

Mathews, Donald; Shaw, Virginia; and Bohnen, Melra. "Hip Flexibility of College Women as Related to Length of Body Segments." *Child Development Abstracts and Bibliography* 32 (June–August, 1958):85.

Mathews, Donald K.; Shaw, Virginia; and Woods, John B. "Hip Flexibility of Elementary School Boys as Related to Body Segments." *Research Quarterly* 30 (October, 1959):297–302.

Meyers, Earle J. "Effect of Selected Exercise Variables on Ligament Stability and Flexibility of the Knee." *Research Quarterly* 42 (1971):411–22.

Morton, Dudley, J. *The Human Foot.* New York: Columbia University Press, 1935.

Murray, M. P., et al. "Function of the Triceps Surae During Gait: Compensatory Mechanisms for Unilateral Loss." *Journal of Bone and Joint Surgery* 60 (June, 1978):473–476.

O'Connell, A. L. "Electromyographic Study of Certain Leg Muscles During Movements of the Free Foot and During Standing." *American Journal of Physical Medicine* 37 (December, 1958):289–301.

Rarick, L., and Thompson, J. "Roentgenographic Measures of Leg Muscle Size and Ankle Extensor Strength." *Research Quarterly* 27 (October, 1956):321.

Ricci, Benjamin, and Karpovich, Peter V. "Effect of Height of Heel Upon the Foot." *Research Quarterly* 35 (October, 1964):385–88.

Sheffield, F. J., et al. "Electromyographic Study of the Muscles of the Foot in Normal Walking." *American Journal of Physical Medicine* 35 (1956):223–36.

Stern, Jack T. "Anatomical and Functional Specializations of the Human Gluteus Maximus." *American Journal of Physical Anthropology* 36 (1972):315–39.

Sutherland, David H. "An Electromyographic Study of the Planter Flexors of the Ankle in Normal Walking on the Level." *Journal of Bone and Joint Surgery* 48-A (January, 1966):66–71.

Wheatley, M. D., and Johnke, W. D. "Electromyographic Study of the Superficial Thigh and Hip Muscles in Normal Individuals." *Archives of Physical Medicine* 32 (1951):508–15.

Woodman, R. M., et al. "Testing Hip Extensor Strength when Passive Hip Extension is Restricted." *Physical Therapy* 58 (July, 1978):882.

6

Spinal Segment Myology

The purposes of this chapter are to present the bones, skeletal landmarks, major joints, and the muscles which constitute the articulations within the spinal segments. The spinal segments presented are the pelvic girdle, lumbar-thoracic spine, and the cervical spine-skull.

Pelvic Girdle

The pelvic girdle is a solid ring of bone consisting of the fused ilium, ischium, and pubis bones. Posteriorly, the pelvic girdle is fused to the sacrum of the spine. The prominent skeletal landmarks of the pelvic girdle are shown in figures 5.21, 5.22 and 5.23. The hip joint muscles which attach to the pelvic girdle were presented in figure 5.30.

The motions of the pelvis are functionally related to hip and spinal column movements. However, pelvic motions are observable as distinct, albeit slight, motions through the three traditional planes. Therefore, there are six pelvic girdle motions: (1) left transverse rotation, (2) right transverse rotation, (3) anterior rotation, (4) posterior rotation, (5) left lateral rotation, and (6) right lateral rotation.

Describing muscle function in this area depends upon what the performer is using as a base of support. The three major variations are (1) standing with the weight distributed bilaterally, (2) standing while shifting the weight from one foot to the other, and (3) hanging by the hands from a piece of apparatus.

Muscles which attach to the pelvic girdle cause or control pelvic motions. In this area of the body, the focal point for motion is within the pelvic segment instead of a joint per se. Therefore, one must apply the spatial relationship concept differently in order to deduce muscle functions. The critical spatial relationships are those of the muscles to the pelvic girdle. The major muscles which move the pelvic girdle also move the hip joint and lumbar-thoracic spine. In studying pelvic motion, as one example, the posterior spatial relationship of the gluteus maximus to that segment becomes an essential factor when determining posterior pelvic rotation while the weight is borne on both feet. All the hip joint muscles described in Chapter 5 and the lumbar-thoracic muscles presented in this chapter should also be studied by considering their spatial relationships to the pelvic girdle. These spatial relationships of muscles to the pelvic girdle are outlined in figure 6.1.

Pelvic Girdle Muscles	Spatial Relationship(s)
Hip and Pelvic Girdle	
Gluteus Maximus	Posterior and Lateral
Tensor Fasciae Latae	Anterior and Lateral
Gluteus Minimus	Anterior and Lateral
Gluteus Medius	Lateral
Rectus Femoris	Anterior
Iliopsoas	Anterior
Pectineus	Anterior
Semitendinosus	Posterior
Semimembranosus	Posterior
Biceps Femoris	Posterior
Lumbar-Thoracic Spine and Pelvic Girdle	
Rectus Abdominis	Anterior
Erector Spinae	Posterior
External Oblique	Anterior and Lateral
Internal Oblique	Anterior and Lateral
Quadratus Lumborum	Lateral

Figure 6.1 Major spatial relationships of hip joint and lumbar-thoracic spine musculature to the pelvic girdle. Muscles which attach to the pelvic girdle cause the motions of that segment.

The muscles listed in figure 6.1 with their spatial relationships to the lumbar-thoracic spine and pelvic girdle are illustrated and described in figures 6.6 through 6.10.

All the pelvic girdle rotations are functional in a variety of skills, notably in lateral pelvic girdle rotation by the gluteus medius during locomotor skills such as walking, jogging, running, and sprinting. Rotations occur at each stride (unilateral weight bearing), e.g., when the body weight is borne through the left lower limb segment, right lateral pelvic rotation is performed by the concentrically contracting left gluteus medius. This "lateral lifting" of the pelvic girdle allows the right hip joint musculature to contract and flex the hip joint. The right lateral rotation of the pelvic girdle gives the right lower limb enough clearance from the ground to allow it to move through the anteroposterior plane of motion at the hip in pendular fashion. If human beings were not able to laterally rotate the pelvic girdle while walking, the walking gait would literally be a shuffle! Manifestly, this would have a significant negative impact on athletic performance.

Space does not permit a thorough discussion of all pelvic motions. However, it should be noted that transverse pelvic rotation is critical in determining the outcome of most ballistic skills involving throwing, kicking, and striking. The timing, extent, velocity, and direction of transverse pelvic rotation during the final weight shifting (unilateral weight bearing) procedure during the "preparation phase" of a ballistic skill largely determine the success or failure of the throw, kick, or striking action. This pelvic motion will be discussed later as an essential element of the *Serape Effect* (see Chapter 7).

Figure 6.2 outlines the functions of the pelvic girdle muscles when there are varying bases of support. For attachments of these muscles, see figures 5.30 and 6.10.

I. Pelvic Girdle Musculature

   A. Transverse pelvic rotation—LEFT
      1. Hanging position
         a. LEFT EXTERNAL OBLIQUE
         b. RIGHT INTERNAL OBLIQUE
      2. Unilateral weight bearing
         a. LEFT EXTERNAL OBLIQUE
         b. RIGHT INTERNAL OBLIQUE
         c. RIGHT ERECTOR SPINAE
      3. Bilateral weight bearing
         a. LEFT GLUTEUS MAXIMUS
         b. RIGHT TENSOR FASCIAE LATAE
         c. RIGHT GLUTEUS MINIMUS
         d. RIGHT GLUTEUS MEDIUS (ANTERIOR FIBERS)

   B. Anterior pelvic rotation
      1. Hanging position
         a. ERECTOR SPINAE
      2. Bilateral weight bearing
         a. RECTUS FEMORIS
         b. ILIOPSOAS
         c. PECTINEUS

   C. Posterior pelvic rotation
      1. Hanging position
         a. RECTUS ABDOMINIS
      2. Bilateral weight bearing
         a. GLUTEUS MAXIMUS
         b. SEMITENDINOSUS
         c. SEMIMEMBRANOSUS
         d. BICEPS FEMORIS (LONG HEAD)
         e. RECTUS ABDOMINIS
      3. Prone position
         a. Same MMI as bilateral

   D. Lateral pelvic rotation—RIGHT
      1. Hanging position
         a. RIGHT ERECTOR SPINAE
         b. RIGHT RECTUS ABDOMINIS
         c. RIGHT EXTERNAL OBLIQUE
         d. RIGHT INTERNAL OBLIQUE
         e. Right quadratus lumborum
      2. Unilateral weight bearing
         a. LEFT GLUTEUS MEDIUS

Figure 6.2 Pelvic girdle musculature. The muscles most involved for the motions against resistance are shown in uppercase letters. Muscles supplying supplemental force during concentric contraction are in lowercase letters.

Spinal Segment Myology

The spinal column consists of seven vertebrae in the cervical (neck) area, twelve in the thoracic region, and five in the lumbar portion of the spine. Movement within and between these twenty-four vertebrae is extensive; however, movement between any two vertebrae is relatively small. Most of the motion occurs at the cervical and lumbar areas of the spine. The thoracic area, including the rib cage, is a relatively stable or fixed portion of the spinal column. In addition to the twenty-four vertebrae in the cervical, thoracic, and lumbar areas, five fixed vertebrae compose the sacrum. Four partially movable vertebrae comprise the coccyx. Thus, the spinal column consists of thirty-three vertebrae (fig. 6.3).

Spinal Column

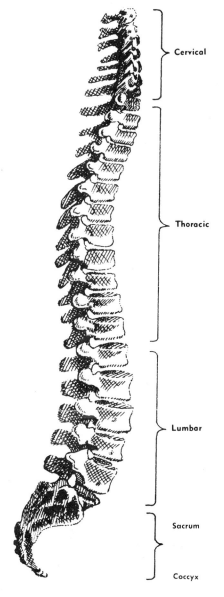

Figure 6.3 Spinal column—lateral view.

Cervical

Thoracic

Lumbar

Sacrum

Coccyx

Basic Concepts of Anatomic Kinesiology

The twelve thoracic vertebrae and the rib cage to which they are attached can be considered as one functional unit. This unit has a limited degree of movement potential, with the exception of rib movement required in breathing. From a kinesiologic standpoint, this unit can be thought of as a constant volumetric structure because the rib cage, unlike the lungs, does not change its size appreciably during movement.

Functioning with the thoracic spine-rib cage segment to make up what is commonly known as the "trunk" is the pelvic girdle. The pelvic girdle, like the thoracic rib cage area, can also be thought of as a functional unit. These two units of the trunk are joined together by the five lumbar vertebrae. The lumbar vertebrae can be regarded as a pivotal support structure linking the two units together.

The seven cervical vertebrae function as another pivotal support structure. These vertebrae serve as a connective link between the rib cage and the head. This area, forming the neck, is moved by muscles connecting the rib cage to the head. The head, like the rib cage, can be thought of as nonchanging volumetric structure. Thus it can be seen that in the so-called trunk area there are three nonchanging volumetric units connected by two pivotal support structures. As a result, virtually all movements of the spinal column take place within two supportive areas known as the lumbar and cervical portions of the spine.

Figure 6.4 Skeletal landmarks of the vertebrae. Note the size differences between the three vertebrae from the three spinal column areas.

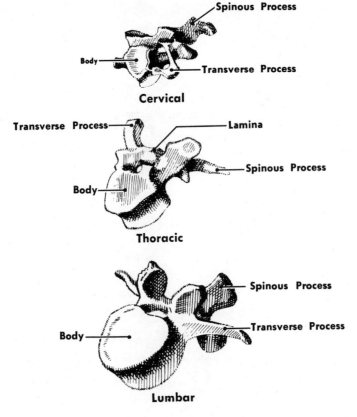

Spinous Process

Body

Transverse Process

Cervical

Transverse Process

Lamina

Body

Spinous Process

Thoracic

Spinous Process

Transverse Process

Body

Lumbar

Spinal Segment Myology

In this part of the body the segments to be analyzed are: (1) pelvic girdle, (2) pelvic girdle and lumbar spine, (3) lumbar-thoracic spine and pelvic girdle, (4) lumbar-thoracic spine, and (5) cervical spine and head. Observation of the center of mass (pelvic girdle and lumbar spine) is valuable as the starting point for most biomechanic analyses. It is also very useful to make critical observations of the relationships of these and other body segments to each other during the sequential phases of any sport skill.

The functional skeletal landmarks of the individual vertebrae and the rib cage are shown in figures 6.4 and 6.5 respectively. The spinous processes of the vertebrae can be marked prior to filming a performance such as diving. This helps define the extent of spinal column motions and indicates accurately the posterior aspect of the cardinal anterposterior plane. The anteriorly located manubrium, sternum, and xiphoid process can serve the same purpose for determining the cardinal anteroposterior plane of motion.

Motions involving the spinal column are described for the lumbar-thoracic and cervical areas. The motions possible in both areas are flexion, extension, and hyperextension through the cardinal anteroposterior plane, rotation (left and right) through the transverse plane, and lateral flexion (left and right) through

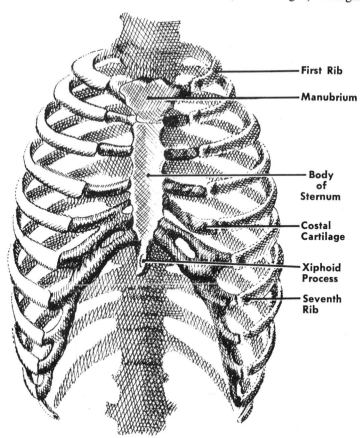

Figure 6.5 Skeletal landmarks of the rib cage—anterior view.

First Rib

Manubrium

Body of Sternum

Costal Cartilage

Xiphoid Process

Seventh Rib

the lateral plane of motion. The two moving areas of the spinal column may also be circumducted. The return motion from lateral flexion to the anatomic position is known as reduction.

An understanding of the spatial relationships of the muscles which move the spinal column is basic to comprehending their functions. Also, to facilitate learning, some of the spinal column muscles are discussed in terms of their motion functions as a group rather than individually. These muscle groups and the muscles which constitute the group are as follows:

ERECTOR SPINAE (Lumbar)
    Iliocostalis thoracis
    Iliocostalis lumborum
    Longissimus thoracis
    Spinalis thoracis
DEEP POSTERIOR SPINAL GROUP
    Intertransversarii
    Interspinales
    Rotatores
    Multifidus
ERECTOR SPINAE (Cervical)
    Iliocostalis cervicis
    Longissimus cervicis
    Longissimus capitis
    Spinalis cervicis
ABDOMINALS
    Rectus abdominis
    External oblique
    Internal oblique

Figure 6.6 shows a superior, transverse section of the spinal column through a lumbar vertebra. The spatial relationships of the abdominals and erector spinae to the spinal column are clearly seen in a cross section drawing. The reader should also note the positions of the muscles in relation to each other.

The oblique muscles have diagonal, lateral-to-anterior lines of pull. They are positioned around the lumbar-thoracic spine to perform strong rotary motions and flexions through transverse and cardinal anteroposterior planes of motion. Furthermore, their lateral spatial relationships help make them the muscles most involved for lateral flexion. The transverse plane functions of the oblique muscles involve them in the *Serape Effect* to be discussed in Chapter 7.

A diagram outline of the rectus abdominis muscles is shown in figure 6.7. These muscles have anterior spatial relationships to the lumbar-thoracic spine. They are relatively long muscles lying in and parallel to the cardinal anteroposterior plane of motion. This provides them with substantial leverage for flexion against gravity as one example. Their length could be detrimental to their strength potential. However, the tendinous inscriptions tend to compartmentalize these

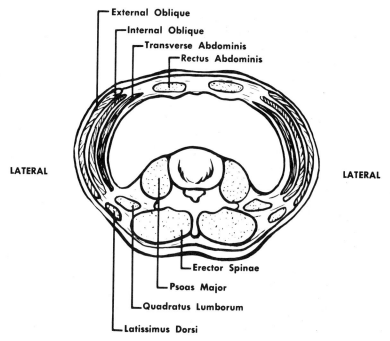

ANTERIOR

External Oblique
Internal Oblique
Transverse Abdominis
Rectus Abdominis

LATERAL

LATERAL

Erector Spinae
Psoas Major
Quadratus Lumborum
Latissimus Dorsi

POSTERIOR

Figure 6.6
Transverse section of the spinal column through a lumbar vertebra showing the spatial relationships of trunk musculature to the spinal column.

muscles. This configuration gives these muscles considerable strength potential by allowing their contractile energy to be utilized rather than diffused, as would be the case if they extended uninterrupted from the pubis to the rib cage.

Figure 6.8 provides lateral views of the external oblique, internal oblique, and transverse abdominis muscles. It should be noted that the deep transverse abdominis muscle does not work with the other abdominal muscles to move the lumbar-thoracic spine. However, it compresses the visceral contents. In addition, it is vital for such functions as defecation, urination, and parturition. All the abdominal musculature plus the erector spinae are essential to the health and well-being of the individual because they compress the viscera.

Figure 6.9 shows the erector spinae muscle group in a diagrammatic fashion. This complex muscle group actually has three subdivisions; the individual muscles which constitute this group are in bilateral pairs throughout each of these subdivisions, from the posterior ilium and sacrum to the mastoid processes. The erector spinae group is considered as a unit for its major functions, extension and hyperextension. These two motions through the anteroposterior plane require equal and bilateral contraction. Other motion capabilities of the erector spinae require unilateral contractions.

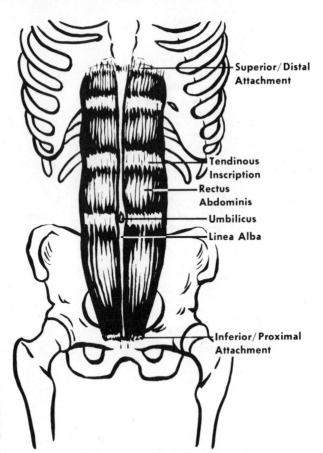

Figure 6.7 Rectus abdominis muscles— anterior spatial relationships to the lumbar-thoracic spine and pelvic girdle.

Labels in figure:
Superior/Distal Attachment
Tendinous Inscription
Rectus Abdominis
Umbilicus
Linea Alba
Inferior/Proximal Attachment

Figure 6.9 includes examples of three of the muscles in each of the three erector spinae subdivisions: (1) iliocostalis lumborum, (2) longissimus thoracis, and (3) spinalis cervicis. As noted, these muscles are arranged in bilateral pairs over the posterior length of the spinal column. All of the iliocostalis muscles from inferior to superior lie lateral to the spinal column. The longissimus muscle pairs are positioned between the iliocostalis and spinalis muscle pairs. The spinalis muscles within the erector spinae group lie virtually within the cardinal anteroposterior plane of motion.

The lines of pull and concentric contraction functions for the spinal column musculature are outlined in figure 6.10. The reversal of muscle function concept should be recalled while studying such a figure. For example, while doing a "sit-up" (flexion against resistances) exercise, the abdominals are the muscles most involved. Moreover, if the lumbar-thoracic spine is lowered (extended) slowly to the mat under muscular control, the abdominals continue to be the muscles most involved for controlling this extension eccentrically. The erector spinae would be the contralateral muscles in that situation, and the two pairs of oblique muscles would be the guiding muscles. They provide a force-counterforce to each other, which negates their rotary functions through the transverse plane of motion.

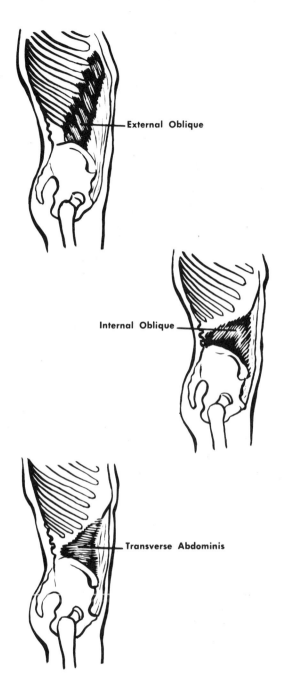

External Oblique

Internal Oblique

Transverse Abdominis

Figure 6.8
Abdominal musculature
with lateral to anterior
spatial relationships to
the lumbar-thoracic spine
and pelvic girdle.

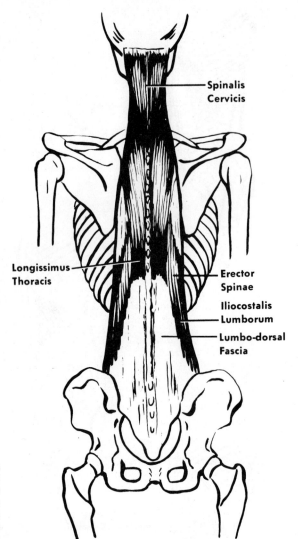

Spinalis
Cervicis

Longissimus
Thoracis

Erector
Spinae

Iliocostalis
Lumborum

Lumbo-dorsal
Fascia

Figure 6.9 The posterior spatial relationships of the erector spinae musculature to the lumbar-thoracic spine and pelvic girdle.

Muscles	Proximal Attachment	Distal Attachment
I. Lumbar-Thoracic Spine Musculature		
A. Flexors		
1. RECTUS ABDOMINIS	Pubic crest, superior	Anterior on the cartilages of ribs 5-7 and the xiphoid process
2. EXTERNAL OBLIQUE	Inferior border of the lower eight ribs	Anterior half of the iliac crest, pubic crest, and linea alba
3. INTERNAL OBLIQUE	Anterior two-thirds of the iliac crest, lateral half of the inguinal ligament and lumbodorsal fascia	Cartilage of the lower four ribs of the linea alba
4. Psoas	Twelfth thoracic and all lumbar vertebrae	Lesser trochanter of femur
B. Extensors		
1. ERECTOR SPINAE	Thoracolumbar fasciae; posterior lumbar, thoracic and lower cervical vertebrae; posterior ribs	Posterior ribs; posterior cervical and thoracic vertebrae and mastoid processes
2. Deep posterior spinal group	Posterior sacrum and processes of all vertebrae	Spinous and transverse processes and laminae of vertebrae superior to proximal attachments
3. Semispinalis thoracis	Transverse processes of thoracic vertebrae 6-10	Spinous processes of upper four thoracic and lower two cervical vertebrae
C. Rotators		
1. INTERNAL OBLIQUE	Anterior two-thirds of the iliac crest, lateral half of the inguinal ligament and lumbodorsal fascia	Cartilage of the lower four ribs and the linea alba
2. EXTERNAL OBLIQUE	Inferior border of the lower eight ribs	Anterior half of the iliac crest, pubic crest, and linea alba
3. ERECTOR SPINAE	Thoracolumbar fasciae; posterior lumbar, thoracic and lower cervical vertebrae; posterior ribs	Posterior ribs; posterior cervical and thoracic vertebrae and mastoid processes
4. Semispinalis thoracis	Transverse processes of thoracic vertebrae 6-10	Spinous processes of upper four thoracic and lower two cervical vertebrae
5. Deep posterior spinal group	Posterior sacrum and processes of all vertebrae	Spinous and transverse processes and laminae of vertebrae superior to proximal attachments

Figure 6.10 Spinal column musculature. The muscles most involved for the motions against resistance are shown in uppercase letters. Muscles supplying supplemental force during concentric contraction are in lowercase letters.

Figure 6.10—*Continued*

Muscles	Proximal Attachment	Distal Attachment
D. Lateral Flexors		
1. RECTUS ABDOMINIS	Pubic crest, superior	Anterior on the cartilages of ribs 5-7 and the xiphoid process
2. INTERNAL OBLIQUE	Anterior two-thirds of the iliac crest, lateral half of the inguinal ligament and lumbodorsal fascia	Cartilage of the lower four ribs and the linea alba
3. EXTERNAL OBLIQUE	Inferior border of the lower eight ribs	Anterior half of the iliac crest, pubic crest, and linea alba
4. ERECTOR SPINAE	Thoracolumbar fasciae; posterior lumbar, thoracic and lower cervical vertebrae; posterior ribs	Posterior ribs; posterior cervical and thoracic vertebrae and mastoid processes
5. Semispinalis thoracis	Transverse processes of thoracic vertebrae 6-10	Spinous processes of upper four thoracic and lower two cervical vertebrae
6. Quadratus lumborum	Iliac crest and iliolumbar ligament	Inferior aspect of the twelfth rib and transverse processes of the upper four lumbar vertebrae
II. Cervical Spine Musculature		
A. Flexors		
1. STERNOCLEIDOMASTOID	*Sternum head:* Anterior manubrium	Mastoid process and superior nuchal line
B. Extensors		
1. ERECTOR SPINAE	Thoracolumbar fasciae; posterior lumbar, thoracic and lower cervical vertebrae; posterior ribs	Posterior ribs; posterior cervical and thoracic vertebrae and mastoid processes
2. Deep posterior spinal group	Posterior sacrum and processes of all vertebrae	Spinous and transverse processes and laminae of vertebrae superior to proximal attachments
3. Splenius capitis	Ligamentum nuchae, spinous processes of the seventh cervical and upper three thoracic vertebrae	Mastoid process and occipital bone
4. Semispinalis cervicis	Transverse processes of upper five thoracic vertebrae	Spinous processes of first five cervical vertebrae
5. Semispinalis capitis	Articular processes of 4-6 cervical vertebrae; transverse processes of the seventh cervical and superior six thoracic vertebrae	Occipital area between the superior and inferior nuchal lines

Figure 6.10—*Continued*

Muscles	Proximal Attachment	Distal Attachment
C. Rotators		
1. STERNOCLEI-DOMASTOID	*Sternum head:* Anterior manubrium	Mastoid process and superior nuchal line
2. ERECTOR SPINAE	Thoracolumbar fasciae; posterior lumbar, thoracic and lower cervical vertebrae; posterior ribs	Posterior ribs; posterior cervical and thoracic vertebrae and mastoid processes
3. Splenius capitis	Ligamentum nuchae, spinous processes of the seventh cervical and upper three thoracic vertebrae	Mastoid process and occipital bone
4. Splenius cervicis	Spinous processes of thoracic vertebrae 3-6	Transverse processes cervical vertebrae two and three
D. Lateral Flexors		
1. STERNOCLEI-DOMASTOID	*Sternum head:* Anterior manubrium	Mastoid process and superior nuchal line
2. ERECTOR SPINAE	Thoracolumbar fasciae; posterior lumbar, thoracic and lower cervical vertebrae; posterior ribs	Posterior ribs; posterior cervical and thoracic vertebrae and mastoid processes
3. Three Scaleni	First two ribs	Transverse processes of cervical vertebrae
4. Splenius capitis	Ligamentum nuchae, spinous processes of the seventh cervical and upper three thoracic vertebrae	Mastoid process and occipital bone
5. Splenius cervicis	Spinous processes of thoracic vertebrae 3-6	Transverse processes cervical vertebrae two and three
6. Semispinalis cervicis	Transverse processes of upper five thoracic vertebrae	Spinous processes of first five cervical vertebrae
7. Semispinalis capitis	Articular processes of 4-6 cervical vertebrae; transverse processes of the seventh cervical and superior six thoracic vertebrae	Occipital area between the superior and inferior nuchal lines

Several skeletal landmarks of the skull are shown in figures 6.11 and 6.12. The cardinal anteroposterior plane passes through the middle of the occipital protuberance, maxilla, and mandible. Muscles attaching to the trunk also attach to the skull at the mastoid processes, nuchal line, and the occipital protuberance. The mastoid process is a prominent skeletal landmark slightly posterior to the cardinal lateral plane of motion.

The Skull

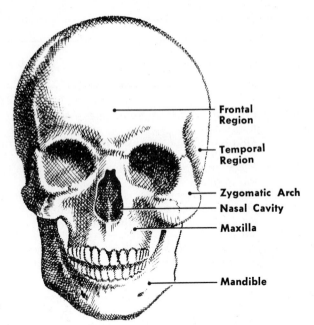

Frontal
Region

Temporal
Region

Zygomatic Arch

Nasal Cavity

Maxilla

Mandible

Figure 6.11 The skull—anterior view.

Figure 6.12 The skull—posterior view.

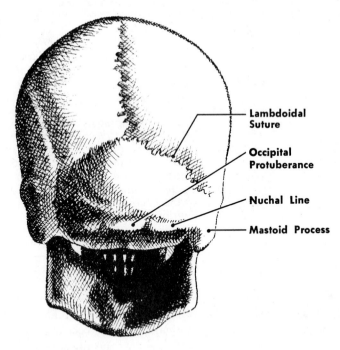

Lambdoidal
Suture

Occipital
Protuberance

Nuchal Line

Mastoid Process

Benteler, A. M. "Change in Tone of the Spinal Muscle in Man." *Sechnor Physiological Journal of the U.S.S.R.* 47 (1961):393–98.

Flint, M. M. "Effect of Increasing Back and Abdominal Muscle Strength on Low Back Pain." *Research Quarterly* 29 (May, 1958):160–71.

Gough, Joseph G., and Koepke, George H. "Electromyographic Determination of Motor Root Levels in Erector Spinae Muscles." *Archives of Physical Medicine and Rehabilitation* 47 (January, 1966):9–11.

Gutin, Bernard, and Lipetz, Stanley. "An Electromyographic Investigation of the Rectus Abdominis in Abdominal Exercises." *Research Quarterly* 42 (1971):256–63.

Halpren, A. A., et al. "Sit-up Exercises: An Electromyographic Study." *Clinical Orthopaedics and Related Research* 145 (November–December, 1979):172–178.

Keagy, Robert D.; Brumlick, Joel; and Bergan, John J. "Direct Electromyography of the Psoas Major Muscle in Man." *Journal of Bone and Joint Surgery* 48-A (October, 1966):1377–82.

Kottke, Frederic J., and Mundale, Martin O. "Range of Mobility of the Cervical Spine." *Archives of Physical Medicine and Rehabilitation* 40 (September, 1959):379–82.

La Ban, Myron M., et al. "Electromyographic Study of Function of Iliopsoas Muscle." *Archives of Physical Medicine and Rehabilitation* 46 (October, 1965):676–79.

Michele, Arthur A. *Iliopsoas.* Springfield: Charles C Thomas, Publisher, 1962.

Partridge, Miriam J., and Walters, C. Etta. "Participation of the Abdominal Muscle in Various Movements of the Trunk in Man." *Physical Therapy Review* 39 (December, 1959):791–800.

Rab, G. T. "Muscle Forces in the Posterior Thoracic Spine." *Clinical Orthopaedics and Related Research* 139 (March–April, 1979):28–32.

Rab, G. T., et al. "Muscle Force Analysis of the Lumbar Spine." *Orthopedic Clinics of North America* 8 (January, 1977):193–199.

Wilson, Gary L., Capen, Edward K. and Stubbs, Nancy B. "A Fine-Wire Electromyographic Investigation of the Gluteus Minimus and Gluteus Medius Muscles." *Research Quarterly* 47 (December, 1976):824–828.

Recommended Reading

7

Upper Limb Segment Myology

The purposes of this chapter are to present the bones, skeletal landmarks, major joints, and muscles which move the upper limb segments. The areas covered are the shoulder girdle, shoulder joint, elbow joint, radio-ulnar joint, wrist, and internal joints of the hand. In addition, a specialized motion concept for highly skilled *ballistic* performances (the Serape Effect) is presented in this chapter.

Shoulder Girdle

The shoulder girdle consists of the two clavicles and two scapulae shown in figure 7.1. The pivotal point for shoulder girdle motions is the sternoclavicular joint. The skeletal landmarks on the shoulder girdle bones are shown in figures 7.2, 7.3 and 7.4. The acromion process is a recognized landmark used to define the lateral plane of motion. It is also used to determine the extent of elevation, protraction, and retraction occurring within the shoulder girdle.

There are six motions possible within the shoulder girdle: (1) abduction (protraction), (2) adduction (retraction), (3) downward rotation, (4) upward rotation, (5) elevation, and (6) depression. The rotations of the shoulder girdle contribute to shoulder joint motions, because the glenoid fossa, which holds the head of the humerus, must be rotated upward or downward to accommodate similar motions through the traditional and diagonal planes at the shoulder joint. If the muscles of the shoulder girdle could not perform these functions, it would (as examples) be impossible to fully flex, diagonally abduct or adduct the shoulder joint.

The shoulder girdle is not a bony mass similar to the pelvic girdle, and does not have the structural stability of the pelvic girdle. Anteriorly, the clavicles articulate with the sternum. Laterally, they articulate with the scapulae (figure 7.1). Posteriorly, the two vertebral borders of the scapulae do not articulate with each other. Furthermore, there is not a bone to bone relationship between

Figure 7.1   Shoulder girdle—superior view.

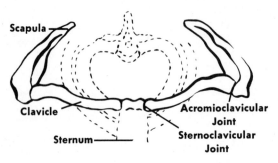

Scapula

Clavicle

Sternum

Acromioclavicular Joint

Sternoclavicular Joint

126

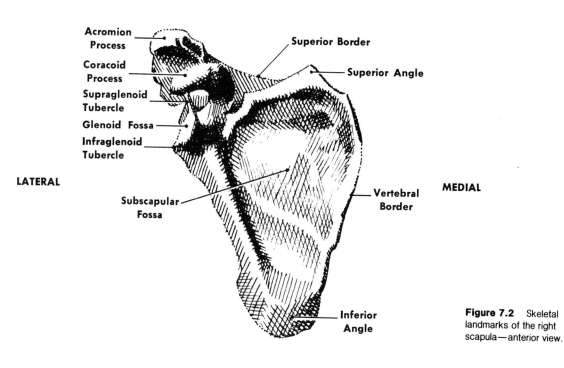

Acromion
Process

Coracoid
Process

Supraglenoid
Tubercle

Glenoid Fossa

Infraglenoid
Tubercle

Superior Border

Superior Angle

LATERAL

MEDIAL

Subscapular
Fossa

Vertebral
Border

Inferior
Angle

Figure 7.2 Skeletal
landmarks of the right
scapula—anterior view.

Figure 7.3 Skeletal
landmarks of the right
scapula—posterior view.

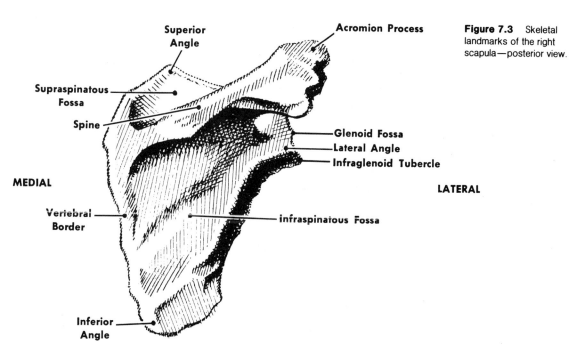

Superior
Angle

Acromion Process

Supraspinatous
Fossa

Spine

Glenoid Fossa
Lateral Angle
Infraglenoid Tubercle

MEDIAL

LATERAL

Vertebral
Border

Infraspinatous Fossa

Inferior
Angle

Basic Concepts of Anatomic Kinesiology   127

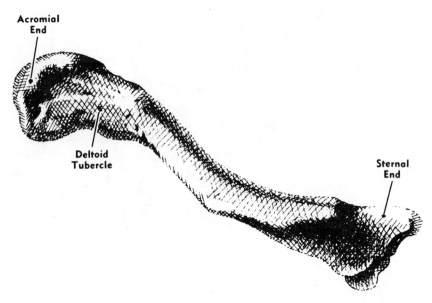

Acromial
End

Deltoid
Tubercle

Sternal
End

Figure 7.4 Right
clavicle—anterior view.

the scapulae and thoracic vertebrae. The posterior continuation of the shoulder
girdle from the vertebral borders of the scapulae to the thoracic spine is effected
by the lines of pull of the rhomboid muscles. This can be observed in figure 7.5.
The rhomboids lie beneath the superficial trapezius muscle which is shown in
figure 7.6.

The application of the spatial relationship concept to the shoulder girdle is
the same as that which was presented in Chapter 6 in relation to the pelvic girdle.
The spatial relationship reflects the location of the muscle to the shoulder girdle.
As an example, the superior spatial relationship of the levator scapulae (figure
7.5) places it in an ideal leverage position to elevate the shoulder girdle when it
contracts concentrically.

As can be noted in figures 7.7 and 7.8, there is effective collaboration among
the shoulder girdle muscles to produce the desired motion. Whenever the upper
limb is moved upward through any plane, the shoulder girdle must be rotated
upward. The glenoid fossa must accommodate the head of the humerus at all
times to maintain the integrity of the shoulder joint. To bring about upward rotation of the shoulder girdle when the shoulder joint is moved more than 60 degrees, the fibers of the serratus anterior and trapezius muscles must cooperate
as they contract. This is shown in figure 7.7.; and downward rotation cooperative
action against resistance is shown in figure 7.8 involving the rhomboids and pectoralis minor muscles.

The lines of pull and motion functions for shoulder girdle muscles are included in figure 7.9.

Upper Limb Segment Myology

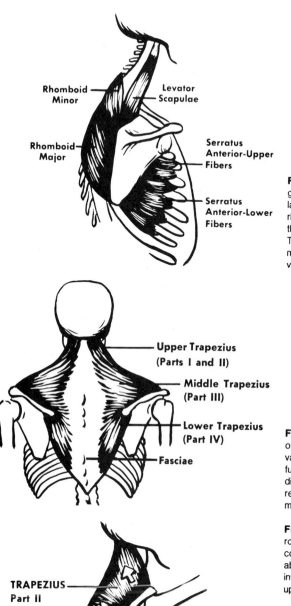

Rhomboid Minor

Levator Scapulae

Rhomboid Major

Serratus Anterior-Upper Fibers

Serratus Anterior-Lower Fibers

Figure 7.5 Shoulder girdle musculature— lateral view. The rhomboids lie beneath the superficial trapezius. The anterior pectoralis minor is not seen in this view.

Upper Trapezius (Parts I and II)

Middle Trapezius (Part III)

Lower Trapezius (Part IV)

Fasciae

Figure 7.6 The parts of the trapezius have a variety of shoulder girdle functions due to their different spatial relationships to that moving segment.

Figure 7.7 Upward rotation of the scapula concurrent with diagonal abduction. (Muscles most involved appear in uppercase letters.)

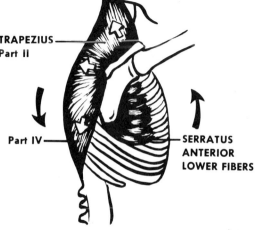

TRAPEZIUS Part II

Part IV

SERRATUS ANTERIOR LOWER FIBERS

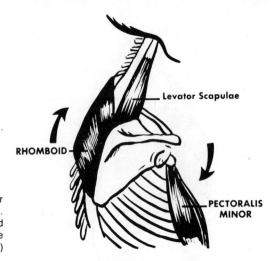

Figure 7.8 Scapular downward rotators. (Muscles most involved appear in uppercase letters.)

RHOMBOID

Levator Scapulae

PECTORALIS MINOR

Muscles	Proximal Attachment	Distal Attachment
I. Shoulder Girdle Musculature		
A. Abductors		
1. PECTORALIS MINOR	Upper, anterior aspects of ribs 3-5	Coracoid process of the scapula
2. SERRATUS ANTERIOR	Outer and lateral aspects of ribs 1-8	Anterior, vertebral scapula border
B. Adductors		
1. TRAPEZIUS a. (MIDDLE FIBERS)	Occipital protuberance; ligamentum nuchae and spinous process of the seventh cervical and all thoracic vertebrae	Scapular spine, superior border
2. RHOMBOID	Seventh cervical to the fifth thoracic spinous processes	Posterior, vertebral scapula border
C. Downward Rotators		
1. RHOMBOID	Seventh cervical to the fifth thoracic spinous processes	Posterior, vertebral scapula border
2. PECTORALIS MINOR	Upper, anterior aspects of ribs 3-5	Coracoid process of the scapula

Figure 7.9 Shoulder girdle musculature. Muscles most involved for the motions against resistance are shown in uppercase letters.

Upper Limb Segment Myology

Figure 7.9— *Continued*

Muscles	Proximal Attachment	Distal Attachment
D. Upward Rotators		
1. TRAPEZIUS (ALL PARTS)		
a. UPPER FIBERS	Occipital protuberance; ligamentum nuchae and spinous process of the seventh cervical and all thoracic vertebrae	Posterior border of the lateral third of the clavicle and acromion process
b. LOWER FIBERS	Occipital protuberance; ligamentum nuchae and spinous process of the seventh cervical and all thoracic vertebrae	Medial aspect of the scapular spine
2. SERRATUS ANTERIOR (LOWER FIBERS)	Outer and lateral aspects of ribs 1-8	Anterior, vertebral scapula border
E. Elevators		
1. LEVATOR SCAPULAE	Transverse processes of the first four cervical vertebrae	Superior angle of the scapula
2. TRAPEZIUS (UPPER FIBERS)	Occipital protuberance; ligamentum nuchae and spinous process of the seventh cervical and all thoracic vertebrae	Posterior border of the lateral third of the clavicle and acromion process
3. RHOMBOID	Seventh cervical to the fifth thoracic spinous processes	Posterior, vertebral scapula border
F. Depressors		
1. TRAPEZIUS (LOWER FIBERS)	Occipital protuberance; ligamentum nuchae and spinous process of the seventh cervical and all thoracic vertebrae	Medial aspect of the scapular spine
2. PECTORALIS MINOR	Upper, anterior aspects of ribs 3-5	Coracoid process of the scapula

The Serape Effect

The serape is a brightly colored woolen blanket worn as an outer garment by some people who live in Mexico and other Latin-American countries. It is designed to hang around the shoulders and cross diagonally on the anterior aspect of the trunk of the wearer. This is analogous to the direction of pull of a series of four pairs of muscles in the same general region covered by a serape. The four pairs of muscles are: (1) rhomboids, (2) serratus anterior, (3) external obliques, and (4) internal obliques. (Figure 7.10)

Synchronized contractions by these four pairs of muscles cause interrelated motions within the pelvic girdle, lumbar-thoracic spine, and shoulder girdle-joint during throwing, striking, and kicking ballistic skills. These motions produce the Serape Effect. (It is recognized that external forces and other muscles also contribute some force for the critical segment motions which result in the Serape Effect.) The Serape Effect is a synthesis of several principles of anatomic kinesiology and biomechanics (Logan and Wallis, 1960).

The importance of The Serape Effect lies in the fact that performers who use it during a ballistic skill are almost always more skilled than athletes who do not use the motions in proper timed sequence or who minimize the ranges of motion of the segments. (The Serape Effect is a maximum force concept; therefore, it applies only to throwing, striking, kicking, and other ballistic skills.)

Figure 7.10 The serape musculature of a right handed discus thrower. The pelvic girdle and lumbar-thoracic segments rotate prior to release.

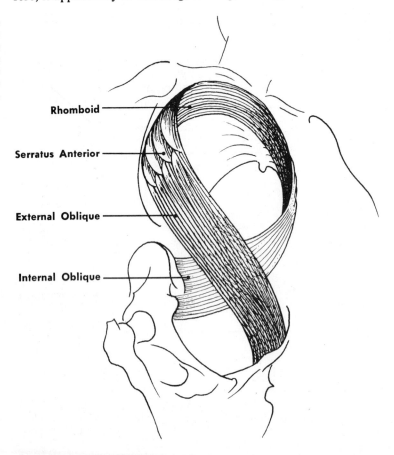

Rhomboid

Serratus Anterior

External Oblique

Internal Oblique

Upper Limb Segment Myology

The critical moving segments which produce the Serape Effect are the transverse plane rotations of the pelvic girdle and lumbar-thoracic spine. The transverse rotation of the lumbar-thoracic spine during throwing and striking skills utilizing the upper limbs is assisted by shoulder girdle-joint motions within the limb used in the skills. The timing of the pelvic girdle and lumbar-thoracic spine transverse rotations is critical. They must be observed during the *preparatory phase* of the ballistic skill immediately prior to the start of the motion phase. Furthermore, these two major segments must rotate in *opposite directions* (torque-countertorque) during the preparatory phase of the skill. As one example, in throwing skills where the upper limb is moved through a high diagonal plane, a left-handed performer must have right transverse pelvic girdle rotation concurrent with left transverse rotation of the lumbar-thoracic spine during the weight shift portion of the preparation phase of the throw (figure 7.11). If this sequence of motions within these segments has occurred, it will produce the Serape Effect.

Figure 7.11 Rich Rivera, California State University-Long Beach pitcher, demonstrates a classic example of The Serape Effect. The right transverse lumbar-thoracic spine rotators and left shoulder joint diagonal adductors have been placed on maximum stretch due to the concurrent torque-countertorque of the pelvic girdle and lumbar-thoracic spine segments plus left shoulder diagonal abduction during his preparatory motion for the pitch. He is just starting into the motion phase. *Photo by Bruce Hazelton. Reprinted with permission of the Sports Information Bureau, California State University, Long Beach.*

Basic Concepts of Anatomic Kinesiology

The *effect* of the perfectly timed torque-countertorque of the pelvis and lumbar-thoracic spine during the preparatory phase of a ballistic skill is to place the muscles to be used during the critical motion phase on maximum stretch (muscle length-force principle). *This is the Serape Effect.* The *effect* of the two body segments rotating in opposite directions concurrently through the transverse plane, stretches the abdominal and shoulder girdle-joint muscles. The force exerted by a muscle is in direct proportion to its length-tension at the time of contraction (muscle length-force ratio). The "stretched muscles" during the preparatory phase become the "force muscles" during the motion phase.

In throwing, striking, and kicking skills where maximal to near maximal forces and velocities of limbs, sport objects, and sport implements are desired, the Serape Effect adds significantly to the summation of force. For example, there is an efficient transfer of force from the large, stretched muscles of the trunk, once they contract to initiate the motion phase, to the smaller muscle mass at the shoulder, which moves the throwing limb. From a biomechanic standpoint, this is an effective way to summate and transfer force.

In order to fully describe the Serape Effect, the functional relationship of the "serape muscles" on either side of the body should be considered. Beginning with the rhomboids, which have a downward and lateral direction and attach proximally to the spinal column and distally to the vertebral border of the scapula, there is a functional completion of these muscles in the serratus anterior. The serratus anterior also attaches at the vertebral border of the scapula. The

Figure 7.12 Serape Muscles—posterior view.

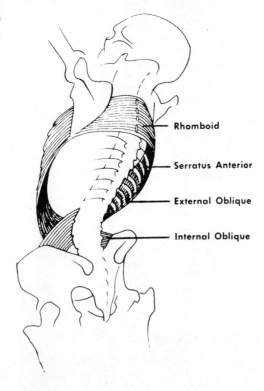

— Rhomboid

— Serratus Anterior

— External Oblique

— Internal Oblique

Upper Limb Segment Myology

serratus anterior continues diagonally and downward as it attaches to the rib cage laterally and anteriorly. These two pairs of muscles work together on the vertebral border to move the scapula. Therefore, the "serape muscles" provide dynamic stability as well as movement of the scapula (figure 7.12).

Continuing in a circular downward and diagonal direction on the rib cage is the external oblique on one side which functionally, but not literally, continues into the internal oblique on the opposite side. The internal obliques terminate at the pelvis. When the bilateral pairs of these four muscles are considered, there are two diagonals crossing in front of the body and working in conjunction with each other; this is a "muscular serape" wrapping diagonally around the trunk (figure 7.13).

Figure 7.11 provides an excellent example of a performer who utilizes the Serape Effect. As noted, the motions which produce the Serape Effect must be observed in the preparatory phase of a ballistic skill. The left handed pitcher in figure 7.11 has just completed that phase, and he is now beginning to take advantage of the effect of stretched muscles during the start of the motion phase of the pitch.

The pelvic girdle is rotated right through the transverse plane concurrent with left lumbar-thoracic rotation in figure 7.11. These motions occurred during the unilateral weight shift from the left to the right leg in the preparatory phase

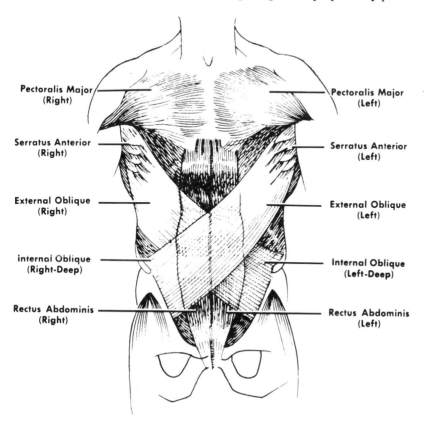

Pectoralis Major
(Right)

Serratus Anterior
(Right)

External Oblique
(Right)

internal Oblique
(Right-Deep)

Rectus Abdominis
(Right)

Pectoralis Major
(Left)

Serratus Anterior
(Left)

External Oblique
(Left)

Internal Oblique
(Left-Deep)

Rectus Abdominis
(Left)

Figure 7.13 The anterior muscular serape; there is a literal interaction between the serratus anterior and external oblique muscles. The interaction between the external oblique of one side is functional, not literal, with the internal oblique muscle of the opposite side.

of the skill. The lumbar-thoracic rotation was assisted by the diagonal abduction momentum of the left upper limb at the shoulder girdle-joint segment. The effect of these motions was to place the right transverse plane rotators of the lumbar-thoracic spine on maximum stretch. The Serape muscles involved for the latter motion are the left external oblique and right internal oblique muscles. These were reciprocally innervated contralateral muscles during the torque-counter-torque of the pelvic and spinal segments in the preparatory phase of the pitch.

As the pitcher in figure 7.11 starts his motion phase, the pelvic girdle is dynamically stabilized throughout the completion of the pitch (DeMille, 1962). The lumbar-thoracic spine is rotated to the right by the stretched muscles noted above and by the right erector spinae. The left shoulder girdle-joint musculature contracts to adduct diagonally and to bring the left upper limb through a high diagonal plane of motion. The reader should observe figure 7.11 closely and try to visualize the stretch placed on the left pectoralis major, anterior deltoid, coracobrachialis, and biceps brachii (short head) by the action of the left shoulder girdle-joint diagonal abduction plus the pelvic and spine rotations during the preparatory phase for the pitch. Those muscles with anterior spatial relationships to the shoulder joint are the important diagonal adductors used in the motion phase. They must be placed on maximum stretch to functionally summate force within the total kinetic chain of this skill.

The sequence of motions by the pitcher in figure 7.11 shows effective utilization of the timing and ranges of motion at the segments and joints involved to produce the Serape Effect. His only technique problem in figure 7.11 is related to the placement of the right foot during the unilateral weight shift of the preparatory phase; his line of thrust should be toward the hitter. There appears to be lateral rotation of the knee causing a lateral displacement of the foot. This could dissipate some of his summated forces in a tangent away from the home plate area. It could also be a control factor in this skill. For those reasons, a coach should modify the skill by eliminating lateral rotation of the right knee if that is a *consistent* motion problem.

In summary, the Serape Effect is an essential motion concept for ballistic sport skills. Performers who execute the pelvic girdle and spinal column rotations in the proper timed sequence concurrently with limb motions are more apt to produce a greater summation of force. Athletes who can control these motions for the desired outcomes within the strategies of their games are better skilled than their counterparts.

Shoulder (Glenohumeral) Joint

Scientifically, the shoulder joint is known as the glenohumeral joint. The reason for this is that the glenoid fossa of the scapula and the humerus join to make the articulation. This articulation is shown in figure 7.14. This literal joining together of the shoulder joint with the shoulder girdle necessitates a functional motion relationship between the two areas, i.e., shoulder girdle motions must be synchronized with shoulder joint motions in order to maintain the proper relationship between the glenoid fossa and humeral head. This relationship protects the integrity of the shoulder joint.

The important skeletal landmarks of the humerus are shown in three views in figure 7.15. The ones shown are primarily significant as muscle attachments.

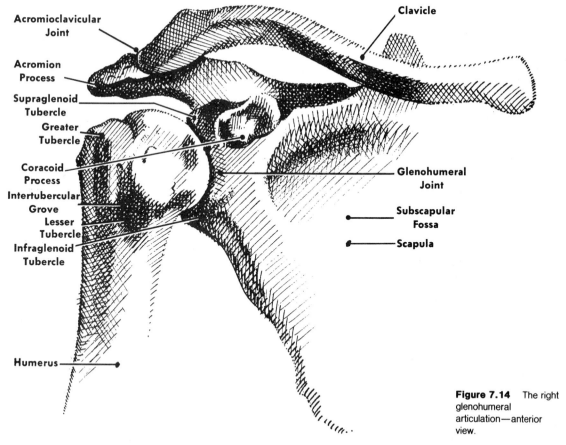

Acromioclavicular Joint

Acromion Process

Supraglenoid Tubercle

Greater Tubercle

Coracoid Process

Intertubercular Grove

Lesser Tubercle

Infraglenoid Tubercle

Humerus

Clavicle

Glenohumeral Joint

Subscapular Fossa

Scapula

Figure 7.14 The right glenohumeral articulation—anterior view.

The greater tubercle, deltoid tuberosity, and lateral epicondyle can at times be used as superficial landmarks for cinematographic analyses utilizing lateral views.

The shoulder joint is multiaxial. Motions can be made through five planes:

Plane	*Motions Possible*
1. Anteroposterior	Flexion, extension, hyperextension (figure 7.16)
2. Lateral	Abduction, adduction (figure 7.17)
3. Transverse	Horizontal abduction, horizontal adduction, lateral rotation, medial rotation
4. High diagonal	Diagonal abduction, diagonal adduction
5. Low diagonal	Diagonal abduction, diagonal adduction

Circumduction is also possible through 180° of diagonal abduction and 180° of diagonal adduction.

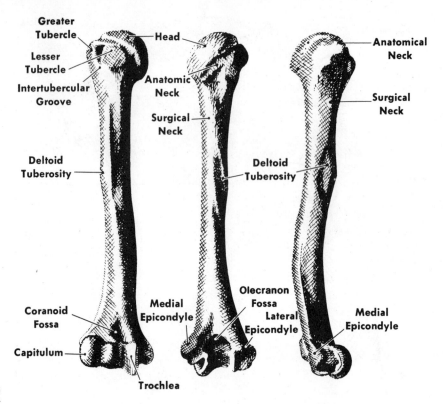

Figure 7.15 Skeletal landmarks of the right humerus.

ANTERIOR   **POSTERIOR**   **MEDIAL**

Greater Tubercle
Lesser Tubercle
Intertubercular Groove
Head
Anatomic Neck
Surgical Neck
Deltoid Tuberosity
Deltoid Tuberosity
Coranoid Fossa
Medial Epicondyle
Olecranon Fossa
Lateral Epicondyle
Capitulum
Trochlea
Anatomical Neck
Surgical Neck
Medial Epicondyle

Figure 7.16
Conceptualization and determination of the extent of shoulder joint flexion-extension through an anteroposterior plane of motion. The total upper limb segment has been flexed 90 degrees.

Figure 7.17
Conceptualization and determination of the extent of shoulder joint abduction-adduction through the lateral plane of motion. The total upper limb segment has been abducted 90 degrees. For this segment to be abducted 180-degrees, it must also be laterally rotated.

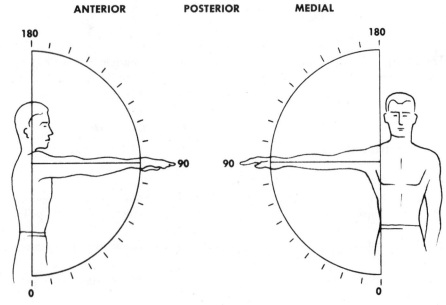

Upper Limb Segment Myology

The complexity of the shoulder joint makes determination of muscle spatial relationships to the joint seem slightly abstract at times. Nevertheless, basic criteria for spatial relationships of muscles to joints apply. The observer must realize that this joint is highly mobile in space itself, and there is an on-going relationship of the humerus segment to the center of the shoulder joint and the trunk segment. The attachments of the muscles on the humerus do not change, but the position of the humerus in space changes frequently; this is what appears to change muscle functions in some shoulder joint motions.

Depending upon where the humerus is located in space, leverages will change for certain muscles. The pectoralis major (figure 7.18) provides an example of this. This versatile muscle attaches to the humerus on the intertubercular groove. Its upper fibers from the clavicle flex the shoulder against resistance. Once the humerus segment is moved more than 100 degrees in flexion, the lower fibers are stretched significantly. If needed, they can contract concentrically and forcefully move the humerus into extension against resistance. The pectoralis major also helps the lateral deltoid and supraspinatus muscles abduct the humerus from 90 to 180 degrees in the range of motion. Again, the attachments do not change, but their position in space is considerably altered in shoulder joint motions. This should be kept in mind while studying all shoulder joint motions and muscles.

Figure 7.19 is provided to show the rotator cuff muscles which attach to the humeral head. These four useful muscles help maintain the integrity of the shoulder joint by keeping the humeral head firmly positioned in the glenoid fossa of the scapula. If an athlete in throwing and striking sports damages the tendons of these muscles, it can lead to considerable diminution of skilled performance.

Observation of figure 7.19 will show the following spatial relationships of the four tendons of the rotator cuff to the center of the shoulder joint:

Muscles	*Spatial Relationships*
Subscapularis	Anterior
Supraspinatus	Superior and lateral
Infraspinatus	Posterior
Teres Minor	Posterior

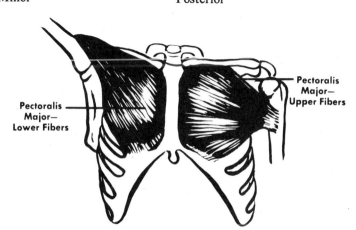

Figure 7.18 The motion functions of the pectoralis major change depending upon the position of the humerus segment in space during shoulder joint motions.

Pectoralis Major— Lower Fibers

Pectoralis Major— Upper Fibers

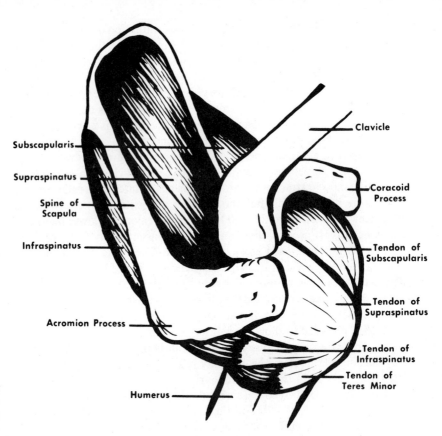

Figure 7.19 Rotator cuff muscles—superior view of the right shoulder.

Subscapularis

Supraspinatus

Spine of Scapula

Infraspinatus

Acromion Process

Humerus

Clavicle

Coracoid Process

Tendon of Subscapularis

Tendon of Supraspinatus

Tendon of Infraspinatus

Tendon of Teres Minor

The deltoid muscle is a superficial muscle of the shoulder (figure 7.20). It is relatively large with very good leverage, especially through the lateral plane of motion. However, this muscle has considerable motion versatility by reason of its anterior, lateral, and posterior spatial relationships to the center of the shoulder joint.

The latissimus dorsi is another powerful and versatile muscle of the shoulder joint (figure 7.21). It performs its motion functions with the teres major muscle (figure 7.22). These muscles have medial spatial relationships to the shoulder joint for adduction against resistance. (Shoulder adduction is the key motion in the "lat pull" weight training exercise.) They also contribute great force for medial rotation (figure 7.21) if the humerus has been laterally rotated and then diagonally abducted effectively through a high diagonal plane range of motion. These motion patterns are seen extensively in throwing and striking skills.

The muscles most involved for the shoulder joint motions against resistance are outlined in figure 7.23, which illustrates the concentric contraction functions of these muscles. The reader is reminded once again to apply the reversal of muscle function concept to explain or describe the eccentric contraction control of motion. For example, if the total upper limb segment is being lowered (extended)

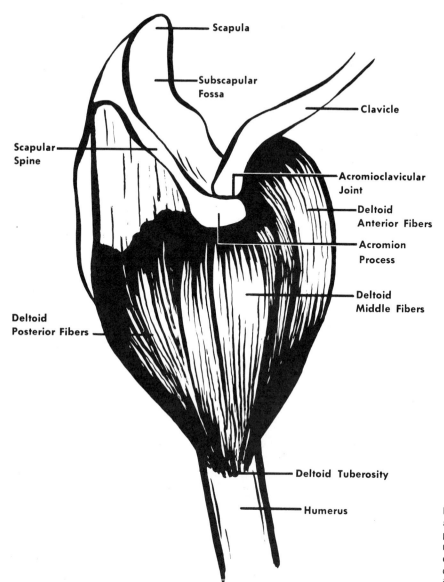

Scapula

Subscapular Fossa

Clavicle

Scapular Spine

Acromioclavicular Joint

Deltoid Anterior Fibers

Acromion Process

Deltoid Middle Fibers

Deltoid Posterior Fibers

Deltoid Tuberosity

Humerus

Figure 7.20 The anterior, lateral and posterior spatial relationships of the deltoid muscle to the center of the shoulder joint.

slowly through an anteroposterior plane of motion, the flexors listed in figure 7.23 will contract eccentrically to control shoulder extension with gravity. Therefore, the muscles most involved for that type of extension would be the anterior deltoid and upper pectoralis major. The reversal of muscle function applies to each motion in this manner.

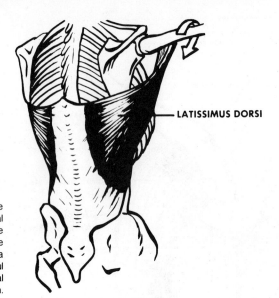

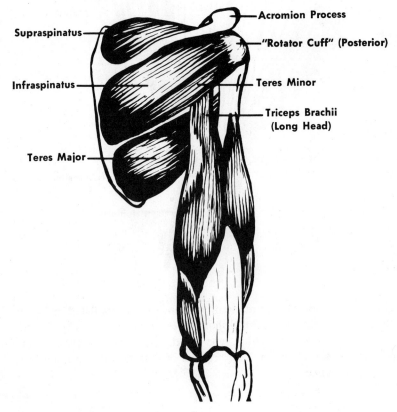

LATISSIMUS DORSI

Figure 7.21 The medial spatial relationship of the latissimus dorsi to the shoulder joint makes it a potentially powerful muscle for medial rotation and adduction.

Figure 7.22 Posterior view of the right shoulder joint musculature.

Supraspinatus

Infraspinatus

Teres Major

Acromion Process

"Rotator Cuff" (Posterior)

Teres Minor

Triceps Brachii (Long Head)

Muscles	Proximal Attachment	Distal Attachment
I. Shoulder Joint Musculature		
A. Flexors		
1. DELTOID (ANTERIOR FIBERS)	Anterior aspect of the lateral one-third of the clavicle, acromion and scapular spine	Deltoid tuberosity of the humerus
2. PECTORALIS MAJOR (UPPER FIBERS)	Anterior and medial half of the clavicle	Lateral to the intertubercular groove
3. Coracobrachialis	Coracoid process	Medial to the deltoid tuberosity
4. Biceps brachii (Short head)	Coracoid process	Radial tuberosity
B. Extensors		
1. PECTORALIS MAJOR (LOWER FIBERS)	Anterior sternum and cartilages of ribs four, five and six	Lateral to the intertubercular groove
2. LATISSIMUS DORSI	Spinous processes of thoracic vertebrae 7-12; all lumbar and sacral vertebrae; posterior lateral and iliac crest; four lower ribs posteriorly	Parallel to the pectoralis major tendon
3. TERES MAJOR	Inferior angle of the scapula posterior	Inferior to lesser tubercle of the humerus; medial to the latissimus dorsi
4. Deltoid (Posterior fibers)	Lower margin of scapular spine	Deltoid tuberosity of the humerus
5. Triceps brachii (Long head)	Infraglenoid tubercle of the scapula	Olecranon process—ulna

Figure 7.23
Shoulder joint musculature. The muscles most involved for the motions against resistance are shown in uppercase letters. Muscles supplying supplemental force during concentric contraction are in lowercase letters.

Figure 7.23—*Continued*

	Muscles	Proximal Attachment	Distal Attachment
C.	Abductors		
1.	DELTOID (Mid)	Lateral clavicle and acromion process	Deltoid tuberosity of the humerus
2.	SUPRASPI-NATUS	Supraspinatus fossa of the scapula	Greater tubercle of the humerus
3.	PECTORALIS MAJOR (UPPER FIBERS)	Anterior and medial half of the clavicle	Lateral to the intertubercular groove
D.	Adductors		
1.	LATISSIMUS DORSI	Spinous processes of thoracic vertebrae 7-12; all lumbar and sacral vertebrae; posterior lateral and iliac crest; four lower ribs posteriorly	Parallel to the pectoralis major tendon
2.	TERES MAJOR	Inferior angle of the scapula posterior	Inferior to lesser tubercle of the humerus; medial to the latissimus dorsi
3.	PECTORALIS MAJOR (LOWER FIBERS)	Anterior sternum and cartilages of ribs four, five and six	Lateral to the intertubercular groove
4.	Triceps brachii (Long head)	Infraglenoid tubercle of the scapula	Olecranon process—ulna
E.	Medial Rotators		
1.	LATISSIMUS DORSI	Spinous processes of thoracic vertebrae 7-12; all lumbar and sacral vertebrae; posterior lateral and iliac crest; four lower ribs posteriorly	Parallel to the pectoralis major tendon
2.	TERES MAJOR	Inferior angle of the scapula posterior	Inferior to lesser tubercle of the humerus; medial to the latissimus dorsi
3.	PECTORALIS MAJOR	Anterior clavicle and sternum	Lateral to the intertubercular groove
4.	SUBSCAPU-LARIS	Medial two-thirds of the subscapular fossa	Lesser tubercle of the humerus
F.	Lateral Rotators		
1.	INFRASPINA-TUS	Upper two-thirds of the infraspinatus fossa—posterior	Posterior aspect of the greater tubercle of the humerus
2.	TERES MINOR	Auxiliary border of the upper two-thirds of the scapula	Posterior aspect of the greater tubercle of the humerus

Figure 7.23—*Continued*

Muscles	Proximal Attachment	Distal Attachment
G. Horizontal Abductors		
1. DELTOID (POSTERIOR AND MIDDLE FIBERS)	Scapular spine, lateral clavicle and acromion process	Deltoid tuberosity of the humerus
2. INFRASPINATUS	Upper two-thirds of the infraspinatus fossa—posterior	Posterior aspect of the greater tubercle of the humerus
3. TERES MINOR	Auxiliary border of the upper two-thirds of the scapula	Posterior aspect of the greater tubercle of the humerus
H. Horizontal Adductors		
1. DELTOID (ANTERIOR FIBERS	Anterior aspect of the lateral one-third of the clavicle, acromion and scapular spine	Deltoid tuberosity of the humerus
2. PECTORALIS MAJOR	Anterior clavicle and sternum	Lateral to the intertubercular groove
3. CORACO-BRACHIALIS	Coracoid process	Medial to the deltoid tuberosity
4. Biceps brachii (Short head)	Coracoid process	Radial tuberosity
I. Diagonal Abductors		
1. DELTOID (POSTERIOR FIBERS)	Lower margin of scapular spine	Deltoid tuberosity of the humerus
2. INFRASPINATUS	Upper two-thirds of the infraspinatus fossa—posterior	Posterior aspect of the greater tubercle of the humerus
3. TERES MINOR	Auxiliary border of the upper two-thirds of the scapula	Posterior aspect of the greater tubercle of the humerus
4. TRICEPS BRACHII (LONG HEAD)	Infraglenoid tubercle of the scapula	Olecranon process—ulna
J. Low Diagonal Adductors		
1. DELTOID (ANTERIOR FIBERS)	Anterior aspect of the lateral one-third of the clavicle, acromion and scapular spine	Deltoid tuberosity of the humerus
2. CORACO-BRACHIALIS	Coracoid process	Medial to the deltoid tuberosity
3. BICEPS BRACHII (SHORT HEAD)	Coracoid process	Radial tuberosity
4. PECTORALIS MAJOR (UPPER FIBERS)	Anterior and medial half of the clavicle	Lateral to the intertubercular groove

Basic Concepts of Anatomic Kinesiology   **145**

Figure 7.23—*Continued*

Muscles	Proximal Attachment	Distal Attachment
K. High Diagonal Adductors		
1. DELTOID (ANTERIOR FIBERS)	Anterior aspect of the lateral one-third of the clavicle, acromion and scapular spine	Deltoid tuberosity of the humerus
2. CORACO-BRACHIALIS	Coracoid process	Medial to the deltoid tuberosity
3. BICEPS BRACHII (SHORT HEAD)	Coracoid process	Radial tuberosity
4. PECTORALIS MAJOR (LOWER FIBERS)	Anterior sternum and cartilages of ribs four, five and six	Lateral to the intertubercular groove

Elbow Joint

The elbow articulation is composed of three bones: the humerus, ulna, and radius. On the medial aspect, the humerus articulates with the ulna, and on the lateral side the humerus articulates with the radius. The bony articulation between the humerus and ulna tends to be stable. On the other hand, the articulation between the humerus and the radius lacks bony stability. The hinge arrangement of the elbow joint occurs at the articulation between the humerus and ulna. This deep articulation provides considerable bony stability. It also is a limiting factor as far as motion is concerned, because it only allows flexion and extension to occur within the elbow joint. This means the elbow motions from the anatomic position are limited to an anteroposterior plane of motion only. Skeletal landmarks around the elbow joint are shown in figure 7.24. The superficial olecranon process can be useful in posterior views of a performer on film to locate the joint center.

There are three major elbow joint muscles which contract concentrically to cause flexion against resistance. These are the biceps brachii, brachialis, and brachioradialis muscles. They have anterior spatial relationships to the elbow joint, as seen in figures 7.25 and 7.26. If elbow extension is performed slowly with gravity, these same muscles control that motion by contracting eccentrically.

The extension force for moving the elbow against resistance is provided by the posterior triceps brachii muscle (figure 7.27). This muscle has excellent leverage within the anteroposterior plane of motion at the elbow joint. The fibers from the three heads of the muscle converge into a common tendon which applies their contractile forces on the olecranon process of the ulna. This utilization of force by means of one tendon is analogous to the quadriceps femoris and triceps surae muscle groups at the knee and ankle respectively.

The lines of pull for the muscles of the elbow, along with their motion functions, are outlined in figure 7.28.

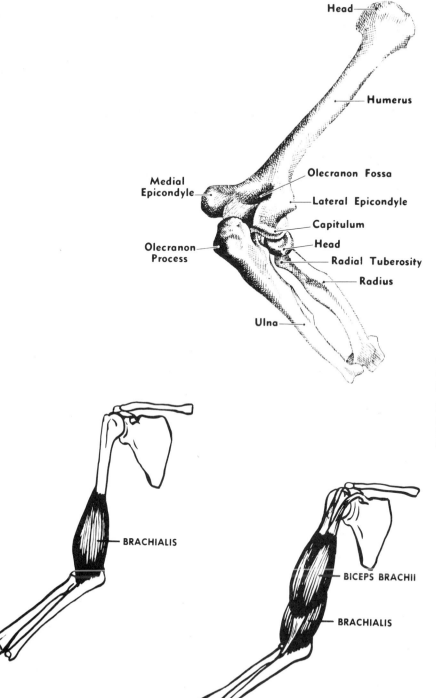

Head

Humerus

Olecranon Fossa

Medial
Epicondyle

Lateral Epicondyle

Capitulum

Head

Olecranon
Process

Radial Tuberosity

Radius

Ulna

Figure 7.24 Skeletal landmarks of the right elbow joint in a flexed position—diagonal view.

Figure 7.25 Anterior spatial relationships of two of the elbow flexors against resistance.

BRACHIALIS

BICEPS BRACHII

BRACHIALIS

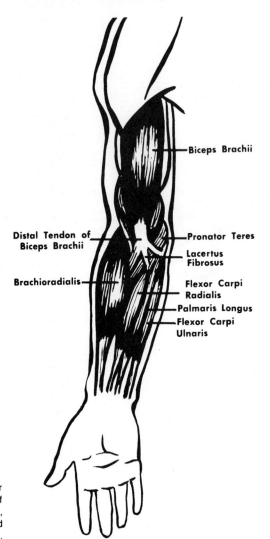

Biceps Brachii

Distal Tendon of
Biceps Brachii

Brachioradialis

Pronator Teres

Lacertus
Fibrosus

Flexor Carpi
Radialis

Palmaris Longus

Flexor Carpi
Ulnaris

Figure 7.26 Anterior
spatial relationships of
some elbow, radio-ulnar,
wrist, and extrinsic hand
musculature.

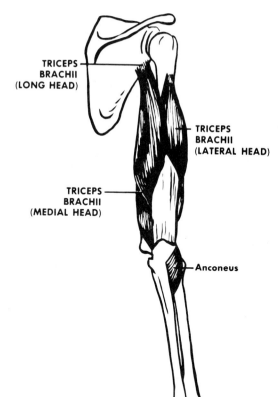

TRICEPS
BRACHII
(LONG HEAD)

TRICEPS
BRACHII
(LATERAL HEAD)

TRICEPS
BRACHII
(MEDIAL HEAD)

Anconeus

Figure 7.27 The posterior spatial relationship of the triceps brachii muscle to the elbow joint.

Muscles	Proximal Attachment	Distal Attachment
I. Elbow Joint Musculature		
A. Flexors		
1. BICEPS BRACHII	*Short head:* Coracoid process *Long head:* Supraglenoid tubercle	Radial tuberosity
2. BRACHIALIS	Anterior and lower half of the humerus	Tuberosity of the ulna and the coronoid process
3. BRACHIORA-DIALIS	Upper two-thirds of the lateral supracondyloid ridge of the humerus	Lateral aspect of the styloid process of the radius
B. Extensor		
1. TRICEPS BRACHII (THREE HEADS)	*Long head:* Infraglenoid tubercle *Lateral head:* Superior half of the humerus to the greater tubercle *Medial head:* Medial-posterior two-thirds of the humerus	Olecranon process

Figure 7.28 Elbow joint musculature. Muscles most involved for the motions against resistance appear in uppercase letters.

Basic Concepts of Anatomic Kinesiology

Radio-Ulnar Joint

The bones involved at the radio-ulnar joint are the radius and ulna. These two bones articulate with each other proximally and distally. They literally rotate around each other during pronation and supination. The radius and ulna articulate proximally when the head of the radius contacts the radial notch of the ulna. Distally, the head of the ulna articulates with the ulnar notch of the radius. This is a relatively unstable bony arrangement. The main skeletal landmarks of the radio-ulnar joint are shown in figure 7.29.

Pronation and supination are the only two motions possible at the radio-ulnar joint. These motions, as well as shoulder joint and wrist movements, are necessary in the sequence of motions needed to impart trajectory modifying spins to

Figure 7.29 Right radius and ulna—anterior and posterior views.

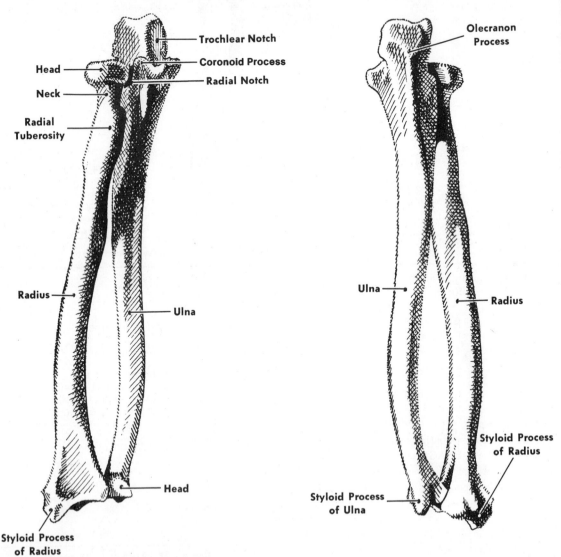

Trochlear Notch

Head

Coronoid Process

Radial Notch

Neck

Radial Tuberosity

Olecranon Process

Radius

Ulna

Ulna

Radius

Head

Styloid Process of Radius

Styloid Process of Ulna

Styloid Process of Radius

Upper Limb Segment Myology

Muscles	Proximal Attachment	Distal Attachment
I. Radio-Ulnar Joint Musculature		
A. Pronators		
1. PRONATOR TERES	*Humeral Head:* Medial epicondyle of the humerus *Ulnar Head:* Medial coronoid process of the ulna	Middle and lateral aspect of the radius
2. PRONATOR QUADRATUS	Lower one-fourth of the anterior ulna	Lower one-fourth of the anterior radius
3. BRACHIORA-DIALIS	Upper two-thirds of the lateral supracondyloid ridge of the humerus	Lateral aspect of the styloid process of the radius
B. Supinators		
1. SUPINATOR	Lateral epicondyle of the humerus	Lateral and upper third of the radius
2. BICEPS BRACHII	*Short Head:* Coracoid process *Long Head:* Supraglenoid tubercle	Radial tuberosity
3. BRACHIORA-DIALIS	Upper two-thirds of the lateral supracondyloid ridge of the humerus	Lateral aspect of the styloid process of the radius

sport objects. While making noncinematographic and basic cinematographic analyses, it is very essential not to confuse radio-ulnar pronation and supination with shoulder joint medial and lateral rotations. If the observations are made of the hand segment only, it is easy to miscalculate which joint was responsible for moving the hand through the transverse plane of motion.

Figure 7.30 Radio-ulnar joint musculature. Muscles most involved for the motions against resistance appear in uppercase letters.

Figure 7.26 shows the biceps brachii, pronator teres, and brachioradialis muscles. These are functioning radio-ulnar joint muscles. The brachioradialis is unique because it contributes force for both pronation and supination. After initiation of elbow flexion against resistance, the brachioradialis will contract to pronate or supinate to the midposition. This is due to its lateral spatial relationship to the radio-ulnar joint. The biceps brachii is the strongest supinator.

The lines of pull and functions of the radio-ulnar musculature are outlined in figure 7.30.

Wrist Joint

The wrist joint involves articulation of the distal end of the radius with the navicular, lunate, and triangular bones. The latter three bones are a part of the eight carpal bones shown in figures 7.31 and 7.32. The remaining carpals are the hamate, pisiform, capitate, greater multangular and lesser multangular. The greater multangular, lesser multangular, capitate, and hamate articulate with the metacarpals to form the carpometacarpal joints.

The wrist is capable of six motions: (1) flexion, (2) extension, (3) hyperextension, (4) radial flexion, (5) ulnar flexion, and (6) circumduction. These motions must be clearly defined by the teacher-coach in terms of the objective of

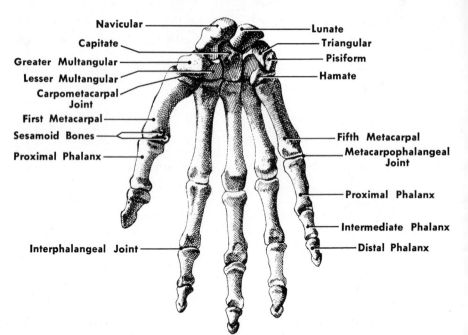

Navicular —
Capitate —
Greater Multangular —
Lesser Multangular —
Carpometacarpal
Joint —
First Metacarpal —
Sesamoid Bones —
Proximal Phalanx —

— Lunate
— Triangular
— Pisiform
— Hamate

— Fifth Metacarpal
— Metacarpophalangeal
Joint

— Proximal Phalanx

— Intermediate Phalanx
— Distal Phalanx

Interphalangeal Joint —

Figure 7.31 Palmar
surface—right hand.

Figure 7.32 Dorsal
surface—right hand.

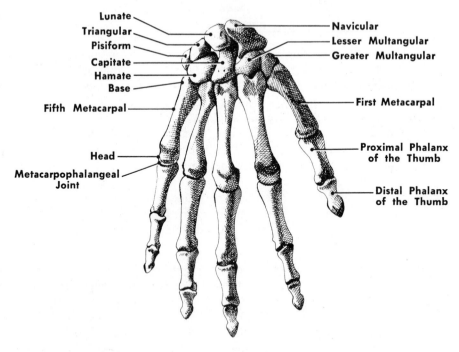

Lunate —
Triangular —
Pisiform —
Capitate —
Hamate —
Base —
Fifth Metacarpal —

— Navicular
— Lesser Multangular
— Greater Multangular

— First Metacarpal

— Proximal Phalanx
of the Thumb

Head —
Metacarpophalangeal
Joint —

— Distal Phalanx
of the Thumb

Upper Limb Segment Myology

the skill being performed. Cliches such as "bending the wrist" and "snapping the wrist" are meaningless, because they do not indicate direction. Terms such as flexion and ulnar flexion, however, are specific in terms of direction.

The football player in figure 7.33, an All-American and All-Pro, demonstrates a common human trait when a camera is used for producing publicity photographs and in filming noncompetitive situations. What is recorded on such film has only a minor relationship to what actually occurs in competition. As an example, this excellent athlete rarely fumbled a football throughout his entire university and professional careers. The grip on the ball in figure 7.33 is very tenuous with the extremely flexed left wrist, slight flexion of the distal interphalangeal joints on one end of the ball, and mild pressure on the ball exerted against the left radio-ulnar joint. A tackler seeing a football back holding a ball in this manner and with his center of mass high would separate him from the ball and his senses! Obviously, the grip and body position shown in figure 7.33 are not the ones commonly used by the athlete in competition. *This example reinforces the principle that film for analysis should be taken in competitive situations.*

From a bony standpoint, the wrist is a stable structure. The three carpal bones are received into the radius in a deep ovoid structure. This articulation allows flexion, extension, radial flexion, and ulnar flexion. The latter two move-

Figure 7.33 Frank Gifford, former USC halfback, demonstrates a ball grip he rarely used in competition. *Courtesy USC Athletic News Service.*

ments can be considered as abduction and adduction movements; consequently, circumduction is also possible at the wrist joint.

The wrist is stable by reason of the arrangement of the ligaments surrounding the joint. There is anterior, posterior, lateral, and medial ligamentous support. Anterior ligamentous support is maintained by the volar radiocarpal ligament. Posterior support is offered by the dorsal radiocarpal ligament. The ulnar collateral ligament provides support medially, and the radiocarpal ligament maintains support laterally.

The wrist joint is amply provided with muscular support. There are fifteen *extrinsic muscles* of the hand located on the anterior, posterior, medial, and lateral aspects of the wrist joint. The spatial relationships of these extrinsic muscles to the wrist joint are shown in figure 7.34. This figure shows a cross section of the left wrist, with the palm facing upward. The tendons of the muscles are included where they cross the joint enroute to their distal attachments. Figure 7.34 can be utilized to conceptualize the spatial relationships of the muscles which cross the reader's right wrist by observing the anterior aspect of the joint. The discussion of wrist and hand motions is restricted to these muscles only; the intrinsic muscles of the hand are not included.

Muscles in figure 7.34 with anterior spatial relationships to the axis of the wrist joint are flexors when they contract concentrically. Conversely, the posterior muscles extend and hyperextend the wrist joint when they contract against

Figure 7.34 Spatial relationships of the extrinsic muscles of the hand to the wrist joint—left wrist seen in cross section with the palmar surface rotated upward due to radio-ulnar supination.

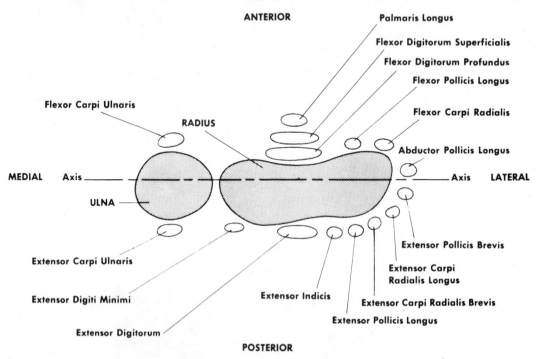

Upper Limb Segment Myology

resistance. The medial muscles on the ulna are the ulnar flexors; the five flexors with the greatest lateral spatial relationships are the radial flexors against resistance.

The muscles that cause the motions of the wrist against resistance are listed in figure 7.35. Generally, the flexors attach in the area of the medial epicondyle of the humerus, and their tendons have an anterior spatial relationship to the wrist. The extensors general line of pull is from the lateral epicondyle of the humerus, and their tendons have a posterior spatial relationship to the wrist joint.

Figure 7.35 Wrist joint musculature. The muscles most involved for the motions against resistance are shown in uppercase letters.

Muscles	Proximal Attachment	Distal Attachment
I. Wrist Joint Musculature		
A. Flexors		
1. FLEXOR CARPI ULNARIS	*Humeral head:* Medial epicondyle of the humerus *Ulnar head:* Medial olecranon and upper two-thirds of the ulna	Pisiform, hamate and fifth metacarpal—palmar surface
2. PALMARIS LONGUS	Medial epicondyle of the humerus	Anular ligament and palmar aponeurosis
3. FLEXOR DIGITORUM SUPER- FICIALIS	Medial epicondyle of the humerus and the radial tuberosity	Via four tendons into each side of the medial phalanx of the four fingers
4. FLEXOR DIGITORUM PROFUNDUS	Upper anterior and medial ulna	Via four tendons to the distal phalanx of the four fingers
5. FLEXOR POLLICIS LONGUS	Volar surface of the radius below the radial tuberosity and medial epicondyle of the humerus	Base of the distal phalanx of the thumb
6. FLEXOR CARPI RADIALIS	Medial epicondyle of the humerus	Bases of the first and second metacarpals
B. Extensors		
1. EXTENSOR CARPI ULNARIS	Lateral epicondyle of the humerus	Ulnar side of the base of the fifth metacarpal
2. EXTENSOR DIGITI MINIMI	Common extensor tendon of the extensor digitorum	Dorsal aspect of the proximal phalanx of the little finger
3. EXTENSOR DIGITORUM	Lateral epicondyle of the humerus	Via four tendons to the dorsal surface of the distal phalanx of each finger
4. EXTENSOR INDICIS	Dorsal aspect of the ulna	Via the extensor digitorum tendon to the index finger
5. EXTENSOR POLLICIS LONGUS	Middle third of the dorsal aspect of the ulna	Base of the distal phalanx of the thumb

Figure 7.35—*Continued*

Muscles	Proximal Attachment	Distal Attachment
6. EXTENSOR CARPI RADIALIS BREVIS	Lateral epicondyle of the humerus	Radial and dorsal aspects of the base of the third metacarpal
7. EXTENSOR CARPI RADIALIS LONGUS	Lower third of the lateral supracondylar ridge of the humerus	Radial and dorsal aspects of the base of the second metacarpal
C. Radial Flexors		
1. FLEXOR CARPI RADIALIS	Medial epicondyle of the humerus	Bases of the first and second metacarpals
2. EXTENSOR CARPI RADIALIS LONGUS	Lower third of the lateral supracondylar ridge of the humerus	Radial and dorsal aspects of the base of the second metacarpal
3. EXTENSOR CARPI RADIALIS BREVIS	Lateral epicondyle of the humerus	Radial and dorsal aspects of the base of the third metacarpal
4. EXTENSOR POLLICIS BREVIS	Dorsal aspect of the radius	Base of the proximal phalanx of the thumb
5. ABDUCTOR POLLICIS LONGUS	Lateral and dorsal aspect of the ulna, middle third of the dorsal aspect of the radius	Radial aspect of the base of the first metacarpal
D. Ulnar Flexors		
1. FLEXOR CARPI ULNARIS	*Humeral Head:* Medial epicondyle of the humerus *Ulnar Head:* Medial olecranon and upper two-thirds of the ulna	Pisiform, hamate and fifth metacarpal—palmar surface
2. EXTENSOR CARPI ULNARIS	Lateral epicondyle of the humerus	Ulnar side of the base of the fifth metacarpal

Metacarpophalangeal and Interphalangeal Joints

The major skeletal landmarks of the hand are shown in four views in figures 7.31, 7.32, 7.36 and 7.37. Figure 7.36 shows the metacarpophalangeal and interphalangeal joints.

The motions of the metacarpophalangeal joints include flexion, extension, abduction, adduction, and circumduction. The interphalangeal joints are limited to flexion and extension only because of their structure.

The hand, like the foot, is equipped with two sets of muscles. One group has proximal attachments on the upper limb and distal attachments within the hand.

These are known as *extrinsic* hand muscles. The other group of muscles attach entirely within the hand, and they are known as *intrinsic* hand muscles. The listing of muscles in figure 7.38 is limited to the extrinsic muscles. It should be understood, however, that the muscles most involved for abduction and adduction of the metacarpophalangeal joints are intrinsic. These are the abductor digiti minimi manus, opponens digiti minimi, abductor pollicis brevis, opponens pollicis, adductor pollicis, interossei palmares, and interossei dorsales manus.

The multiarticular extrinsic muscles of the hand that flex and extend the metacarpophalangeal and interphalangeal joints when they undergo concentric contraction are listed in figure 7.38.

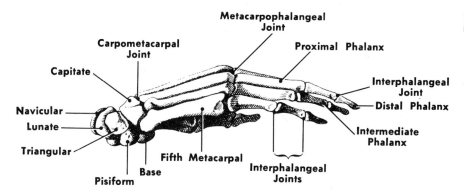

Figure 7.36 Right hand—ulnar view.

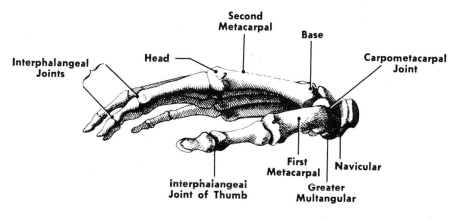

Figure 7.37 Right hand—radial view.

Muscles	Proximal Attachment	Distal Attachment
I. Metacarpophalangeal and Interphalangeal Musculature (extrinsic)		
A. Flexors		
1. FLEXOR DIGITORUM PROFUNDUS	Upper anterior and medial ulna	Via four tendons to the distal phalanx of the four fingers
2. FLEXOR DIGITORUM SUPERFICIALIS	Medial epicondyle of the humerus and the radial tuberosity	Via four tendons into each side of the medial phalanx of the four fingers
3. FLEXOR POLLICIS LONGUS	Volar surface of the radius below the radial tuberosity and medial epicondyle of the humerus	Base of the distal phalanx of the thumb
B. Extensors		
1. EXTENSOR DIGITORUM	Lateral epicondyle of the humerus	Via four tendons to the dorsal surface of the distal phalanx of each finger
2. EXTENSOR INDICIS	Dorsal aspect of the ulna	Via the extensor digitorum tendon to the index finger
3. EXTENSOR DIGITI MINIMI	Common extensor tendon of the extensor digitorum	Dorsal aspect of the proximal phalanx of the little finger
4. EXTENSOR POLLICIS LONGUS	Middle third of the dorsal aspect of the ulna	Base of the distal phalanx of the thumb
5. EXTENSOR POLLICIS BREVIS	Dorsal surface of the radius and interosseous membrane	Base of the first phalanx of the thumb

Figure 7.38
Metacarpophalangeal and interphalangeal musculature (extrinsic). Muscles most involved for causing the motions against resistance appear in uppercase letters.

Recommended Reading

Bankoff, Dalla, et al. "Simultaneous EMG of Latissimus Dorsi and Sternocostal Part of Pectoralis Major Muscles During the Crawl Stroke." *Electromyography and Clinical Neurophysiology* 18 (June–August, 1978):289–295.

Buller, N. P., et al. "Recording of Isometric Contractions of Human Biceps Brachii Muscle." *Journal of Physiology* 277 (April, 1978):11–12.

Dempster, Wilfrid T. "Mechanisms of Shoulder Movement." *Archives of Physical Medicine and Rehabilitation* 46 (January, 1965):49–70.

DeSousa, O. Machado, et al. "Electromyographic Study of the Brachioradialis Muscle." *Anatomical Record* 139 (1961):125–31.

Diviley, Rex L., and Meyer, Paul W. "Baseball Shoulder." *Journal of the American Medical Association* 171 (November, 1959):1959–61.

Duvall, E. N. "Critical Analysis of Divergent Views of Movement at the Shoulder Joint." *Archives of Physical Medicine* 36 (1955):149–54.

Inman, Verne T.; Saunders, J. B.; and Abbott, Leroy C. "Observations on the Function of the Shoulder Joint." *Journal of Bone and Joint Surgery* 26 (1944):1–30.

Ismail, H. M., et al. "Isometric Tension Development in a Human Skeletal Muscle in Relation to its Working Range of Movement: The Length-Tension Relation of Biceps Brachii Muscle." *Experimental Neurology* 62 (December, 1978):595–604.

Ketchem, L. D., et al. "The Determination of Moments of Extension of the Wrist Generated by Muscles of the Forearm." *Journal of Hand Surgery* 3 (May, 1978):205–210.

Logan, Gene A. "Movement in Art." *Quest* 2 (April, 1965):42–45.

————. *Adaptations of Muscular Activity.* Belmont, CA: Wadsworth Publishing Company, 1964.

Logan, Gene A., and McKinney, Wayne C. "The Serape Effect." *Journal of Health, Physical Education, Recreation* 41 (February, 1970):79–80.

Logan, Gene A., and Wallis, Earl L. "Recent Findings in Learning and Performance." Southern Section Meeting, California Association for Health, Physical Education and Recreation, October, 1960, at Pasadena City College.

Poppen, N. K., et al. "Forces at the Glenohumeral Joint in Abduction." *Clinical Orthopaedics and Related Research* 135 (September, 1978):165–170.

Provins, K. A., and Salter, N. "Maximum Torque Exerted About the Elbow Joint." *Journal of Applied Physiology* 7 (1955):393–98.

Ramsey, Robert W., et al. "An Analysis of Alternating Movements of the Human Arm." *Federation Proceedings* 19 (March, 1960):254.

Rasch, Philip J. "Effect of the Position of Forearm on Strength of Elbow Flexion." *Research Quarterly* 27 (1956):333–37.

Sullivan, P. E., et al. "Electromyographic Activity of Shoulder Muscles During Unilateral Upper Extremity Proprioceptive Neuromuscular Facilitation Patterns." *Physical Therapy* 60 (March, 1980):283–288.

Taylor, Craig L., and Schwartz, Robert J. "The Anatomy and Mechanics of the Human Hand." *Artificial Limbs* 2 (1955):22–35.

Travill, A. A. "Study of the Extensor Apparatus of the Forearm." *Anatomical Review* 144 (1962):373–76.

Travill, Anthony, and Basmajian, John V. "Electromyography of the Supinators of the Forearm." *Anatomical Record* 139 (1960):557–60.

Van Linge, B., and Mulder, J. O. "Function of the Supraspinatus Muscle and Its Relation to the Supraspinatus Syndrome." *Journal of Bone and Joint Surgery* 45-B (1963):750–54.

Weathersby, Hal T., et al. "The Kinesiology of Muscles of the Thumb: An Electromyographic Study." *Archives of Physical Medicine and Rehabilitation* 44 (June, 1963):321–26.

Whitley, Jim D., and Smith, Leon E. "Measurement of Strength of Adduction of the Arm in Various Positions." *Archives of Physical Medicine and Rehabilitation* 45 (July, 1964):326–28.

Wiedenbauer, M. M., and Mortensen, O. A. "An Electromyographic Study of the Trapezius Muscle." *American Journal of Physical Medicine* 31 (1952):363–73.

Yamshon, L. J., and Bierman, W. "Kinesiologic Electromyography, I. The Trapezius." *Archives of Physical Medicine* 24 (1948):647–51.

———. "Kinesiologic Electromyography, III. The Deltoid." *Archives of Physical Medicine* 30 (1949):286–89.

3

Physics of Sport

8

Understanding Basic Physics

Part 3, "Physics of Sport," is designed to acquaint the student with the terms and relationships used in the physical description of human body motion. The remainder of the text will then use these concepts from physics in their applications to the analysis of sport. The basic purpose of physics is to provide knowledge and ability to describe precisely the interactions that occur between objects and bodies in the universe. Mechanics is that part of physics dealing with the description of motion and with the forces and interactions which affect motion. Part 3 is written for the student with limited background in physics and mathematics.

It may help the student at this point to note the special nature of chapter 11 of the text. Following the basic mathematic-physics development of the terminology and relationships of mechanics, chapter 11 reviews the material in a descriptive fashion by providing a nonmathematic emphasis on the concepts that are most important in biomechanic applications. Thus, the material presented in that chapter serves as a *glossary* of terms and, in many cases, a condensed description of specific physical relationships. It may prove useful in some situations for students to become familiar with that chapter before working on material in chapters 8, 9, and 10.

Within the literature of biomechanics and physics, confusion sometimes arises regarding the use of the major terms that categorize the study of motion. The student should be aware of an author's basic definitions of such terms in order to compare statements which appear in textbooks and professional journals. As indicated in chapter 1, *mechanics* is the most general term used to describe the study of the behavior of objects or fluids under gravitational and contact forces. *Kinematics* technically refers to the study of motion without reference to the forces causing motion. *Dynamics* is the term used for the study of relationships that exist between forces and motion. *Kinetics* refers to the description of a system in terms of the motions of its components.

In the same sense that each sport in physical education has a specialized vocabulary to describe motions and techniques, the field of biomechanics also utilizes a unique vocabulary in describing various aspects of the relationship between forces and motion. This specific vocabulary puzzles students at times, because they may be unfamiliar with the terms or with the precise meaning scientists give to these terms. In science, every effort is made to keep terms unambiguous so a given statement can have only one possible interpretation. In order to do this, terms are usually used in a much more restricted sense than they are in everyday life. In a few cases, scientific terms can be used in a more general sense.

An example of the more restricted usage is the scientific definition of the word *power*. In everyday usage this term connotes everything from muscular strength to social authority. *To the scientist utilizing biomechanics, however, the term "power" means precisely the total amount of work accomplished in one unit of time.* For example, a 220-pound man running up a ten-foot flight of stairs in four seconds would be exerting one horsepower. In order to exert two horsepower, he would have to run up the same flight of stairs, accomplishing the same amount of work, in two seconds.

On the other hand, it is found that the term *acceleration* is used in a much more generalized sense in scientific description of motion than during everyday usage of the word. Most people who use the term *acceleration* imply that an increase in speed of a moving object is occurring. In a biomechanic description of motion, however, *acceleration* refers to any change of motion, whether it represents a speeding up, a slowing down, or simply a change in direction of a moving body. For example, a football back in broken field running may give himself a positive acceleration at times to escape a pursuing player. He may next provide a negative acceleration (change of pace), more commonly called deceleration, to allow blocking to occur in front of him. Following this, he may provide an acceleration changing the direction of motion in order to follow his blocking.

One purpose of this chapter is to define many of the basic terms of biomechanics as they are used in the scientific sense. Examples of the use of these concepts are provided, and in some cases numerical calculations will indicate the relation between mechanical quantities.

In order to describe the effect of forces upon motion of bodies, a precise set of terms and relationships to describe motion itself is needed. Motion is generally described in terms of velocities and accelerations.

Description of Linear Motion

Familiarity with a number of terms is necessary to understand correctly the descriptions of objects in space. *Position* simply refers to the location of a body in space. This is given with reference to other bodies and a directional coordinate system. *Distance* refers to the measure of space between two positions. *Length* is the special distance between two points, often the end points, on a body. In describing motion, the term *displacement* normally means the difference between the beginning and final positions of a motion. The terms *displacement, change of position,* and *distance covered* are used interchangeably in biomechanics literature.

Velocity

Velocity is defined as the change of position of a body per unit of time. When an object is moving, it is constantly changing position. Its velocity is measured by determining how much this position changes within a given time span.

In sport applications, there are two ways to determine an average velocity. A fixed distance for a motion or event may be set (90 feet, 100 yards, 5000 meters) and the time necessary to move this distance may be measured. Or a fixed time may be used (12 minutes, 1 hour), and the distance covered during this period is then determined. In either case, the mathematical calculation of the velocity is the same:

$$\text{Velocity} = \frac{\text{Distance}}{\text{Time}} = \frac{d}{t}$$

When communicating scientific quantities, it is necessary to use universally accepted units. Distances in the English system are commonly expressed in feet, yards, or miles. The Metric system describes distances of interest in sport in centimeters, meters, or kilometers. In both systems the commonly accepted unit of time is the second. Thus, the quantities of velocity would normally be expressed in such units as feet per second (ft/sec), meters per second (m/sec), or miles per hour (mi/hr). These units may be converted from one to the other by the use of multiplying factors (see Appendix A for an extensive conversion table):

$$1 \text{ ft/sec} = .305 \text{ m/sec} = .682 \text{ mi/hr}$$

$$1 \text{ m/sec} = 3.28 \text{ ft/sec} = 2.24 \text{ mi/hr}$$

As an example, the highest average velocity achieved by a runner in N.C.A.A. championship track-and-field competition occurs in the 220-yard dash. A performer who can run this distance in 20 seconds flat would be considered a potential champion in most college or university meets. In terms of velocity, this represents a distance covered of 660 feet in 20 seconds.

$$V = \frac{660 \text{ ft}}{20 \text{ sec}} = 33 \text{ ft/sec}$$

$$33 \text{ ft/sec} = 22.5 \text{ mi/hr}$$

It should be noted that this 33 ft/sec is an *average velocity* over the distance and not the maximum velocity achieved by the performer at a specific point during the race. This type of velocity calculation does not take into account the fact that the performer is actually starting from a velocity of zero. This, of course, decreases the sprinter's average velocity. Theoretically, a maximum velocity for the human body during a running event would seem to be about 40 ft/sec or approximately 27 mi/hr.

In sprinting conditions in sport, a performer covers between three and four feet in 0.1 second. Races up to 220 yards or 200 meters in length often end with several participants within 0.1 second of winning. Since hand timing is only accurate to this interval of time, electronic timing has become necessary for determining winners and places of performers in sprinting events.

In field sports such as baseball, football, and soccer, many plays are decided by distances of a foot or less. This represents a time interval much less than the tenth of a second which is basic in human response. Therefore, strategies or techniques which can gain an extra tenth of a second often insure success in base stealing and outmaneuvering a defensive alignment in these sports. Or a loss of this same interval often can lead to failure.

In another example, the present world record for running 30,000 meters is 1 hour, 29 minutes, 18.8 seconds. The average velocity for this feat would be

$$V = \frac{30,000 \text{ meters}}{(1 \times 3600) + (29 \times 60) + 18.8 \text{ sec}} = \frac{30,000}{5358.8} = 5.60 \text{ m/sec}$$

When converted into familiar English units, this means the performer ran nearly 18 2/3 miles at an average speed of 12.50 miles per hour.

Figure 8.1 indicates the average velocities attained in record performances at several metric distances. Data are also included for the marathon, 1 hour run, and 24 hour run/walk. Examination of these numbers will show not only how human speed changes as the distance of races increases, but also will give a feeling for the comparison of these speeds in the three most commonly expressed sets of units.

Velocities encountered in sport motions vary from a few feet per second in slow precision movements to nearly 150 ft/sec in ballistic movements of body parts. The head of a golf club may reach a speed over 200 ft/sec at the point of impact.

It should be emphasized that within the area of biomechanics there is a scientific distinction between the terms *velocity* and *speed*. Velocity is known as a vector quantity. *A vector quantity represents both magnitude and direction.* The

Figure 8.1 Average speed versus distance for selected running events from 100 meters through the 24 hour run.

Distance	Time	Speed		
	Hr:Min:Sec	Meters/sec	Feet/sec	Miles/hr
100 meters	9.95	10.05	32.97	22.48
200 meters	19.72	10.14	33.27	22.68
400 meters	43.86	9.12	29.92	20.40
800 meters	1:41.72	7.86	25.79	17.58
1500 meters	3:31.36	7.10	23.29	15.88
1 mile	3:48.20	7.05	23.14	15.78
5 kilometers	13:08.40	6.34	20.80	14.18
10 kilometers	27:22.40	6.09	19.98	13.62
13 miles 24 yards	1:00:00.00	5.82	19.09	13.02
25 kilometers	1:13:55.80	5.64	18.50	12.61
30 kilometers	1:29:18.80	5.60	18.37	12.53
26 miles 385 yards Marathon	2:10:00.00	5.42	17.79	12.13
167 miles 440 yards	24:00:00.00	3.12	10.22	6.97

units expressed above actually express the magnitude of the velocity. This is referred to as *speed*. In order to define a velocity completely, speed must be indicated together with the direction in which the motion is taking place. Examples of describing vector quantities are (1) a baseball moving at 100 ft/sec into a head wind and (2) a motion in which a performer moves his hand upward at 50 ft/sec as a result of shoulder flexion. In these simple examples, the amount of motion and a direction in which motion occurs have been specified; both are necessary for velocity calculations.

There are times when the direction of motion is not exceptionally important. For example, in the previous question regarding average velocity for a man performing a 220-yard dash, the reference was actually to speed and not to the total components of velocity. The direction of motion in this case is clearly defined in terms of the rules governing the race and the way running lanes are marked on the track. In the case of distance races performed on a 440-yard track, the path (direction) is a closed loop. In fact, from the previous definition of velocity, the average velocity over a closed loop must add up to zero because the performer ends at the same point he started. In this case, it is actually the speed, i.e., the total linear distance covered divided by the time, which is important.

Since velocity is a vector quantity, there are certain properties of vectors considered important in analyzing velocities. The fact that a vector has direction as well as magnitude means that it can be described geometrically. For example, a vector quantity in a given direction can be expressed in terms of components in other directions. A ball thrown or batted into the air has a component of velocity in some horizontal direction. It also possesses a component of velocity in a vertical direction, i.e., either up or down. These are referred to as the horizontal and vertical components, respectively, of the velocity vector (fig. 8.2).

This relationship between the direction of the velocity vector and the observed interactions will be discussed in detail in Chapter 9, where some important trajectories which occur during sport events are considered.

Figure 8.2 Horizontal and vertical components of a velocity vector.

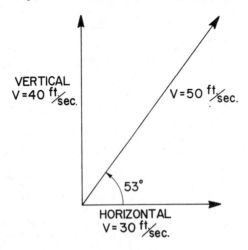

VERTICAL
$V = 40 \, \frac{ft}{sec.}$

$V = 50 \, \frac{ft}{sec.}$

53°

HORIZONTAL
$V = 30 \, \frac{ft}{sec.}$

Understanding Basic Physics

The directional factor of velocity makes calculations in terms of *vector addition* possible. For example, a series of motions affecting the same body or object will often add together to give a large total velocity vector. The execution of the spike in volleyball requires such a combination of motions. Along with the leap for height, the spiker also moves the whole body toward or along the net. This motion is normally added to a medial rotation of the shoulder, a high diagonal motion of the arm, and a final elbow extension, with flexion of the wrist when contacting the ball. The vector sum of these motions gives the hand a total velocity vector in the proper direction to propel the ball downward across the net at a high velocity (figure 8.3).

Figure 8.3(b) is a line drawing of the spiker, indicating the body velocity toward and along the net (B), the shoulder motion resulting from rotation of the upper body by the Serape Effect (S), the arm motion from high diagonal adduction of the shoulder (A), and additional hand velocity from the wrist flexion (W). These four vector velocities are added in Figure 8.3(c) to indicate the final velocity of the hand as it strikes the ball.

The final velocity of most sport objects is the sum of several previous velocities developed sequentially in moving joints. This is an example of the stereotype of perfect mechanics discussed earlier; it is the procedure which allows a set of individual movements to sum together most effectively, thus combining to pro-

Figure 8.3
Summation of velocities during a volleyball spike by Terry Place of USC. *Photo courtesy USC Athletic News Service.*

(a)

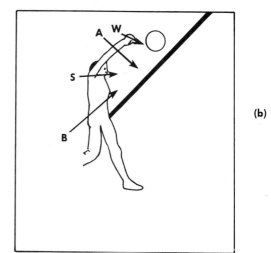

(b)

(c)

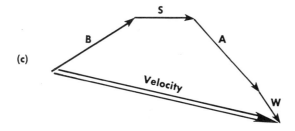

vide an optimum result. A deviation from ideal form in the volleyball spike would cancel some part of this summation, affecting the speed and/or accuracy of the ball adversely. This same principle applies in the case of a pitched baseball, tennis serve, shot put, long jump, and many other skills of a ballistic nature.

Acceleration As mentioned previously, the term *acceleration* is used to connote any change of velocity occurring during motion. *Acceleration is defined as the change of velocity occurring per unit time.*

$$\text{Acceleration} = \frac{\text{Change of velocity}}{\text{Time}} = \frac{V_{(\text{final})} - V_{(\text{initial})}}{\text{Time}} = \frac{V_f - V_i}{t}$$

These changes of velocity may result in an increase in speed or magnitude of the velocity; they may result in a decrease in speed or magnitude of the velocity; or they may simply result in a change of direction. Acceleration is also expressed in terms of precise scientific units. These will be the units of velocity divided again by units of time normally expressed in seconds. Quantities of acceleration are usually expressed in terms of feet per second per second (ft/sec²) or meters per second (m/sec²).

Acceleration may be either positive or negative. As an example of positive acceleration, a shot-put performer in the linear move across the ring may change the motion of the shot from near rest to a final velocity of 40 ft/sec. If he or she performs this motion in one second, the acceleration is—

$$a = \frac{40 - 0}{1} = 40 \text{ ft/sec}^2$$

Again, this represents an *average acceleration* for the motion of the shot across the circle, and not necessarily the acceleration occurring at any specific point in the maneuver.

An example of negative acceleration is given by a trampoline performer stopping motion at the end of a rebound by flexions of knee and hip joints. As the performer falls to the trampoline from a height of about six feet above the bed, he or she attains a velocity of 20 ft/sec. If the motion is stopped within one-third of a second of contact with the trampoline bed, the acceleration is—

$$a = \frac{0 - 20}{1/3} = -60 \text{ ft/sec}^2$$

The negative sign in this case indicates that the direction of acceleration was opposite to the initial velocity; consequently, the velocity is decreased to zero. In a motion such as this, the magnitude of the acceleration depends principally on the time used in stopping the motion. Motions stopped in very short periods of time lead to very high negative accelerations.

The purpose of the follow-through in many sport motions is to allow a reasonable time for stopping the motions used in a throwing or stiking performance. Longer times lead to smaller negative accelerations; consequently, there is less force required by muscles at joints. This reduction of the danger of damage to

Figure 8.4 Use of follow-through in a golf drive by Scott Simpson of the USC golf team. *Photo courtesy USC Athletic News Service.*

joints following a throw or the swing of a sport implement allows the performer to concentrate on fast, smooth motions during the action. For example, the high velocities of the head of a golf club are possible because the good golfer uses a long follow-through after contact with the ball (fig. 8.4).

Still a third type of acceleration is seen in the case of a baseball runner taking a wide turn at second base at constant speed (figs. 8.5 and 8.6). Calculating the acceleration in this case is more complex, since geometrical relations must be used to determine the change of velocity and time. If the speed is 20 ft/sec, for instance, the dimensions of the triangle reveal that the change of velocity is slightly more than 28 ft/sec. It can be shown that for a circular path traversed at constant speed, this acceleration is calculated by the equation $a = V^2/r$ where r is the radius of the circle of curvature. If a radius of 16 feet is used in the foregoing example of a runner rounding second base at 20 ft/sec, the acceleration is—

$$a = \frac{V^2}{r} = \frac{20 \times 20}{16} = 25 \text{ ft/sec}^2$$

In this special case of a change of direction at constant speed, the term *radial acceleration* is used. It can be seen that either an increase in speed or a decrease in radius leads to a larger acceleration. Large radial accelerations are often observed in athletic skills requiring sharp turns at high speed.

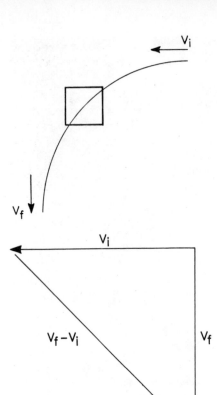

Figure 8.5 Spatial path of the base runner.

Figure 8.6 Vector change of velocity by base runner.

It will be noted from these examples that acceleration, like velocity, is a vector quantity, i.e., both the direction of the acceleration and its magnitude are involved. There is no special term for the magnitude of acceleration analogous to the term *speed* used to describe the magnitude of velocity. The direction of acceleration is of utmost importance. This is especially true as it relates to the direction of velocity. An acceleration in the same direction as the velocity produces an increase in speed; an acceleration in the opposite direction produces a decrease in speed; and an acceleration perpendicular to the velocity produces a change of direction.

For these reasons, consideration is given to the components of the acceleration vector parallel and perpendicular to the velocity vector. The component of acceleration parallel to the velocity is normally called the *tangential acceleration*. The tangential acceleration produces an increase or decrease of speed. The component perpendicular to the velocity is called the *radial acceleration*. The radial acceleration changes the direction of motion (figure 8.7).

In the example of the shot put, acceleration was in the same direction as movement of the shot. Therefore, the shot was increasing in speed. In the example of the trampoline performer, acceleration was opposite to the direction the performer was moving through a series of flexions at the major joints. Consequently, the performer slowed to a stop. In the example of the base runner, acceleration was perpendicular to the direction of movement. This was the causative factor for the change in direction with no change of speed.

One significant form of acceleration frequently observed in sport events is acceleration due to gravity. Disregarding the effects of air resistance, all falling bodies will accelerate toward the center of the earth with an acceleration of 32 ft/sec² or 9.8 m/sec². Because of the significance of this acceleration, other accelerations are often expressed in comparison with *gravitational acceleration. An acceleration of 32 ft/sec² is sometimes called one g*. If, for example, during the start of a 50-yard dash, a performer were to experience an acceleration of 64 ft/sec², it could be stated that this acceleration was equal to two *g*'s. In the earlier shot put example, the average acceleration was 1.25 *g* and in the baserunner rounding second base it was about 0.8 *g*. It should be noted that the term *g* as used here refers to an acceleration and not to a force. Gravitational force will be discussed more fully in the next section of this chapter.

Everyone is familiar with the concept of force; force is usually thought of in terms of "pushes and pulls." From the viewpoint of biomechanic analysis, the concept of force is especially important because it represents the way in which various bodies or parts of bodies interact with each other.

There are two effects produced by forces acting on bodies. First, the action of force may distort the shape of the body being acted upon. This phenomenon will be discussed when the concept of energy is presented. The second effect is that application of an unbalanced force to a body or body part results in an acceleration of the body or body segment. The term *unbalanced* is used in this context to represent those forces not being counteracted by any other opposing force on the body. Many times a performer has forces acting on the body but producing no apparent effect, simply because there is some counteracting force. For example, when an individual is standing virtually motionless, there is a large force of air pressure tending to push this person backward, but there is no acceleration impelling the body backward body because there is an equal and opposite counteracting force from air pushing against the back.

Concepts of Force and Mass

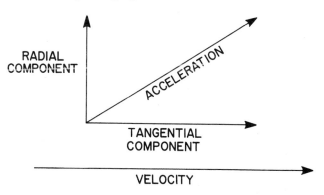

Figure 8.7 Radial and tangential components of acceleration.

This relation between force and acceleration is one of the most compelling relationships to be found in biomechanics. Not only are forces always associated with changes of motion of bodies, but the amount of acceleration is always directly proportional to the force. A force twice as great will produce an acceleration twice as great. A force only one-half as great will produce only one-half as much change of motion per unit time. Also, acceleration is always in the same direction as the force with which it is associated. This basic relationship allows detection and measurement of the effects of forces by observing changes in the motion of bodies or body parts.

This proportional relationship between force and acceleration introduces another very basic concept in mechanics. *The ratio between a force and the acceleration it produces is called the mass of the body. Mass is equal to force divided by acceleration; or, force is equal to mass times acceleration:*

$$\text{Mass} = \frac{\text{Force}}{\text{Acceleration}} = \frac{F}{a} \text{ or } F = ma$$

This is sometimes referred to as Newton's Second Law of Motion. A body having a large mass will require a large force to produce a given acceleration. For a body with a small mass, only a small force is required to produce the same acceleration.

There is a difference between the use of the terms *mass* and *weight*. Oftentimes these two terms are used synonymously and incorrectly. In the scientific sense, each has a specialized meaning. As indicated, *mass is the ratio of the force acting on a body to the acceleration it produces. Weight is the force with which the earth attracts a body gravitationally.* Thus, weight is a force dependent not only on the mass of the body but on the basis of gravitational attraction. For example, if a body were taken to the moon, where the gravitational force is one-sixth that of the earth, its mass would remain the same. Its weight would decrease by a factor of six, since the force of gravity changes by that amount.

The "force equals mass times acceleration" relationship can be expressed as "weight equals the mass times gravitational acceleration $(w = mg)$." Inversely, "mass equals weight divided by gravitational acceleration $\left(m = \dfrac{w}{g} \right)$" where g is the symbol for gravitational acceleration. In making calculations involving forces, masses, and accelerations, the mass of a body is often replaced by its weight divided by g. For example, the mass of a 16-pound shot is given by—

$$\text{Mass} = \frac{\text{Weight}}{g} = \frac{16}{32} = \frac{1}{2} \text{ mass unit}$$

In the English system of units, the normal measure of a body is its weight in pounds. Thus, to express its mass, one needs to use the factor $\dfrac{w}{g}$. (The actual unit of mass in this system is the *slug,* and a mass of 1 slug weighs 32 pounds at the surface of the earth. This unit is used in many engineering texts, but seldom in biological or biomechanic literature.)

In the metric system of units, the normal measure of a body is its mass in kilograms. To express body weight, the factor *mg* needs to be used. (The actual unit of weight in the metric system is the *newton,* but this term is seldom seen in sport literature. A mass of one kilogram weighs 9.8 newtons at the surface of the earth.)

In the earlier example of the trampoline performer, it was found that she was stopped with an upward acceleration of 60 ft/sec². If we choose a nominal weight of 120 pounds for the performer, we find that:

$$F = ma = \frac{120}{32} \times 60 = 225 \text{ pounds}$$

This indicates that a net force of 225 pounds must act on the performer to stop her in the indicated time. The force exerted by the trampoline must be this much more than her weight, giving:

$$F \text{ (trampoline)} = 225 + 120 = 345 \text{ pounds upward.}$$

In the shotput example, a force of:

$$F = ma = \frac{16}{32} \times 40 = 20 \text{ pounds}$$

would be needed to provide the indicated acceleration. This is in addition to a 16 pound force to support the weight of the shot. Since the weight and the necessary acceleration are at an angle of 45° to each other, these two forces must be added as vectors. This calculation is given in Appendix *C* as an example of the use of a calculator to perform biomechanics procedures.

A second major relationship involving forces in mechanics is called the *Law of Action and Reaction. This is sometimes referred to as Newton's Third Law of Motion.* This law states simply that when an object exerts a force on a second object, the second object exerts an equal and opposite force on the first. Forces never occur singly or in isolation, but always in a force-counterforce situation. One object cannot affect the motion or condition of a second object without itself also being affected.

In the context of biomechanics applied to sport, it should be noted that a sport object may be the whole human body or it may represent only certain body segments. The performer may also use a sport implement and/or a sport object as discussed earlier. Thus, the force with which a tennis racket strikes the ball is also equal and opposite to the force of the ball against the racket, producing elastic distortion of the strings.

The Law of Action and Reaction is of great importance in the study of human motions because of the forces and counterforces various body parts exert on one another. For example, this force interaction is noticed during a runner's arm motion. As the left shoulder joint is being diagonally adducted, the right shoulder joint is being diagonally abducted. This force-counterforce relationship of the arms helps maintain the body in an equilibrium position from which the legs can function most effectively to accelerate the runner. There is also a similar relationship in the leg motions during running involving anteroposterior plane motions.

Force, like velocity and acceleration, is a vector quantity; it has both magnitude and direction. As stated above, the direction of a single operating force is the same as the direction of the acceleration it produces. As in all other vector quantities, components of a force can be identified. At times, a single diagonal force vector will be replaced by a horizontal force vector plus a vertical force component. Also, force vectors may be added at a given point to produce a resultant force that has magnitude and a direction given by summing the individual force vectors. The proper sequential production of these individualized forces at joints to produce an optimum result is what the physical educator sometimes refers to as *timing* or *coordination* in a wide variety of sport and dance skills.

In anatomic kinesiology the term *summation of internal forces* is sometimes used to denote sequential motion within joints of a performer. From the viewpoint of biomechanics, this is termed a *summation of motions,* since an optimal motion either of the sport implement, such as a golf club or tennis racket, or at the body segment responsible for the final motion of the skill, such as the hand in handball, is actually produced. The difference in this terminology occurs because the internal forces produced by the muscle contractions of the performer result in an optimum series of motions whose velocities are additive. In a kinesthetic sense, what the performer experiences is actually a summation of internal forces. To the observing physical educator making a biomechanic analysis, this appears as a sequential set of additive motions executed by the performer to produce a given skill. From the standpoint of biomechanics, objectivity and precision of motion description are attained by mathematically summing the components of motion as viewed cinematographically.

This quantitative manipulation of the data from biomechanic observations is in the realm of intermediate cinematographic analysis and biomechanic research. Examples of this level of analysis will be found in Chapter 14 of the text and in Appendix C illustrating the use of electronic calculators to simplify such procedures.

Pressure and Frictional Forces

Within the scope of biomechanics, *the term "pressure" refers to the force applied per unit area across a surface.* A moderate force applied over a large area may produce a very low pressure. On the other hand, this same moderate force on a very small area of a body or implement may result in pressures high enough to cause tissue damage or trauma to body parts. Boxing gloves are designed to spread the force of a blow over a large area; this avoids large pressures. Without the gloves, the knuckles striking in regions such as the orbital ring of the eye tend to apply the force of the blow to an extremely small area. Such a blow would produce intense pressures resulting in traumatic laceration and contusions of the skin and underlying tissue.

One type of force extremely significant in biomechanic analysis is *friction*. This force occurs when two surfaces are in contact, and where motion is either present or tending to occur along the contact surface. If motion is taking place, the opposing force is known as *kinetic friction*. This tends to produce heat along the surface. If no motion is taking place, the opposing force is known as *static friction*. Static friction cancels out any component of applied force tending to cause motion.

The amount of frictional force available depends on the characteristics of the surfaces and on the force pressing them together. This force holding two surfaces in contact is known as the *normal force*. Thus, all forces acting between surfaces can be represented by components perpendicular to the surface (normal forces) and components parallel to the surface (frictional forces). This is illustrated in figure 8.8.

The ratio between the normal force and the available frictional force is known as the *coefficient of friction:*

$$\text{Coefficient of friction} = \frac{\text{Frictional force available}}{\text{Normal force}}$$

The coefficient of friction between two surfaces depends on the smoothness of the surfaces, dryness or lubrication, composition of the surface materials, and, to a lesser extent, the total area of contact and the relative velocity between the surfaces. The coefficient of friction is higher in static situations than when movement is present. As a result, higher frictional forces are available for changing motion before slippage or skidding occurs between surfaces of contact.

Friction is often considered as an undesirable force leading to loss of motion or to other problems in athletic events. But friction is also a very necessary force for skilled performance. Without it, athletic events as we know them would not be possible. For example, friction is needed in order to hold onto an object or implement. Performers need friction in order to exert forces parallel to the ground, floor, or other performers. In some instances, the problem is to obtain more frictional force, not less. The latter is particularly true during periods of high acceleration as seen in athletic and dance skills. Thus, fast starts, quick stops, or

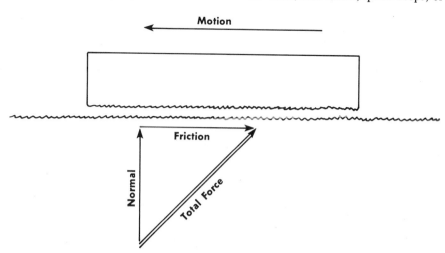

Figure 8.8 The figure illustrates a sport object such as a hockey puck sliding across a rough surface. Normal force is the force supporting the weight of the sport object. Frictional force is the force which tends to retard the motion.

rapid changes of direction of motion are skills that require considerable frictional force.

Performers use materials such as rosin or pine tar in order to increase the coefficient of friction while handling sport implements. This is a common practice in baseball; the hitter uses pine tar on his hands or hitting gloves to increase the coefficient of friction between the hands and bat during the very high, positive acceleration of the bat at the time of the swing. Without friction between hands and bat at this point, control of the bat would be extremely difficult, and without bat control, hitting averages tend to be low. Therefore, friction is a vital factor in this particular skill.

As a numerical example of frictional calculations, consider the following situation. A performer is attempting to move laterally from rest on a floor with a coefficient of friction of 0.5. What is the maximum acceleration that he or she can obtain if body weight is 160 pounds?

If it is assumed that weight provides the entire normal force, then—

$$\text{Frictional force} = \text{Coefficient of friction} \times \text{Normal force}$$

$$= \quad 0.5 \quad \times \quad 160$$

$$= 80 \text{ pounds lateral force}$$

Using the relation between force and acceleration—

$$\text{Force} = \frac{\text{Weight}}{g} \times \text{Acceleration}$$

$$80 = \frac{160}{32} \times a$$

$$\text{Acceleration} = \frac{80}{5} = 16 \text{ ft/sec}^2$$

An increase in the coefficient of friction to 0.8 would allow an increase in acceleration as follows:

$$\text{Frictional force} = 0.8 \times 160 = 128 \text{ pounds}$$

$$= 5 \text{ mass units} \times \text{Acceleration}$$

$$\text{Acceleration} = \frac{128}{5} = 25.6 \text{ ft/sec}^2$$

In actual performance, it is possible for a short period of time to push off against the floor and exert a normal component of force greater than the weight of the body. This allows frictional forces and accelerations even greater than those calculated above. This, however, also lifts the body. The leg will eventually reach full extension at the hip and knee, resulting in lifting the foot from the floor, and no lateral forces will then be possible.

Certain quantities are important in analysis because they are conserved. Within an isolated system, a constant amount of such quantities will be present throughout any given performance or interaction. Because of this, such properties are valuable for accounting purposes. If the change in a conserved property within one part of the system is known, this must be offset by an opposite change of the same quantity in some other part of the system. The two most important conserved quantities in biomechanics are *energy* and *momentum*.

Energy is a term which has many connotations when found in common usage. For the scientist, however, the term *energy* has a precise meaning, and this is directly related to the specific definition of the term *work*. In biomechanics, *work is accomplished by the operation of a force through a distance in the direction of the force.*

An example follows, computing the amount of work performed during weight training. During a bench-press workout, the average college football lineman should be able to press 240 pounds easily through a distance of approximately three feet. Since the direction of motion is exactly the same as the direction of force, the total amount of work performed is 240 pounds times three feet, or 720 foot-pounds of work during each repetition. Work = force × distance = $240 \times 3 = 720$ foot-pounds. Since this lightweight load would be handled for ten repetitions, 7,200 foot-pounds of work would be performed at the bench-press station. If three sets were performed, this would amount to a total of 21,600 foot-pounds.

The unit of work and energy in the English system is the product of the unit of distance, the *foot,* times the unit of force, the *pound.* This unit is called the *foot-pound.* In the metric system, the unit of work and energy is the *joule,* which is the product of the newton and the meter. Thus, one joule of work is performed when a force of one newton is exerted through a distance of one meter. Conversions between these units are given in Appendix A.

Another unit of work often used in biomechanics is the *Calorie.* Although originally defined as a unit of heat, the importance of this unit in food conversion within the human body has caused its wide acceptance in many areas of physical education. *The Calorie is equal to 3087.4 foot-pounds.* In the preceding example, the 21,600 foot-pounds of work is equivalent to seven calories. It should be noted here that the human body is far less than 100 percent efficient in converting food energy into work. To do seven calories of external work might require many more calories of fuel consumption.

From the point of view of mechanics, if the student-athlete simply holds this 240 pounds in the air during a repetition, no further work is being performed. This is where the big difference lies between the scientist's concept of work and the common connotation of work. *The essential act in performing work is maintaining a motion against a resisting force.* For example, if during a relay race at a picnic a 120-pound woman carries another 120-pound woman over a fifty-yard distance, the total amount of work accomplished would be zero unless she met with some force resistive to the horizontal motion. Since in carrying the woman she is exerting the force upward but is moving horizontally, there is no component of motion in the direction of the force. From the standpoint of biomechanics no work has been performed. From a *physiologic* standpoint, however, energy is being expended in both of the cited examples, through all aspects of the weight-training exercise and the run.

The conservation properties regarding work and energy are derived from the fact that every force has an equal and opposite counterforce. When there is a motion in the direction of one force, causing that force to do work, the opposite counterforce is having motion against it. Therefore, it is receiving the same quantity of work. *The term "energy" is used to represent the ability of a body to perform work.* Thus, one body doing work and losing energy implies that a second body receives work and gains that amount of energy. In this way, the total amount of energy within any system remains the same.

To return to the example of the bench-press exercise, when the athlete has pressed the weight upward, he has performed work. Therefore, he has 720 foot-pounds less energy. The weights, on the other hand, have been lifted higher in the air and are capable of doing work as they return to their original position. This capability to perform work is called *potential energy*. Therefore, the 720 foot-pounds lost by the man are counterbalanced by a 720 foot-pound gain of potential energy by the weights themselves. In this way, energy is conserved within the system.

When the man returns the weights to their original position through flexion at the elbow and horizontal abduction of the shoulder joint, a new phenomenon occurs. The weights now have lost the energy, and the 720 foot-pounds of work were transferred to the man. Unfortunately, the human system is incapable of mechanically storing the energy received. Within the human being, this energy is transformed to the form of heat and serves only to elevate the temperature of the body and its surroundings. This energy is lost mechanically to the system and is dissipated as heat energy.

Within the context of sports, the principal energy interactions are the change from chemical energy within the body into mechanical energy, and the change from received mechanical energy into heat within the human body. It should be noted that transformation of energy into the form of heat does not represent a total loss, since the generation of heat within the body triggers many other physiologic processes. These include an increased rate of perspiration, heart rate, and mechanical (accelerated) breathing rate, which facilitate biochemical respiratory actions. It is precisely this property of energy being transformed from one form to another that makes it such a useful and necessary element in the field of biomechanics.

Figure 8.9 illustrates the critical function of energy considerations in the analysis of human performance. This curve, taken in part from the data of figure 8.1, shows the average velocity for races of various distances. The distances are plotted on a logarithmic scale, and they range from 50 meters to the marathon.

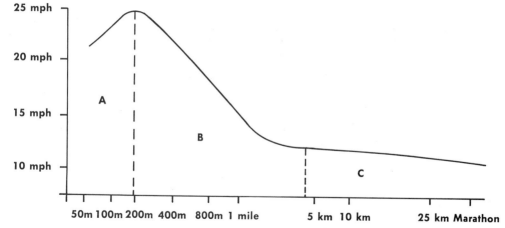

For sprints (A), the average velocity increases with the distance, to a maximum near 200 meters. At these distances, performers are limited in speed only by their step rate and stride length combination. The necessary energy for a 30 second maximum output is stored at the muscle level in the body. The principal differences are determined by the fact that the race starts from zero velocity. This is an example of *anaerobic performance*. Anaerobic means "not requiring oxygen."

In middle distance races (B), the body is unable to furnish oxygen for muscle energy as fast as it is being consumed. This is known as an "oxygen deficit" performance. The rate of decrease of available energy needs to be controlled to allow the performer to finish the race before oxygen levels in the blood become low enough to impair coordination and strength. As a result, races requiring longer times are necessarily run at a slower pace.

For long distances (C), much of the race is run at a pace at which the oxygen processing can furnish energy at the same rate it is being consumed. This is an aerobic or "oxygen equilibrium" situation. The average velocity is highly dependent on the speed which can be maintained under "oxygen equilibrium" conditions. Only during small segments of the race will the "oxygen deficit" allow a faster pace. Endurance training helps to increase this equilibrium speed. Only for distances requiring more than an hour to run do other components of the energy producing system contribute significantly.

Within the human body, the conservation of energy is illustrated by the relation between work, food consumption, and body fat. Energy input into the body is in the form of food and in oxygen to burn it. This energy may be used in the basic life processes or may be expended during work. Most of the energy for basic metabolism is dissipated in the form of heat. If the heat loss plus the work performed does not equal the input energy, the body stores the excess in the form of fat. At times, some of this fat may then be burned to provide a greater expenditure of energy over input.

There are two types of mechanical energy of concern within this text. *The first of these is kinetic energy or energy due to motion.* Any moving object has

Figure 8.9 Average velocity vs. race distance for a wide range of expert running performances.

the ability to do work simply because it is moving. In slowing this object to a stop, some force must be exerted against the motion. As a consequence, the object does the work, and the body exerting force receives the energy. For example, after striking the ball in a tennis serve, the arm and racket have a great amount of kinetic energy. The muscles within the racket arm and shoulder absorb this energy by eccentric contraction. This contraction serves the purposes of stopping the total motion, preparing for subsequent motions at these joints by placing critical muscles on stretch, and protecting the elbow joint against hyperextension.

The numerical calculation of kinetic energy uses the equation—

$$\text{Kinetic energy} = \tfrac{1}{2} \times \text{Mass} \times (\text{Velocity})^2 = \tfrac{1}{2}mv^2$$

Thus, a bowling ball weighing 16 pounds and moving at 20 ft/sec has a kinetic energy given by—

$$\text{Kinetic energy} = \tfrac{1}{2}mv^2 = \tfrac{1}{2}\frac{16}{32}(20)^2 = 100 \text{ ft-lb of kinetic energy}$$

and a 160-pound runner moving at 30 ft/sec has a kinetic energy of—

$$\text{Kinetic energy} = \tfrac{1}{2}\frac{160}{32}(30)^2 = 2{,}250 \text{ ft-lb}$$

As an example of kinetic energy calculations within the metric system, consider the case of a 100-kilogram hockey player skating at 15 meters per second.

$$\text{Kinetic Energy} = \tfrac{1}{2}mv^2 = \tfrac{1}{2}(100)(15)^2 = 11{,}250 \text{ joules}$$

It can be seen that kinetic energy increases dramatically as either the mass of the body or the speed increases.

The other principal type of mechanical energy mentioned previously is *potential energy*. This is the ability to work which an object possesses because of its position within the gravitational field or because of an elastic change of the object's shape. An example of potential energy is observed in the approach of a diver to his final step on the diving board. This step is calculated to impart an optimal amount of potential energy into the bending of the board so the board will return this energy to the diver and provide an optimal angle for the trajectory of the dive.

Numerical calculation of the gravitational potential energy is given by—

$$\text{Gravitational potential energy} = \text{Weight} \times \text{Distance raised or lowered}$$

$$= W \times h$$

In the previous example of the bench press, the potential energy gained by the weight equals—

$$\text{P.E. (grav.)} = W \times h = 240 \text{ pounds} \times 3 \text{ feet} = 720 \text{ ft-lb}$$

As another example, a 190-pound pole-vaulter raising his center of gravity 16 feet acquires a gravitational potential energy of—

$$\text{P.E. (grav.)} = W \times h = 190 \times 16 = 3{,}040 \text{ ft-lb}$$

A 60-kilogram diver at the top of a 5-meter tower has a gravitational potential energy of—

$$\text{P.E. (grav.)} = mg \times h = 60 \times 9.8 \times 5 = 2{,}940 \text{ joules}$$

This must be dissipated after entering the water.

Calculation of the elastic potential energy stored in a system is more complex than the previous calculations. If the system behaves perfectly linearly, i.e., if the distortion is directly proportional to the applied force, the system is said to obey Hooke's Law. In this case:

$$\text{Force} = K \times \text{Distortion, where } K \text{ is the elastic constant}$$

$$\text{P.E. (elas.)} = \tfrac{1}{2}(K)\,(\text{distortion})^2$$

Unfortunately, there are few cases of interest in biomechanics where this simple relation holds. In the compression of balls, the bending of sport implements, and the distortions of human body parts, there is a nonlinear relation between force and distortion. In these cases, a graph of the force vs. distortion needs to be made, and the energy can be calculated from this curve.

In many sport events there is a long chain of energy transformations from one type to another before the conclusion of the performance. For example, a polevaulter uses the running approach to produce a large amount of kinetic energy. During the early part of the vault immediately following the pole plant, much of this energy is stored in the bending of the pole. The good vaulter then moves himself into a position to enable the pole to deliver this potential energy to his body in the form of kinetic energy. This aids motion in an upward direction. Kinetic energy is transformed into gravitational potential energy as the vaulter goes high enough to clear the bar. Following clearance of the bar, gravitational potential energy is then transformed into kinetic energy of the fall. The purpose of the padding in the pole-vault pit is to absorb this kinetic energy in a harmless form of work and to finally dissipate it as heat (fig. 8.10).

Figure 8.10 Jeff Davis, Southwest Missouri State University pole-vaulter, nearing the point of maximum conversion to gravitational potential energy. *Photo courtesy SMSU Public Information Office.*

The subject of elasticity within biomechanics involves the interplay between kinetic and potential energies. During a collision involving an elastic object such as a tennis ball, the kinetic energy that the ball possesses before the collision is stored as elastic potential energy in the deformation of the ball. As the ball returns to its original shape, most of this energy is again converted into kinetic energy. *The term "coefficient of restitution" is used to denote the ratio of the amount of kinetic energy after the collision to the amount of kinetic energy prior to the collision.*

Many times the rules of a sporting event specify either the air pressure or the materials of a sport object in order to limit the coefficient of restitution to an acceptable level. As an example, the United States Handball Association's rules specify that the handball should be made of rubber 1⅞″ in diameter with a ¹⁄₃₂″ variation. It should weigh no more than 2.3 oz. with a variation of 0.2 oz. The rebound from a 70-inch drop at 68°F shall be no more than 42–48 inches.

A handball with the proper coefficient of restitution when dropped from 70 inches should rebound no more than 48 inches. Although not specified in the rules, it is assumed that the handball should be dropped on a nonresilient surface as found on a handball court.

Limitation of the coefficient of restitution for a handball and regulations for the size of the court allow the strategy of the game to fall within the competency of a skilled performer. There are available other balls of the same size as regulation handballs but made of a modern silicone plastic composition. These are sold under the trade name of "Super Ball." The coefficient of restitution of these balls is approximately 0.95. If we were to attempt to play handball with one of these extremely lively balls, the velocities and trajectories of the rebounds would necessitate a complete change of strategy and body motion for the handball performer. It would literally be "a new ball game!" With the present size of the handball court, the reaction and movement time limitations of the performer would make it very difficult to play the game of handball as it is now played.

The sports of handball and racquetball offer an instructive comparison of the changes in strategy which come about when an implement is introduced into a sport. The racquet not only replaces contact with the gloved hand of the performer, but also introduces a new set of rebound coefficients into the collisions with the ball. In addition, the racquet extends the reach of the performer, and this necessitates modifications of the stance and body position when executing "kill shots."

Impulse and Momentum

The other valuable conserved property in biomechanic applications is momentum. Historically, momentum has been known as the "quantity of motion." *Scientifically, it is represented by the product of the mass of a body times the velocity with which that body is moving.* Large momenta are associated with large bodies or bodies moving at high velocity. As an example of a momentum calculation, let us return to the example of the sprinter in the 220-yard dash. If the athlete has a weight of 160 pounds, the momentum can be calculated:

$$\text{Momentum} = \text{Mass} \times \text{Velocity} = mv = \frac{wv}{g} = \frac{160}{32} \times 33 \text{ ft/sec}$$

$$= 165 \text{ momentum units}$$

During recent years, the term *momentum* has received increasing usage in a wide variety of contexts in the field of sport. It has become almost a cliché with sportscasters who use the term *momentum* in a psychological sense to describe a positive expectation of a team on the field. Increasing momentum has also been used as a synonym for positive acceleration. For example, a ski-jumper moving down a slope may be referred to as having positive acceleration, or it can be said that he is gaining momentum.

The latter statement is not inaccurate and represents a very essential relationship between momentum and the previously discussed quantities of acceleration and force. If we consider the concept of change of momentum, the following relation can be shown:

$$\frac{\text{Change of momentum}}{\text{Time}} = \frac{\text{Mass X Change of velocity}}{\text{Time}}$$

$$= \text{Mass X Acceleration} = \text{Force}$$

$$\text{Force} = \frac{\text{Final momentum} - \text{Initial momentum}}{\text{Time}}$$

Forces can be associated both with the production of acceleration and with the changing of momentum. Forces cause accelerations and changes of momentum, and accelerations represent changes of momentum. As an historic footnote, it may be recalled that when Newton first proposed the laws of mechanics, his statement was simply that a force was linearly related to the change of quantity of motion of an object. "Quantity of motion" was Newton's term for what is called momentum in present-day terminology.

Just as work is related to the transfer of energy between objects, there is also a quantity called *impulse,* which describes the transfer of momentum between objects. *Impulse is scientifically defined as the product of force and time.* It can be observed that this product of force and time actually represents the change of momentum of a body. Work represents transfer of energy, and impulse represents change of momentum between bodies.

Neither the English nor the metric system of units has a separately named unit for momentum or impulse. In the English system, either the slug-foot/second or the pound-second can be used. These are equivalent units. Investigators who write biomechanics literature seem to prefer the pound-second. In the metric system, the kilogram-meter/second is more common, but the newton-second is also correct for these quantities.

It should be noted here that only linear momentum has been discussed. There is also a conserved quantity, angular momentum, which is associated with rotational motion. Its units are slightly different, but rotational motion is equally important in describing total motion. Angular momentum will be discussed later in this chapter, following a general discussion of rotational motion.

The fact that forces and counterforces always occur with equal magnitude and opposite direction leads to the concept of conservation of momentum. For every force acting in one direction, there is a force in the opposite direction for the same period of time representing an opposite change of momentum. For every change of momentum occurring in one direction on one body, another body is receiving an equal and opposite change of momentum, so the total momentum of the system remains a constant.

The impulse of a collision within the context of sport action may occur in different ways. The total effect of the impulse is to produce a change of momentum equal to the product of force and time. However, this product may be formed by a large force for a very short time period or by a smaller average force exerted over a longer time period to produce the same total change of momentum. Examples of these two types of impulses are commonly observed in boxing. A jab to the head generally exerts a very large force over a very short contact time, whereas the time of contact for a body hook is much longer. The forces for the two punches may actually be the same. However, the hook requires a longer period of time and a greater distance of motion. Most of the impulse differences between these two punches lie in the type of body tissue with which contact is made. The mandible or jaw area is the most common contact point for the jab. The amount of collision motion is minimal because of the relative hardness of the surface; therefore, the time of contact is very short. In contrast, owing to the relatively softer abdominal area where the hook is often landed, there is a much greater elastic motion. As a consequence, the collision or contact time is greater.

For another example, in a pass-protection block in American football, the emphasis is on short, hard blocks in repetitive sequence by the offensive lineman, whereas during line blocking to "open a hole" for a running play, a sustained block must be used, i.e., contact time for this type of blocking must be longer than for blocking for pass protection.

As a numerical example of impulse-momentum calculations, if a 16 lb. bowling ball is delivered at a velocity of 20 ft/sec, the total momentum of the ball would be:

$$\text{Momentum} = \text{Mass} \times \text{Velocity} = \frac{16}{32} \times 20 = 10 \text{ momentum units}$$

If the time of approach and delivery were 2 seconds, then—

$$\text{Avg. force} \times \text{Time} = \text{Change of momentum} = 10 \text{ momentum units}$$

$$\text{Avg. force} = 10/2 = 5 \text{ pounds}$$

For a second example, consider the case of a softball batter who receives a pitch at 140 ft/sec and hits it back to the pitcher at 200 ft/sec with a total time of contact with the bat of 1/50 sec. In this case, the direction of motion of the ball is reversed. The change of momentum is the numerical sum of the magnitudes of the original and final momenta (weight of a softball is about 6 ounces):

$$\text{Change of momentum} = \text{Mass} \times \text{Change of velocity}$$

$$= \frac{\text{Weight}}{\text{Gravity}} \times \text{Change of velocity}$$

$$= \frac{6/16}{32} \times (140 + 200)$$

$$= 3.98 \simeq 4 \text{ momentum units}$$

$$\text{Avg. force} = \frac{\text{Change of momentum}}{\text{Time}} = \frac{4}{1/50} = 200 \text{ pounds}$$

The concepts of impulse and momentum are extremely important in situations involving collisions of the type presented in the previous example. In chapter 9, various types of collisions that occur in sport will be considered, and the interactions which occur will be analyzed.

Power represents the amount of work performed per unit time. It is related to energy in the same way that momentum is related to force. Power is energy expended or energy gained per unit time. Force represents the change of momentum per unit time. Power can also be considered as the product of force times distance per unit time, or as the product of force times velocity.

Power

The simplest units for expressing power are those of energy per unit time. In the English system, this would be in foot-pounds per second. However, power is more commonly expressed in horsepower. *One horsepower is defined as 550 foot-pounds per second.*

In the metric system, the unit of power is the watt, which is one joule per second. *One horsepower represents 745 watts.* In biomechanics literature, power is often expressed as energy expenditure in Calories per minute. *One horsepower is equal to 10.7 Cal/min.*

The amount of power that can be delivered during a human performance also depends on the time during which it must be sustained. Power is basically limited by the rate of delivery of oxygen to the affected musculature. Several horsepower may be developed for periods of less than a second. Power levels of the order of one horsepower can be maintained for several seconds by a trained athlete. But for sustained or long interval performances, even a trained human body can maintain only a few tenths of a horsepower.

For calculations involving maximum power, the "snatch lift" from the sport of weight lifting serves as a good example. In this event, the performer in one motion lifts the weight from the ground to a position overhead. This involves considerable power (horsepower), especially in T.A.C. and Olympic competitions.

A heavyweight world-class competitor can "snatch" about 300 pounds. If he lifts this weight a total distance of 7 feet during a ¾-second period of time, then—

$$\text{Work} = \text{Force} \times \text{Distance} = 300 \times 7 = 2{,}100 \text{ ft-lb}$$

$$\text{Power} = \frac{\text{Work}}{\text{Time}} = \frac{2{,}100}{\text{¾}} = 2{,}800 \frac{\text{ft-lb}}{\text{sec}} = \frac{2{,}800}{550} = 5.1 \text{ hp}$$

During the time of the lift, the performer expends energy or delivers power at a rate of about five horsepower. Most sport performances require high power expenditures for short periods with rest intervals between. American football is an example of this.

| Rotational Motion | Linear motion, movement from one point in space to another, has been the only consideration in this chapter thus far. In biomechanics, however, rotational motion is often of greater importance than linear motion. Because of the structure of the human musculoskeletal systems, nearly all motions can be represented as rotations about joints. And in many aspects of exercise, sport, and dance, the rotation of a body about some axis is of fundamental importance. |

In many ways, the description of rotational motion parallels that of linear motion. There are rotational positions, displacements, velocities, and accelerations. The basic difference between linear and angular motion occurs in the units in which these must be specified. In rotational motion, rather than having a linear displacement or distance covered, an angle determines the direction.

Angular description differs from linear description in one very basic way. After a certain magnitude of angle has been traversed, the original position is again reproduced. Due to limitations of joint structure and body tissue, however, a rotation will normally be limited to far less than one complete revolution in joint motions. Each articulation of the body has a maximum rotational range of motion about its axis. In biomechanics the term *rotation* as applied to joint motion has a generalized meaning, as compared to the specific anatomic definition of joint rotations described in chapter 2.

Because of the duplication of position at the end of the revolution, a natural way of specifying a unit of angle is the complete revolution. However, other units are also used for the measurement of angles. The most common unit is the degree, or $\frac{1}{360}$ of a revolution. The concept of having 360 degrees in a complete revolution has developed from a historic context, but the number does allow convenient divisions of the circle into equal parts. Normally, ranges of angular motion are specified in joints in terms of the number of degrees through which the joint may be rotated without damage. For example, the average individual can rotate the bony levers about the axis of the elbow during elbow flexion through a range of approximately 150 degrees. A full 360 degrees of motion is not possible due to the structure of the joint, size of the surrounding musculature, and the nature of the noncontractile tissue.

Another unit of angular measure is the radian. *The radian is defined as the angle which includes an arc of a circle equal to the radius of the same circle.* As an example, if an object rotates through an angle of one radian, any point located two feet from the axis of rotation will move a distance of two feet. A point located four feet from the axis of rotation will move a distance of four feet (fig. 8.11). Because the total distance traveled by an object during one complete revolution, or 360 degrees, is 2π times the radius of the circle, it follows that one radian is equal to $\frac{1}{2}\pi$ of 360 degrees. That is—

$$1 \text{ radian} = \frac{360°}{2\pi} = 57.3°$$

$$1 \text{ degree} = \frac{2\pi}{360} = 0.0175 \text{ radian}$$

Understanding Basic Physics

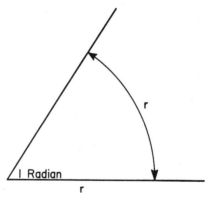

I Radian

r

Figure 8.11 Radian.

The function of the radian will be seen later in the conversion from rotational velocities and accelerations to linear velocities and accelerations.

The development of *flexibility* during conditioning of students actually corresponds to an increase of ranges of motion in each of the applicable joints. This is accomplished by stretching noncontractile tissue such as ligaments and tendons. The importance of flexibility as a biomechanic parameter is often overlooked by physical educators. The increase in range of motion or flexibility allows a performer to exert forces over greater distances and times. This increases velocities, energies, and momenta associated with the performance. Also, the increase of range of motion allows a greater stretch to critically involved muscles. This provides these muscle groups with the potentiality of developing larger forces. For example, a golfer who has an extensive range of motion in the backswing is able to apply a higher velocity to the club head at the point of impact because of the extra distance over which he can provide acceleration. The muscle groups which initiate the torques or motions of the downswing are placed on greater stretch during the backswing; therefore, they are capable of exerting greater force.

It is worth noting at this point the reasons for placing muscles on stretch before beginning a strong concentric contraction. While it is true that man cannot store nearly as much energy in elastic stretching of muscle and connective tissue as some animals and insects, there is still some energy to be gained in this respect. But, there is even more to be gained from the neuromuscular phenomenon known as the *stretch reflex*. Much simplified, this means that when muscle fibers are stretched there is an enhanced firing of the nerve cells near them to cause them to contract. This is a part of the internal safety system that prevents damage to the tissues. By stretching a muscle immediately before a contraction, this extra nerve stimulus can be added to the overt signals to the muscles. Therefore, a much stronger contraction can be obtained than from volitional brain control signals alone.

It should also be noted that the stretch reflex depends on the timing of the stretching immediately before the desired contraction. A slight pause at the end of a backswing in golf, for example, will allow the reflex stimulus to pass and lose its enhancement of the desired motion. In some cases, such as putting in golf, it is necessary to take time for the stretch reflex to dissipate in order to have better control of the desired motion.

Linear velocity was defined as a distance covered per unit time. *Angular velocity is the angle rotated through per unit time.* Units for angular velocity will be described in degrees per second, radians per second, or revolutions per second. Likewise, *angular acceleration is defined in terms of the change of angular velocity per unit of time.* The units for angular acceleration will be degrees per second squared, radians per second squared, or revolutions per second squared.

As an example of these concepts, consider a bowling ball rolling down the lane at 18 ft/sec. Since bowling rules specify that the circumference of the ball is 27 inches, it will roll one revolution each time it travels this distance. Thus, its angular velocity will be:

$$\text{Angular velocity} = \frac{18 \times 12}{27} = 8 \text{ revolutions/sec}$$

$$= 8 \times 360 = 2{,}880 \text{ degrees/sec}$$

$$= 8 \times 2\pi = 50.3 \text{ radians/sec}$$

When released, a bowling ball is usually sliding with no angular velocity. The conversion from sliding to rolling involves angular accelerations—an increase of angular velocity. In the foregoing example, if the conversion from sliding to rolling required 0.7 seconds—

$$\text{Angular acceleration} = \frac{\text{Change of angular velocity}}{\text{Time}}$$

$$= \frac{8 - 0}{0.7} = 11.43 \text{ rev/sec}^2$$

$$= 4{,}114 \text{ degrees/sec}^2 = 71.86 \text{ rad/sec}^2$$

Torque   Force as a quantity is necessary to produce linear acceleration. There is also a quantity called *torque,* which is directly related to the production of angular acceleration. A torque is commonly pictured as a tendency to twist. In many cases, this is the way a torque is produced. In other cases, however, it is produced by a force applied along a line not passing through an axis of rotation; therefore, it causes a rotation of a body or body part. For example, rotation of the hand is the result of torques produced either by supination or pronation motions of the radio-ulnar joint. Hand rotation can also be caused by torques associated with medial and lateral rotations of the shoulder joint. Muscles associated with medial rotation are far more powerful than those associated with lateral rotation at the shoulder or the supination-pronation at the radio-ulnar joint. Therefore, the strongest torques achieved at the hand are produced with the elbow and radio-ulnar joints in stabilized positions and the shoulder joint being medially rotated. This condition is sometimes seen in the sport of arm wrestling (fig. 8.12). A winning effort is associated with the strong torques produced at the shoulder joint by the large muscle masses associated with medial rotation.

Figure 8.12 Torque-countertorque during arm wrestling.

The distance from the line of action of the force to the axis about which rotation takes place is called the lever arm. The magnitude of the torque is determined by the product of the force times the length of the lever arm. A force applied far from the axis of rotation is more effective in producing torque than a force applied near the axis of rotation.

This use of various lever arms in the production of torque is commonly known as *leverage.* A performer working against a resisting force can more effectively operate by utilizing a rotational motion and applying forces at greater distances from the axis of rotation than the resistances. This ratio of lever arms acts as an effective multiplier of one's own force. As a consequence, a force of 100 pounds applied two feet from an axis of rotation is capable of holding or moving a force of 200 pounds applied only one foot from the axis.

Figure 8.13 shows a gymnast establishing a grip to initiate rotation about a vertical axis. Such a move requires a torque exceeding 500 pound-inches to complete the maneuver within the time constraints of the position. If he obtains a hand position separated by about 18 inches, the rotation will require forces of about 30 pounds against the bar:

$$\text{Torque} = \text{force} \times \text{separation} = 30 \times 18 = 540 \text{ pound-inches}$$

If the bar should be gripped with a separation of only 12 inches, however, larger forces would be required:

$$\text{Force} = \text{Torque/separation} = 540/12 = 45 \text{ pounds}$$

This principle explains the necessity of a widened stance in those sport situations where sudden rotations may be a part of the movement strategy. If the feet are too close together, extremely large frictional forces will be necessary to initiate such rotations.

Figure 8.13
Establishment of grip for
effective torque
production by Donald
Jackson of the USC
gymnastic team during a
horizontal bar routine.
*Courtesy USC Athletic
News Service.*

Figure 8.14
Application of leverage
during a wrestling match.
*Courtesy SMSU Public
Information Office.*

Figure 8.14 illustrates another condition where leverage is necessary for a successful sport move. The wrestler underneath needs to rotate his opponent into an upright position in order to move from the kneeling to the standing position. The necessary torque is provided by forceful diagonal abduction of the left shoulder joint concurrent with elevation of the right shoulder girdle. If the lines of action of these forces are separated by a distance of 2 feet, a force of 60 pounds would produce:

$$\text{Torque} = \text{Force} \times \text{lever arm} = 60 \times 2 = 120 \text{ pound-feet}$$

If he allows the hands to slide up the opponent's back so the lever arm is reduced to 6 inches, the same torque would require a force of:

$$\text{Force} = \text{Torque/lever arm} = 120/.5 = 240 \text{ pounds}$$

From this position the lower wrestler would lift the opponent before he could cause the desired rotation. Meanwhile, the dominant wrestler in figure 8.14 has placed his arms in positions to prevent the obtaining of the necessary torque. This production of tactical and compensating torques is an integral part of the intellectual aspect of the sport of wrestling.

Within the human musculoskeletal system, the concept of leverage is used in a different way. Muscle and bone attachments are often arranged as *third class levers*. This means that the lever arm for the input muscular force is much shorter than the lever arm for the output action. This condition is illustrated in figures 8.15 and 8.16 showing the leverage in the Achilles tendon and biceps brachii systems.

Since the lever arm for the Achilles tendon is only 2 inches, while the lever arm to the "ball of the foot" is 7 inches, forces in the tendon will be 3.5 times as great as those borne by the foot. Thus, a 200 pound performer having these measurements would develop a force of:

$$\text{Force} = 200 \times 3.5 = 700 \text{ pounds}$$

in the tendon while standing on the ball of the foot. When landing from a jump or performing a ballistic maneuver, this force would be much larger. It is not uncommon for large performers to be involved in situations where the Achilles tendon must bear a force of more than a ton.

In the biceps brachii contraction, the lever ratio may range from 7:1 to 10:1, depending on the body structure of the performer. This means that the forces experienced, both at the muscle attachment positions and as compression within the elbow joint, are much greater than the weight borne at the hand position. For example, with the dimensions indicated in figure 8.16, a performer doing a "biceps curl" (elbow flexion) using a 30 pound weight would experience the following:

$$\text{Leverage ratio} = 13.5/1.5 = 9 \text{ in favor of the load}$$

$$\text{Biceps brachii tension} = 9 \times 30 = 270 \text{ pounds}$$

$$\text{Elbow compression} = 270 - 30 = 240 \text{ pounds}$$

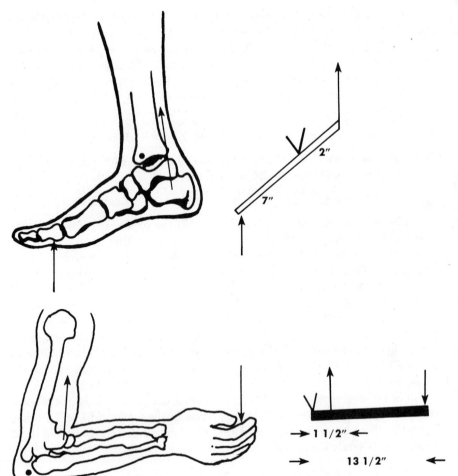

Figure 8.15
Leverage in the Achilles tendon.

Figure 8.16
Leverage in the biceps brachii system.

The actual ratios vary, depending on the angle of the joint and the position of nearby joints in many cases. For example, the biceps brachii ratio is smaller in a position of radio-ulnar supination than in a position of pronation. This is one reason why "chinups" (elbow flexion) with supinated radio-ulnar joints are much easier than with the hands pronated.

This lever condition prevails because we are generally built for mobility and range of motion rather than strength. Thus, a 2 inch contraction of the biceps brachii muscle can produce from 14 to 20 inches of motion of the hand as the elbow flexes. These large lever ratios are particularly valuable at joints responsible for translational motion, the hip and knee, and lead to extreme forces in the muscles associated with such joints.

Third class lever relations are also seen in the use of many sport implements. Bats, rackets, and hockey sticks act as lever extensions. This provides them with greater range and speed from joint rotations. An extreme in this case is the use

Understanding Basic Physics

of a long fly rod for fishing. In this instance, small joint rotations can produce large ranges and speeds of the relatively small sport object (the fly).

Another application of this principle is the widened stance used in many sports requiring quick body rotations. The frictional forces between the feet and the ground required for such movements when the feet are positioned three feet apart are only one-third those required if the feet are placed one foot apart. Within the limits of efficient use of the muscle-joint actions, performers in baseball, tennis, football, wrestling, and other sports use as wide a stance as practical to execute the skills.

In gymnastics, the production of torque is especially critical because of the production and change of rotational motion inherent in gymnastic routines. These torques are produced not only with the feet against mat surfaces but also by means of handholds on apparatus such as bars, rings, and uneven parallel bars. Again, attention to wide spacing in providing large lever arms for effective torque production is of utmost concern.

In the contact strategy at the start of a wrestling match, the wrestler often endeavors to position a hold on the opponent as far from the trunk as conveniently possible. This will enable the first wrestler to turn and control the opponent with a greater torque. Many maneuvers are designed to work against an opponent's extended limb, where torques are greatly hampered by the minimal lever arms obtainable because of the positions of muscle attachment. Likewise, many of the movements in judo and other forms of the martial arts are designed to assure the most efficient use of such leverage.

In many other sport applications, leverage is used in precisely the opposite sense. When applying a rotation against a resistance located much farther from the axis, the performer sacrifices the multiplication of force but gains a multiplication of speed. Many sport implements are designed to use this by effectively extending the performer's own limb distance. One of the most dramatic examples is the use of a fly rod in fishing. The object being thrown—a dry or wet fly—is so light it requires only minimal forces. The long rod is used to impart a high velocity to the fly and line in order effectively to overcome air resistances. The stereotype of perfect mechanics in the use of such sport implements involves the application of a series of torques to produce optimal velocity of some portion or portions of the implement.

In discussing a body or sport object which has linear and rotational motions, it normally is necessary to separate these two motions. This is done by ascribing the linear motion to a fictitious point called the center of gravity. *The center of gravity can be described as that point on or about which a body or object would balance most perfectly.* That is, in any direction from the center of gravity, torque caused by matter on one side exactly balances torque caused by matter on the opposite side. In this context it should be noted that each body or each body segment has its own center of gravity, and this can be determined. At times it becomes necessary to describe motion in terms of the center of gravity for the total body and the individual body segments.

Center of Gravity

To aid in the location of the center of gravity, the human form is sometimes considered to be composed of several segments. Each of these has a relatively fixed weight and internal center of gravity. The 10 major segments are the head, torso, upper arms, lower arms and hands, upper legs, lower legs and feet. If the weight and approximate center of gravity of each of these for an individual performer is known, it is possible to estimate quickly the location of the center of gravity for the total body in a wide variety of positions.

Drawings indicating this procedure are shown in figures 8.17 through 8.20. Each photograph is accompanied by a contour drawing showing the position of the above 10 segments. Segmental centers of gravity are indicated with dots, and the approximate total center of gravity is marked with an X. Typical data and calculation operations for performing this type of analysis are given in Appendix C.

In describing the center of gravity of the total human body, it should be kept constantly in mind that this center of gravity is very much a function of the position of limbs and the motions of the spinal column. There are many times during sport and dance performances when the center of gravity may actually lie outside the body. For example, a diver in a pike position has so much mass distributed in front of the trunk that the center of gravity of the total body actually lies somewhere in the triangle formed by the trunk and limbs.

In figure 8.17, the center of gravity of the body theoretically would lie at the intersection of the three cardinal planes of motion. These are discussed in chapter 3. In figures 8.18, 8.19 and 8.20, the center of gravity of the performers have been shifted to other points, due to the nature of the total motion. In the cases of the gymnast dismounting from the horizontal bar and the high jumper, the center of gravity for each athlete may often lie outside the body.

Figure 8.17 Center of gravity of the body lies at a theoretical point below the umbilicus and between the iliac fossas in the anatomic position.

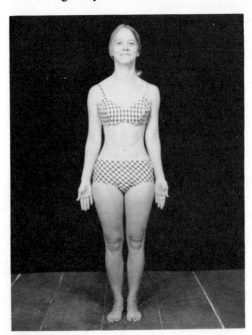 

Understanding Basic Physics

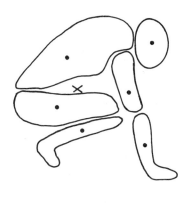

Figure 8.18 Center of gravity of the body shifts due to any change in position of its joints or segments. *Courtesy USC Athletic News Service.*

Figure 8.19 Center of gravity lies outside the body during a horizontal bar dismount by Gareth Burk, USC gymnast. *Courtesy USC Athletic News Service.*

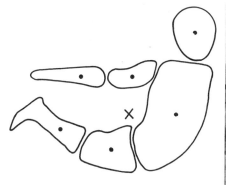

While executing a sport motion, the motion of the center of gravity often follows either a straight line or a parabolic gravitational arc. Meanwhile, the body or body segments are executing rotations about this center of gravity. During biomechanic analysis it sometimes is necessary to describe the linear motion of the center of gravity and various rotations occurring about this point in terms of the motions of the upper and lower limbs and a variety of lumbar-thoracic and cervical spinal motions. In chapter 9, the relationships between the linear motion of the center of gravity and the rotational motion of the various parts about this point will be presented in more detail.

Moment of Inertia

Just as mass has been defined as the ratio between linear force and the observed acceleration, there is also a quantity that applies to a body to describe the ratio of angular torque to the resulting angular acceleration. This quantity is known as the *moment of inertia.* It is a more complex quantity than mass because the moment of inertia depends not only on the total amount of matter, but on the relative positions of each part of a body. *The total moment of inertia of a body or a sport object is calculated by summing the amount of mass contributed by each part times the square of its distance from the axis of rotation:*

$$\text{Moment of Inertia} = \text{Torque/Angular acceleration}$$

$$= \Sigma \, (\text{Mass} \times \text{Radius}^2)$$

For example, the moment of inertia of a football thrown or kicked with spiral rotation is much smaller than the moment of inertia when it is tumbling or rotating end over end following a kickoff or field goal attempt. The reason for this lies in the fact that the total mass tends to be located much closer to the axis of rotation in the case of spiral rotation. Figure 8.21 shows selected moments of inertia.

Understanding Basic Physics

	Weight (Pounds)	Radius (Inches)	Moment of Inertia (Pound-inch²)
BALLS			
Baseball	.31	1.5	.3
Basketball	1.31	4.8	20
Bowling	10–16	4.5	81–130
Field Hockey	.35	1.4	.3
Football—spin	.90	3.3	approx 5
tumble	.90	Ellipse 3.3 × 5.7	approx 10
Golf	.10	.81	.03
Handball	.14	.95	.08
Lacrosse	.32	1.25	.2
Ping-Pong	.006	.75	.002
Shot	16	2.5	40
Soccer	.94	4.4	12
Softball	.40	1.9	.6
Squash	.07	.88	.03
Tennis	.13	1.25	.13
Volleyball	.60	4.2	6.5
Discus—spin	4.4	4.4	approx 40
tumble	4.4	4.4	approx 20
SPORT IMPLEMENTS		(Dimensions)	
Badminton racket	.38	Handle 17	140
		Head 10 × 7	
Baseball bat	2.2	33–36	1900
Golf club—wood	.85	30	800
iron	1.0	28	750
Hockey stick	variable	Handle 40–53	3000+
		Head 2 × 14	
Softball bat	2	31–34	1100
Tennis racket	.85	27 × 9	300

Figure 8.21
Moments of inertia of selected sport objects and implements.

The actual geometrical calculations of moments of inertia of sport objects and the human body are quite complex. However, the moment of inertia about a given axis can be measured by going back to the basic definition and determining the amount of torque necessary to provide a given angular acceleration. Although accurate numerical calculations of the moment of inertia are not always feasible, the use of the concept of a changing moment of inertia quite often allows its utilization through basic levels of biomechanic analysis. For example, when the limbs are brought in close to the longitudinal axis of the body, the moment of inertia and the torque-acceleration relationship are very different than when the limbs are extended. The utilization or the violation of this principle can be readily observed in various gymnastic routines, ice skating rotations, and in the sport of diving.

While it is often quite difficult to calculate numerically the moment of inertia from the geometrical positions of various parts of an object, there is a term sometimes used to indicate the average distance of parts from the axis of rotation. This is called the *radius of gyration.* By definition, the radius of gyration is obtained as follows:

$$\text{Moment of inertia} = \text{Total mass} \times (\text{Radius of gyration})^2$$

$$\text{Radius of gyration} = \sqrt{\frac{\text{Moment of inertia}}{\text{Total mass}}}$$

A typical baseball bat is constructed so the radius of gyration is approximately 30 inches when measured from the knob of the bat. A point at this distance is technically known as the *center of percussion.* It is popularly known as the "sweet spot" of the bat (fig. 8.22). A collision with the revolving bat at this point will impart maximum momentum change to the ball.

The term "sweet spot" comes from the fact that a collision at the center of percussion does not tend to rotate the bat about its own center of mass. Such a collision transmits minimum force to the hands. A collision nearer the handle tends to move the handle backward in the hands, while a collision near the end of the bat imparts a forward motion of the handle within the grip. Either of these provides an inefficient transfer of energy to the ball and a painful force to the hands. On the other hand, a collision at the "sweet spot" allows the bat to reduce velocity and continue its translational motions.

A softball bat is usually shaped with the center of percussion nearer the handle. This leads to a much lower moment of inertia and allows much faster rotational accelerations of the bat. This is necessary because of the shorter interval between the release of a softball pitch and the time it crosses the plate. Thus, the emphasis is more on bat control and bat speed in fast pitch softball than on maximum transfer of momentum by means of a long radius of gyration, as in baseball.

A tennis racket is designed so the balance point is near the throat, but the center of percussion is slightly below the center of the string face (fig. 8.23). This makes possible optimum use of the elasticity of the strings when striking near the center of percussion. Modern designs have concentrated on enlarging this "sweet spot" area to allow more strokes to hit this optimum part of the racket.

Experimentally, the moment of inertia of a man about 6'5" and 200 pounds, rotating about a longitudinal axis with his arms at his side and feet together, is about 1.5 units. If he is rotating about the same axis with arms and legs abducted, the moment of inertia is increased to about 8.5 units. In contrast, the same man will have a moment of inertia about a transverse axis through his center of gravity of about 17 units while he is in an extended position. His radius of gyration in each of these cases is calculated as follows:

$$\text{Basic relationship: Radius of gyration} = \sqrt{\frac{\text{Moment of inertia}}{\text{Total mass}}}$$

A. Longitudinal—Anatomic position

$$K = \sqrt{\frac{1.5}{200/32}} = \sqrt{0.24} = 0.49 \text{ ft} \approx 6 \text{ inches}$$

Understanding Basic Physics

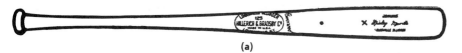

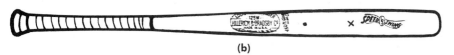

Figure 8.22
Comparison of sizes and shapes of typical baseball (a) and softball (b) bats. • = center of mass x = center of percussion

Figure 8.23
Comparison of position of center-of-mass (·) and center-of-percussion (x) of a tennis racket.

So, the average unit of mass is located about 6 inches from the axis of rotation.

  B. Longitudinal—Arms and legs abducted at shoulder and hip joints

$$K = \sqrt{\frac{8.5}{200/32}} = \sqrt{1.36} = 1.17 \text{ ft} \approx 14 \text{ inches}$$

  C. Transverse—Supine position

$$K = \sqrt{\frac{17}{200/32}} = \sqrt{2.72} = 1.65 \text{ ft.} \approx 20 \text{ inches}$$

From these measurements it can be seen that the moment of inertia and radius of gyration vary greatly according to body position.

Relation between Linear and Angular Motions

The interrelationship between linear and angular motions is one of the most important relationships that must be understood in biomechanics. There is often a tendency during the analytic process to be primarily interested in the linear motion of a specific body part. As examples, the linear velocity of the hand during a throwing motion is conducive to throwing speed, and the linear velocity of a kicker's foot is directly related to the performance outcome. In both cases, however, the final linear velocity achieved is the result of a large number of angular motions at the individual joints of the body. Therefore, the real significance of the interrelationship between angular and linear motions lies in the fact that most human body movements are actually rotational movements at the joints. *There must be a functional integration of linear and angular motions to produce skilled performances.*

This was pointed out in the discussion of the radian. There is a direct relationship between the angle through which a body turns and the distances that the various points on the body move linearly. Since the angular velocity represents the angle turned per unit time, there is a correspondence between the angular and linear velocities of the various parts of the body doing the turning. This relationship also applies to the accelerations and is given in the following equations:

$$\text{Distance} = \text{Angle (radians)} \times \text{Radius (distance from axis)}$$

$$\text{Linear velocity} = \text{Angular velocity} \times \text{Radius}$$

$$\text{Tangential acceleration} = \text{Angular acceleration} \times \text{Radius}$$

$$\text{Radial acceleration} = (\text{Angular velocity})^2 \times \text{Radius}$$

In all the above equations, angular quantities should be expressed in radians, rad/sec, and rad/sec². From this it will be noted that in order to achieve high linear velocities of various parts of the body, the rotations must be made either with high angular velocities or with a large radius of movement (fig. 8.24).

The concept of centripetal and centrifugal forces is indispensable to understanding the relationship between linear and angular motions. There has been considerable confusion regarding these forces in the field of physical education, as well as in many other fields of study involving angular motion. It should be remembered that the individual points on a rotating object are not moving in straight lines but rather in circular paths. As noted during the description of acceleration, this means that their velocities are changing. Moreover, this is a direction change not necessarily associated with change of speed. *This change of direction of motion is called radial acceleration.* Since these points are being accelerated, an external force must be exerted. For example, when a hammer thrower is changing the direction of motion of the hammer during the rotational phase, he is exerting a force inward on the chain in order to provide this change of motion (fig. 8.25). *The force associated with the radial acceleration is called centripetal force.*

Combining the previous expression for radial acceleration with the relationship between force and acceleration:

$$\text{Centripetal force} = \text{Mass} \times \text{Radial acceleration} = \frac{\text{Mass} \times (\text{Velocity})^2}{\text{Radius of curvature}}$$

$$F \text{ (cent)} = \frac{mv^2}{r}$$

In the previous example of a baseball player rounding second base at a speed of 20 ft/sec and a radius of curvature of 16 feet, it was found that the athlete's

Figure 8.24
Integration of angular
velocities by lower limb
joint motions to produce
linear velocity during
running by Jim Ryun, USA
Olympic team. *Courtesy
Visual Track and Field
Techniques, 292 So.
LaCienaga Blvd., Beverly
Hills, CA 90211.*

Figure 8.25
Centripetal force during
hammer throwing by
Romuald Klim of the
USSR. *Courtesy Visual
Track and Field
Techniques, 292 So.
LaCienaga Blvd., Beverly
Hills, CA 90211.*

Understanding Basic Physics

radial acceleration was 25 ft/sec². Assuming he weighs 192 pounds, the centripetal force necessary to accomplish this—the horizontal frictional force between his feet and the ground—is as follows:

$$\text{Force} = \frac{w}{g} \times a \text{ (radial)} = \frac{192}{32} \times 25 = 150 \text{ pounds}$$

If he were to attempt the same maneuver with a radius of curvature of four feet (a tight turn at third base), he would need:

$$\text{Force} = \frac{w}{g} \times \frac{v^2}{r} = \frac{192}{32} \times \frac{20^2}{4} = 600 \text{ pounds}$$

This illustrates the necessity for slowing down immediately prior to executing a very sharp turn.

If a force is being exerted on objects in circular motion in order to keep them traveling along the circular path, they are in turn exerting an outward counterforce. *The counterforce to centripetal force is properly known as centrifugal force.* To return to the example of the hammer thrower, since the performer is exerting an inward force on the hammer, then the hammer itself is exerting an outward force on the performer. In fact, during the performance of this event, the outward or centrifugal force may achieve relatively high magnitude.

If a performer executes two revolutions per second with an effective radius of six feet, he achieves a hammer speed of—

$$\text{Velocity} = \text{Angular velocity} \times \text{Radius} = 2 \times 2\pi \times 6 = 75.4 \text{ ft/sec}$$

To hold the hammer in circular motion requires—

$$\text{Cent. force} = \frac{m \times v^2}{r} = \frac{16}{32} \times \frac{(75.4)^2}{6} = 474 \text{ pounds}$$

If he could increase his rotation to 2.5 revolutions per second while maintaining the same effective radius, the values would be—

$$\text{Velocity} = 2.5 \times 2\pi \times 6 = 94.25 \text{ ft/sec}$$

$$\text{Cent. force} = \frac{16}{32} \times \frac{(94.25)^2}{6} = 740 \text{ pounds}$$

The problem performers face in dealing with centrifugal force is one reason the hammer thrower performs in a cage during track-and-field meets. Either a premature release or loss of angular orientation by the performer could result in the hammer going into an area where spectators or other participants are located.

As another example, the tangential velocity of a foot segment during running or kicking maneuvers is caused partially by a combination of hip flexion and knee extension (fig. 8.26). Maximum rate of hip flexion seems to be about one revolution (2π radians) per second with an extended knee. Therefore, this could contribute about:

$$\text{Velocity} = 2\pi \text{ rad/sec} \times 3 \text{ feet} = 18.85 \text{ feet/sec}$$

to the tangential velocity of the foot.

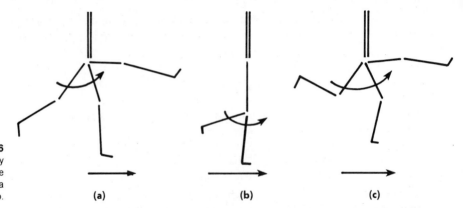

Figure 8.26
Obtaining foot velocity by
(a) hip flexion, (b) knee
extension, and (c) a
combination of the two.

(a) (b) (c)

Knee extension can reach a maximum velocity of about 1.5 revolutions/second with a lever arm of about 16 inches. This could contribute a maximum of:

$$\text{Velocity} = 1.5 \times 2\pi \times 16/12 = 12.55 \text{ feet/sec}$$

Therefore, the maximum tangential foot velocity obtainable by the combination of these motions would be about 31.4 feet per second.

Actually, a kicking or running motion is more complicated than indicated above. Other body motions aid the foot velocity, while neuromuscular factors prevent the performer from obtaining the maximum joint motions concurrently. However, a good kicker should be able to attain a foot velocity near this factor by optimizing his or her technique.

Rotational Kinetic Energy and Power

Just as there is a kinetic energy associated with linear motion of a body, there is also a kinetic energy associated with rotational motion of the body. This rotational kinetic energy is calculated by the formula:

Kinetic energy (rotational) = ½ × (Moment of inertia) × (Angular velocity)²

It can be calculated that the moment of inertia of a softball is about 0.00013 units. If a foul "pop-up" in softball is spinning with 20 revolutions per second, its rotational kinetic energy is—

Kinetic energy = ½ × (Moment of inertia) × (Angular velocity)²

$$= \tfrac{1}{2} \times 0.00013 \times (20 \times 2\pi)^2 = 1 \text{ ft-lb}$$

To stop this spin during a catch, it is necessary to supply a frictional force between the ball surface and the hand or glove. If the slippage during stopping is one-half inch, the force needed is—

$$\text{Force} = \frac{\text{Energy}}{\text{Distance}} = \frac{1}{\tfrac{1}{2} \times \tfrac{1}{12}} = 24 \text{ pounds average frictional force}$$

The fact that rotating bodies are endowed with energy sometimes becomes crucial in sport events. For example, a rolling billiard ball will cause a slightly different type of collision than one that is sliding, since the rotational kinetic energy must be absorbed in the collision. Likewise, when trying to catch a rapidly spinning baseball or softball, the performer must allow for the energy associated with the spin of the ball as well as its linear motion as it approaches the glove. Failure to allow for this spinning motion of the ball sometimes causes the ball to hit the glove and "climb out." In recent years, the engineering and design of baseball gloves have tended to alleviate this problem to a degree by providing greater surface areas.

Just as force times distance per unit time is a measure of linear power developed, rotational power can be expressed as torque times the angle turned per unit time. An example which illustrates some of the concepts of biomechanical power production is observed in the sport of cycling.

Power input into the bicycle is measured in terms of the average torque at the crank axle times the number of radians per second through which the pedals rotate. While amateur cyclists normally turn the pedals over at a rate of 60 to 75 revolutions per minute, trained cyclists can maintain 90 rpm for long periods and may increase to 120 rpm in sprints. Consider the example of cyclists capable of maintaining an average torque of 12 pound-feet at a rotational speed of 90 rpm (1.5 revolutions per second). They would then be developing power at the rate of:

$$\text{Power} = \text{Torque} \times \text{Angular velocity}$$

$$= 12 \times 1.5 \times 2\pi$$

$$= 113 \text{ ft-lb/sec} = 0.2 \text{ Horsepower}$$

An average torque of 12 pound-feet with 8-inch pedal cranks would require an average force of—

$$\text{Torque} = \text{Force} \times \text{Lever arm}$$

$$12 = \text{Force} \times 8/12$$

$$\text{Force} = 144/8 = 18 \text{ pounds against the pedal perpendicular to crank arm}$$

With 6-inch pedal cranks, an average force of 24 pounds would be needed.

For any cyclist there is an optimal speed of pedaling at which he or she can maintain the highest power output. If pedaling is too slow, the forces required for high power are too large. Similarly, high rotational speeds lead to large energy losses in simply moving the legs; this also results in less efficient power production. The purpose of gearing in a racing or touring bicycle is to allow the cyclist to maintain the most efficient pedal rate while meeting a variety of environmental conditions.

The power needed in cycling depends primarily on wind losses and the slope or grade of the roadway. These combine to create a resistive force against which the cyclist works. This force times the velocity of the bicycle represents the power output that must be matched against the power-producing capabilities of the rider. If the cyclist can maintain the most efficient pedaling conditions as road and wind conditions change, higher average speeds or longer endurance in touring will result.

Angular Momentum   As with all the other quantities associated with linear motion, momentum, too, has an angular counterpart. Angular momentum is defined by the following equation:

Angular momentum = Moment of inertia $\times$ Angular velocity

With a weight of about 4.4 pounds and a radius of 4.3 inches, the standard discus has a moment of inertia near 0.1 unit. To find the angular momentum of a discus spinning at five revolutions per second—

Angular momentum = Moment of inertia $\times$ Angular velocity

$$= 0.1 \times 5 \times 2\pi = 3.1 \text{ units of angular momentum}$$

Like linear momentum, angular momentum is a conserved quantity and in the absence of external torques will remain a constant throughout motion. The very significant fact about angular momentum is that the moment of inertia can be varied during motion as the angular momentum is being conserved. This means—since the product of moment of inertia and angular velocity must remain a constant—a performer can decrease the moment of inertia by flexing at the hip, knee, and lumbar spine and provide an automatic increase in angular velocity. This capability is useful in the performance of many athletic or dance maneuvers.

There are many sport actions where the distance of a part of the moving body from the axis of rotation is decreased in order to increase the speed of rotation. An example is flexion of the elbow associated with diagonal abduction and adduction of the shoulder during an overhand throw. This is a common motion in almost all throwing and striking skills involving the high diagonal plane. Flexion of the elbow must occur to increase limb velocity. This can be seen in figure 12.8, frames 22 through 24, where the javelin thrower moves the right elbow from extreme extension to flexion. This increases the velocity of the throwing limb, and hence is essential to obtain maximum distance for the throw.

One of the most notable examples of conservation of angular momentum is the increase of rotational velocity which occurs during the tuck maneuver in diving. An angular momentum is imparted to the diver by his motion as he leaves the board. This angular momentum will be conserved during the entire interval of time in the air. As the tuck position is attained (hip, knee, and lumbar flexions), this reduces the moment of inertia and increases angular velocity. Before entering the water, the diver again extends the joints in order to increase the moment of inertia and reduce angular velocity. The total number of rotations performed before entering the water depends greatly on the degree of decrease of the moment of inertia, the so-called "tightness of the tuck."

There are also examples of this increase of limb velocity through conservation of angular momentum by a knee flexion. Thus, in the running pattern, flexions of the knee and hip joints cause a very rapid swinging of the whole leg forward in preparing for the next step. Knee flexion is also used for the same purpose in both soccer-style and traditional American football kicking patterns in order to attain a higher final foot velocity. In these examples, flexion of the knee draws more of the mass near the axis of rotation taking place at the hip joint. This causes a reduction of the total moment of inertia of the leg about the hip joint and a consequent increase in angular velocity of the moving leg.

Since angular momentum is a vector quantity, it must be conserved in direction as well as magnitude. For this reason, spinning objects have a strong tendency to maintain their axis of spin in the same direction. This principle is illustrated in the stability of flight of a punted football, a discus, or a javelin. Only when they have a strong, stabilizing spin, will these sport objects travel through the air without wobbling.

In this context, it should be noted that each object has certain *principal axes* related to the symmetry of their bodies and about which the spin is most stable. Attempts to spin the object about other axes result in a phenomenon known as *precession;* this appears in biomechanic analysis of the motion as a circular wobble. The stereotype of perfect mechanics for throwing such objects includes the introduction of an optimal spin moving exactly about one of the principal axes.

In competitive diving and gymnastics, a performer often makes complex body and limb motions designed to convert a spin about one principal axis into a spin about a new axis. As examples, the change from turning to twisting motions in diving, or the change from a giant swing to a flying twist dismount from the horizontal bar, involves such procedures. These motions involve quite complex biomechanic interaction.

Two skills which notably offer similar and contrasting elements are cricket bowling and baseball pitching. The ultimate objective of both of these is to deliver the ball to a batter at a great velocity. The rules of the two sports differ considerably, and these variations result in a decided difference in the way the performers are allowed to achieve the ultimate objective of optimal ball velocity. In cricket, the bowler is allowed to run up to the point where he releases the ball; this results in considerable linear velocity of his center of gravity. The baseball pitcher is much more restricted in this respect. The pitcher's total body linear velocity is much less, because he or she must maintain contact with the pitching rubber until the ball is released.

However, the rules of the two sports favor the baseball pitcher in the development of limb velocity. The rules for cricket prescribe that the throwing elbow of the cricket bowler must remain extended at all times during delivery (fig. 8.27). Thus, the bowler cannot gain the added velocity that the baseball pitcher develops by elbow flexion and extension during the pitching process (fig. 8.28). In other words, the baseball pitcher increases the final velocity by decreasing the moment of inertia. The cricket bowler is not allowed to do this. A larger proportion of the bowler's ball velocity is due to linear velocity attained during the approach, and a smaller share is due to the angular velocity of the throwing limb.

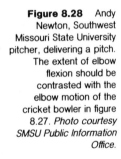
Figure 8.27 Ken Holt, former University of Nottingham cricket player, bowls a medium-paced ball. Compare and contrast this elbow motion with the baseball pitcher's elbow motion in figure 8.28. *Photo by Art Evans.*

Figure 8.28 Andy Newton, Southwest Missouri State University pitcher, delivering a pitch. The extent of elbow flexion should be contrasted with the elbow motion of the cricket bowler in figure 8.27. *Photo courtesy SMSU Public Information Office.*

Cromer, Alan H. *Physics for the Life Sciences.* NY: McGraw Hill Book Co. 1974. (Chapters 1 through 6)

Fuller, Harold Q., et al. *Physics, Including Human Applications.* NY: Harper and Row, Publishers, 1978. (Chapters 3 through 7)

Halliday, David and Resnick, Robert. *Fundamentals of Physics,* Second Edition. NY: John Wiley and Sons, 1980. (Chapters 1 through 13)

Marion, Jerry B. *General Physics with Bioscience Essays.* NY: John Wiley and Sons, 1979. (Chapters 2 through 6)

Tipler, Paul A. *Physics.* NY: Worth Publishers, Inc., 1976. (Chapters 1 through 13)

Recommended
Readings

9

Relationships between Force and Motion

The purpose of this chapter is to demonstrate those interrelationships between force, torque, and motion which are of greatest importance in biomechanic analysis. The chapter is specifically concerned with those situations in which (1) the interaction is between solid surfaces and (2) the interaction with the air is not of great significance. Interactions with fluids, generally water or air, will be considered in the next chapter.

Static and Dynamic Equilibrium

Equilibrium is one of the principal concepts in the biomechanic analysis of sport events. In the scientific sense, *the term "equilibrium" describes those systems existing in a state of zero acceleration*. This includes both linear and rotational motions. In a system without linear acceleration, the condition of equilibrium specifies that each force acting on a body must be opposed by an equal and oppositely directed force on that same body. Under these conditions, the sum of all forces acting on the body will be zero.

Likewise, if the system is in rotational equilibrium, the torques will all be balanced in such a way that the total summation of torque is equal to zero.

The term *static equilibrium* is applied to objects under equilibrium at rest. The term *dynamic equilibrium* refers to those objects moving in a straight line at a constant velocity. For example, a football lineman in his three-point stance (before the ball is put into play) is an example of static equilibrium. A distance runner moving in a straight line at a constant velocity is personifying dynamic equilibrium.

The case of a body under a constant acceleration will also be considered in this chapter. Although this is not a case of true equilibrium, it is of critical importance in many sport and dance events. Therefore, the term *equilibrium under acceleration* is introduced. An example of this type of motion is the extreme forward lean of a sprinter coming out of the blocks at the beginning of a race. This type of motion also occurs any time a performer has an abrupt change of direction, as commonly occurs in court and field sports. The lean accompanying directional changes is especially noticeable during rapid pivoting motions of performers in such sports as soccer, field hockey, basketball, lacrosse, and football (fig. 9.1).

One of the most familiar examples of static equilibrium is the situation normally associated with the term *balance*. *A decisive principle with respect to balance or static equilibrium is that the center of gravity of the body must lie at a point directly above the base of support.* The term *base of support* includes

Figure 9.1
Equilibrium under
acceleration during
directional change by
O. J. Simpson, USC
Heisman Trophy winner.
*Courtesy USC Athletic
News Service.*

the entire area formed by those points at which the body is touching the ground, mat, floor, or sport apparatus. So far as the performer is concerned, the base of support may be one foot, both feet, one hand, both hands, or parts of the back or abdomen touching the mat, ground, or apparatus. In the sports of gymnastics and wrestling, as examples, the base of support is not only widely varied in terms of body segments, but also changes rapidly from one position to another. One need only watch a skilled gymnastic routine to gain an appreciation of varied bases of support.

Associated with the concept of base of support in many sport actions are the techniques involved in a performer's preliminary stance. A wide stance offers a relatively large, stable base of support. This allows a variety of body motions while equilibrium is maintained. A wide stance also furnishes correspondingly greater lever arms to the vertical forces of support, thereby allowing quicker rotational motions of the entire body in beginning or changing a sport action.

Figures 9.2 and 9.3 show the variations possible in the preliminary stance of football linemen. Figure 9.2 illustrates a football stance common nearly fifty years ago. Although the stance is relatively stable, movement from such a stance would be inefficient by today's standards. In contrast, figure 9.3 shows the stances more often used by contemporary offensive and defensive linemen. The wide, level stance of the offensive line (dark jerseys in fig. 9.3) allows motion in many directions. When the stance is taken properly, there is no indication of the impending direction of the play. The stance of the defensive end in figure 9.3, on the other hand, predetermines rapid movement in a specific direction. A comparison of figures 9.2 and 9.3 demonstrates that the stereotype of perfect mechanics for sport skills undergoes change.

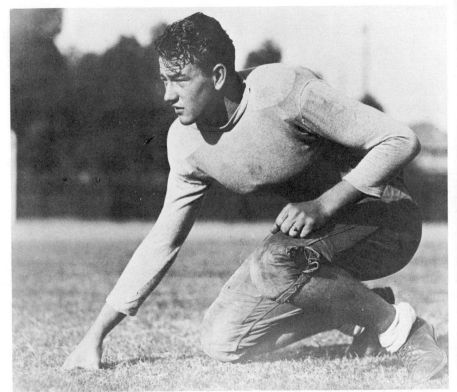

Figure 9.2 Marion M. Morrison, an excellent USC football player from 1926–1928 (better known in movies as John Wayne), demonstrates a football stance of his era that proved inefficient for subsequent movement. *Courtesy USC Athletic News Service.*

Figure 9.3 Contrast between typical offensive and defensive line stances in modern football. *Courtesy SMSU Public Information Office.*

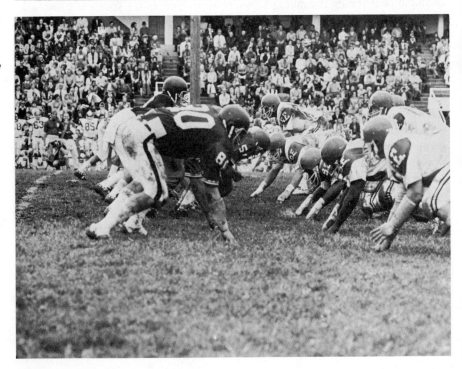

Relationships between Force and Motion

As described in the previous chapter, most situations of contact involve not only those forces pressing the surfaces together—normal forces—but also forces parallel to the surfaces in contact—frictional forces. The importance of frictional forces in maintaining equilibrium should not be overlooked. It is much easier to perform any type of action on a rough surface where slippage is minimized, rather than on a smooth or slick surface where frictional forces are not available. When the center of gravity lies outside the base of support, tending to cause the object to rotate into a falling condition, frictional forces can be used to rotate body segments back into an equilibrium position.

In the case of a departure from an equilibrium condition, a determining reaction of a performer is the reestablishment of a new base of support beneath the body. An excellent example of this is the situation of a "scrambling back" in football who, when he is hit and knocked out of the equilibrium condition, will rapidly move lower limbs and the upper limb not holding the football to new positions beneath the body in order to keep from falling to the ground. What he is actually doing from a biomechanics point of view is reestablishing a base of support beneath the center of gravity, thus enabling him to maintain equilibrium and control volitional motions. He may use movements of the limbs to change complete body motion, making possible a return to the upright running position while moving in another direction.

In contrast, basketball players may lose equilibrium and regain it by means of friction without critical adjustments of the base of support. In order to do this, the frictional forces developed parallel to the floor must be very great. This is the reason for the design, shape, and materials of the basketball shoe. The surfaces of the soles of the shoes and of the floor provide the basketball player with maximum coefficients of friction. The use of frictional forces in this sense requires special flexibility training and the strengthening of noncontractile tissue as well as the muscles most involved in controlling inversion and eversion of the subtalar and transverse tarsal joints.

Because of the common values for weight and moment of inertia of the human body, it requires a period of about one second for a person to fall from slight disequilibrium to a position where recovery is impossible. This is rather long as compared to the few tenths of seconds in which a good athlete can react and perform the necessary compensating movements to reestablish the base of support. Consequently, many athletic movements are performed by purposely moving out of an equilibrium position. Skilled performers know that following the critical maneuver they will once again be able to make the corrections to regain equilibrium.

Examples of this can be found in broken field running in many field sports. A conservative athlete who is concerned with constantly being in an equilibrium position has much more limited movement possibilities than an uninhibited performer whose movements allow short periods of time off balance. As another example, in slalom skiing the turns can be taken much tighter and faster by using disequilibrium followed by compensating shifts of the skis. Basketball players often move far out of balance in order to evade a defensive player and to attempt a shot.

Motions in dynamic equilibrium, where the body is moving rather than remaining static, can be described in a similar context. In fact, the example of the "scrambling back" in football better illustrates dynamic equilibrium than static equilibrium, since he will continue motion after regaining equilibrium. Again, the principal point is that parts of the body in contact with the ground must be in a condition where they can establish torque to counteract any torque due to gravity. If this is not accomplished, falling will be inevitable.

In common locomotor actions such as walking, jogging, and running, linear motion of the body is maintained, in part, by a series of hip, knee, and ankle-joint motions designed to regain equilibrium when it is lost immediately after the body moves over the foot in contact with the ground. The new foot placement always occurs in front of the center of gravity in order to arrest the falling motion of the body. If this were not done, a state of complete disequilibrium would result, and the individual would fall to the ground.

Figure 9.4
Controlling disequilibrium through positive acceleration during the sprint start by Tom Jones, former USA Olympic team member. *Courtesy Visual Track and Field Techniques, 292 So. La Cienaga Blvd., Beverly Hills, CA 90211.*

Relationships between Force and Motion

This concept of arresting disequilibrium and controlling locomotor motion carries over to the earlier described idea of "equilibrium under acceleration." Consider the example of the sprinter during the first four or five yards of the fifty-yard dash. During this time, because of positive acceleration he or she is not in a state of equilibrium (fig. 9.4). That position is also one which could not be maintained in either a static or normal dynamic equilibrium. The center of gravity lies in front of the unilateral weight-bearing foot or base of support. However, the linear motion continues during this state of disequilibrium, due to the fact that positive acceleration is occurring.

From this it can be observed that the lean of a runner during this type of acceleration is reciprocally related to the fact that positive acceleration is occurring. The lean position is a result of the acceleration and not the cause of acceleration, i.e., leaning forward does not automatically insure that a performer will accelerate. For example, during a 440-yard dash a performer cannot hope to lean forward throughout the entire race because he or she cannot positively accelerate over the entire distance. Furthermore, an exaggerated forward lean of the body throughout a race is not an efficient position from a myologic point of view. This arises from the fact that constant lumbar-thoracic flexion elicits an excessive amount of eccentric contraction within antigravity musculature. Therefore, energy consumption is directed toward counteracting the downward pull of gravity rather than moving the athlete linearly. The undesirable effect of a constant forward lean by a performer becomes more obvious as the distance of the race increases. Leans are contraindicated during the running of any distance except at the start, while accelerating, and possibly at the finish line.

Adjustment of body orientation to maintain equilibrium while accelerating also occurs during radial acceleration. When a runner is changing direction, acceleration is inward or toward the center of the curve of directional change. This is observed when a base runner rounds third base trying to score from first base on an extra-base hit. Because of the runner's extreme speed and the acute directional change at the base, it is necessary for the athlete to lean sharply inward in order to maintain equilibrium. This requires some spinal lateral flexion and rotation toward the center of the curve. Furthermore, the horizontal centripetal force necessary to provide this radial acceleration must be supplied by friction between the performer's shoes and the ground. To provide this extra horizontal force, the design of baseball shoes includes cleats. These furnish extremely high coefficients of friction as they dig into the earth during running.

Using this information, it is possible to calculate the expected angles of lean for various situations involving acceleration. A performer leans because the ground or base of support must provide a vertical component of force to support the weight and a horizontal component to support the needed acceleration. The most efficient body position is at an angle which allows the line of action of this support force to lie along the longitudinal axis of the body. In this position the force does not tend to cause rotations leading to imbalance.

For example, a sprinter accelerating at a rate of 16 ft/sec² (½ g) requires a horizontal force equal to one half the body weight. In order for the base of support to furnish this component plus a vertical component equal to the weight, the

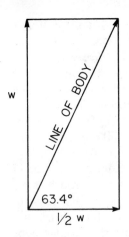

Figure 9.5 Angle of lean during linear acceleration.

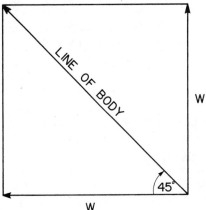

Figure 9.6 Angle of lean during radial acceleration.

body must lean at an angle of 63.4 degrees with the horizontal, or 26.6 degrees with the vertical (Figure 9.5). As acceleration decreases with the approach to maximum velocity, the angle of lean with the vertical decreases to zero.

Likewise, a runner rounding third base at a speed and radius of curvature requiring a radial acceleration of 1 g (32 ft/sec²) must lean inward at a 45 degree angle to allow equal components of vertical and horizontal forces to combine along the longitudinal axis of the body (Figure 9.6). Indoor tracks, where it is common to run a curved path at high velocity, are often banked to lead the runners efficiently into the correct lean.

Trajectories

One of the forces most frequently experienced in its effect on motion is the force of gravity. Since human body motions and the motions of sport objects occur relatively close to the ground, gravitational acceleration is considered a constant factor in biomechanic analysis. Every unsupported object is under the influence of gravitational acceleration or the associated force which is called the *weight* of the object. Gravitational acceleration is always present and is directed down-

Relationships between Force and Motion

ward, regardless of the velocity of the body. When a body is moving straight upward, gravitational acceleration operates against the direction of motion. This causes the speed of the body to decrease.

When the actual velocity of the body is downward, acceleration is downward and, therefore, in the same direction as the velocity. This causes an increase in speed of the body or sport object. *Even at the top of a vertical flight path when an object has a velocity of zero, gravitational acceleration is constantly causing velocity to change and body motion to change direction and move downward.*

Under these conditions of constant acceleration, standard mechanics texts mathematically describe motion in terms of the following three equations:

$$V = V_0 + at$$

$$s = V_0 t + \tfrac{1}{2} at^2$$

$$2as = V^2 - V_0^2$$

The first of these, which comes directly from the definition of acceleration, describes how velocity changes as a function of time. The second equation describes how the displacement changes as a function of time, and the third is a combination of the two, eliminating the time factor and showing the relation between the distance covered and the velocity.

For an object falling from rest under the influence of gravity, these simplify into:

$$V = gt$$

$$h = \tfrac{1}{2} gt^2$$

$$V^2 = 2gh$$

where g has the value of 32 ft/sec² or 9.8 meters/sec² given in the previous chapter, and h, the height, represents the distance fallen.

A condition much more familiar in sport and dance situations is that of a body or object initially moving upward, coming to a stop, then falling back to the ground or performance area under the influence of gravity. Examples are basketball rebounding, dance leaps, and high jumping. In these instances the motion is symmetrical, that is, the time of rising is equal to the time of fall, and the final velocity is equal to the initial velocity. Under these conditions, the total time in the air is twice the time of fall from the maximum height reached. The equations above can be algebraically manipulated to give the following relationships:

$$\text{Max. height} = \frac{(\text{Velocity})^2}{2g}$$

$$\text{Total time} = \frac{2\,(\text{Velocity})}{g}$$

$$\text{Total Time} = \sqrt{\frac{8(\text{Max. height})}{g}}$$

These relationships hold in any sport maneuver where takeoff and landing occur at the same level.

As a numerical example, suppose a coach using basic cinematographic techniques observes that a rebound tumbler reaches a maximum height where the center of mass has been lifted ten feet above the trampoline bed. From this information, how can the coach calculate initial and final velocities and the total amount of time the performer has in the air to execute the required skills?

$$10 = \frac{(\text{Velocity})^2}{64}$$

$$\text{Velocity} = \sqrt{640} = 25.3 \text{ ft/sec}$$

$$\text{Time} = \sqrt{\frac{8 \times 10}{32}} = 1.6 \text{ seconds}$$

Since the upward and downward motions are symmetric, the velocity calculated in the example above represents both the upward speed when leaving the trampoline bed and the downward speed when striking the bed.

In most sport trajectories motion is not purely vertical. It should be recalled that trajectories are a combination of vertical and horizontal motions. A governing principle to be observed here is that these two directions of motion are independent. This makes the task of analysis somewhat easier, since the conditions affecting vertical motion can be used to determine time of flight, while horizontal velocity is used to determine the distance the body or object travels. For example, a hurdler combines vertical and horizontal motions during his trajectory over the hurdle (figure 9.7).

If it is observed that the performer raises his center of gravity 18 inches (1.5 feet) while crossing a hurdle, then:

$$\text{Time} = \sqrt{\frac{8 \times 1.5}{32}} = 0.6 \text{ seconds}$$

$$1.5 = \frac{(\text{Vert. velocity})^2}{64}$$

$$\text{Vert. velocity} = \sqrt{96} = 9.8 \text{ ft/sec}$$

Thus, during takeoff the hurdler must give himself a vertical component of velocity of nearly 10 ft/sec. If his horizontal component of velocity is 22 ft/sec during flight, the total distance covered while crossing the hurdle is—

$$\text{Distance} = \text{Velocity} \times \text{Time} = 22 \times 0.6 = 13.2 \text{ feet}$$

It should be noted in the case of hurdling that the desired result is achieved by maintaining maximum average horizontal velocity. If the hurdler actually jumps the hurdles, giving himself an excessive vertical component, he has a longer

Figure 9.7 Vertical and horizontal motions during a trajectory by Ralph Mann, former member USA Olympic team, over an intermediate hurdle. *Courtesy Visual Track and Field Techniques, 292 So. La Cienaga Blvd., Beverly Hills, CA 90211.*

Figure 9.7—*Continued*

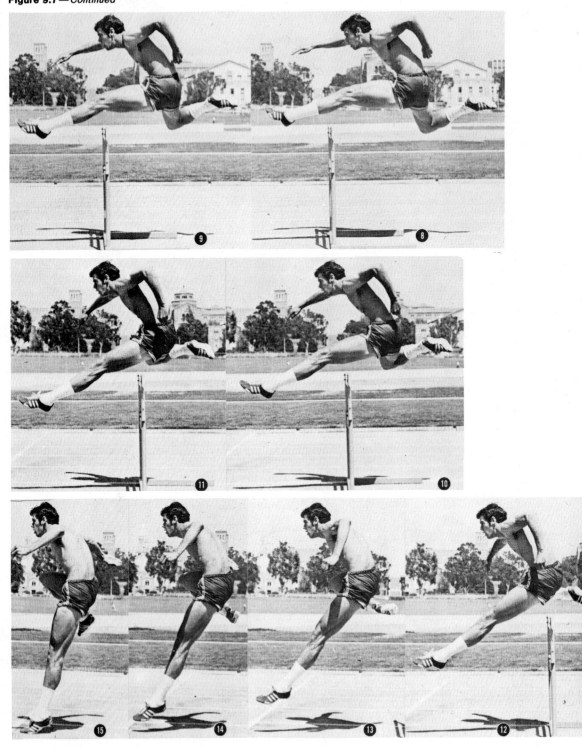

Relationships between Force and Motion

period of time in the air at a correspondingly lower horizontal velocity. Average horizontal velocity is decreased, and it will take the hurdler longer to run the entire race. Therefore, the stereotype of perfect mechanics for hurdling involves the use of a minimum amount of vertical motion of the hurdler's center of mass and the maintenance of a steady, maximum horizontal velocity during the trajectory over the hurdle. This is one reason the upper body is brought low (hip joint and lumbar spine flexions) as the legs are passing over the hurdle. The performer's center of gravity is lifted to a minimum height at the vertical peak of the trajectory.

Consider, for example, a performer whose mechanics lifts his or her center of gravity two feet and whose horizontal velocity is 19 ft/sec:

$$\text{Time of flight} = \sqrt{\frac{8 \times 2}{32}} = 0.7 \text{ seconds}$$

$$\text{Distance covered} = \text{Velocity} \times \text{Time} = 19 \times 0.7 = 13.3 \text{ ft}$$

As compared to the previous performer, this performer is taking 0.1 seconds longer to go essentially the same horizontal distance over the hurdle. Multiply this loss by the number of hurdles in a race and you have the difference between a winning effort and a poor showing.

Under the conditions of a constant horizontal velocity combined with the type of vertical motion described above, the path followed by a body or a sport object conforms to the geometrical curve known as a *parabola*. This is the origin of the term *parabolic trajectory* often used in describing sport motions. The term *trajectory* is a generalized term indicating the actual path followed by a freely moving, nonsupported body. Such factors as "drag" due to air resistance, and lift in a flying object cause the trajectory to deviate from a true parabola. This true parabola is actually a mathematic construct and is probably most closely followed in sports by the paths of shots or hammers after they are put into the air. In these cases, the weight and inertia they have while traveling through the air are much greater than the forces exerted by the air.

The actual shape of the parabola traced by an object moving through the air depends on (1) the velocity with which it leaves its base of support and (2) the angle at which it initially moves into the air. Projection into the air at a low angle gives a larger horizontal component of velocity but a smaller vertical component. Therefore, a low projection angle results in a correspondingly shorter time of flight. Projection at a high angle provides a large vertical component of velocity and a long time of flight, but there is a correspondingly lower horizontal velocity during this time.

Examples of the use of two angles of trajectory in tennis are the low-angle forehand shot on one hand, and the contrasting lob shot on the other (figure 9.8). Using the low trajectory, the amount of time necessary to reach the point where the ball contacts the court is minimal. This has the strategic offensive advantage of giving the opponent minimum time to react. When the lob shot is used, the length of time necessary to reach the contact point on the court becomes greater. The offensive player hitting the lob shot acquires maximum time to recover body position and establish a strategic defensive posture on the court.

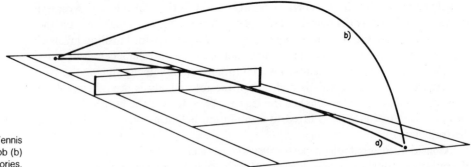

Figure 9.8 Tennis drive (a) and lob (b) trajectories.

Mathematical description of these trajectory motions normally requires the use of trigonometric methods summarized in the Appendix. Basically, the technique is to resolve the motion into horizontal and vertical components of the initial velocity, and then to treat the horizontal and vertical motions independently but cooperatively.

The vertical component of motion is used as in the previous examples to determine the maximum height and time of flight. Of special importance is the relation:

$$\text{Time of flight} = \frac{2 \times \text{Vertical component of initial velocity}}{g}$$

Since, in the absence of air resistance, the horizontal component of velocity is assumed to be constant:

Horizontal range = Horizontal component of initial velocity × Time of flight

In this way, the parameters of interest in biomechanic analysis can be obtained.

For those students familiar with trigonometric notation, the components of initial velocity of a body or sport object projected at an angle of θ with respect to the horizontal are given by:

Vertical component = Initial velocity × sin θ

Horizontal component = Initial velocity × cos θ

Horizontal and vertical components of velocity applicable in these equations can be read directly from film by analyzing the horizontal and vertical positions indicated in each frame. These observations are made during an intermediate cinematographic analysis. The components can also be calculated by measurement of total speed and angle as shown in the film. As an example, if a tennis player strikes the ball near ground level at an angle of 15° above the horizontal and a speed of 60 ft/sec, then:

$$V \text{ (vertical)} = 15.5 \text{ ft/sec}$$

$$V \text{ (horizontal)} = 58 \text{ ft/sec}$$

$$\text{Max height} = \frac{(15.5)^2}{2 \times 32} = 3.75 \text{ ft}$$

$$\text{Time of flight} = \frac{2(15.5)}{32} = 0.97 \text{ sec}$$

$$\text{Range} = 58 \times 0.97 = 56.2 \text{ ft}$$

The ball would barely clear the net and strike the ground 56.2 feet away about one second after the stroke.

If, however, the player strikes the ball with the same speed and an initial angle of 75° with the horizontal:

$$V \text{ (vertical)} = 58 \text{ ft/sec}$$

$$V \text{ (horizontal)} = 15.5 \text{ ft/sec}$$

$$\text{Max height} = \frac{(58)^2}{2 \times 32} = 52.5 \text{ ft.}$$

$$\text{Time of flight} = \frac{2 \times 58}{32} = 3.62 \text{ seconds}$$

$$\text{Range} = 3.62 \times 15.5 = 56.1 \text{ ft}$$

The ball would strike nearly the same point as previously, but 3.6 seconds after the stroke.

The maximum horizontal range across a level surface is obtained with a projection angle of 45°. Although this angle does not give an especially high value for either the horizontal velocity or the time of flight, it leads to the highest value for their product. Surprisingly, a launch angle of 45° is quite rare in biomechanic performances. In some cases—tennis would be an example—a maximum horizontal range would carry the ball beyond the playing area. In others, such as the shot put or baseball hit, the point of release and the point of impact are not at the same altitude. Many objects, such as a golf ball, interact strongly with the air and do not follow a parabolic trajectory. Finally, in trajectories such as the long jump or dance leaps, the human musculature does not allow for a high angle of launch combined with high initial speed.

In the case where the landing point is lower than the point at which the object enters the air, the angle of release for maximum range will be less than 45°. In the case of the shot put, the optimal angle is a function of both velocity of release and the height at which the performer loses contact with the shot. Figures 9.9 and 9.10 illustrate typical values for these parameters for various classes of performers.

In general, the optimal angle decreases with a reduction in velocity or an increase in release height.

	Championship	College	High School	Junior High
Release height	8.5 feet	8.5 feet	8.0 feet	7.0 feet
Velocity	45 ft/sec	40 ft/sec	37 ft/sec	33 ft/sec

Figure 9.9 Release heights and velocities for shot putting.

	Championship	College	High School	Junior High
45°	70.55 ft	57.15 ft	49.5 ft	39.8 ft
44°	70.75	57.35	49.65	40.00
43°	70.90	57.50	49.78	40.10
42°	70.95	57.60	49.88	40.20
41°	70.95	57.65	49.92	40.22
40°	70.90	57.60	49.92	40.25
39°	70.70	57.55	49.88	40.22
38°	70.55	57.40	49.80	40.20

Figure 9.10 Range as a function of angle in shot putting.

The concept of a trajectory being a true mathematic parabola is based on the assumption that the only force acting on the subject in free flight is the force of gravity. In practice, most body and sport objects in flight interact with air. This causes the trajectory to deviate from the true parabolic pattern. The air has two effects on the flights of bodies or objects: (1) *drag,* which is a force exactly opposite to the direction of motion and dependent on the shape of the object and amount of turbulence behind it and (2) in some cases, *lift,* due to a vertical interaction with the air caused by the shape or manner of flight of the object. These two effects combine to trace widely varying paths for different types of objects. Because of the importance of the interaction of bodies or objects with air, the next chapter will discuss the subject of aerodynamics.

An impressive but sometimes unnoticed example of trajectory motion occurs with the human body during jumping activities in sport and dance. In the high jump, long jump, triple jump, and dance leaps, the center of gravity of the performer is following a near-parabolic trajectory during the time of free flight. Parabolic trajectories are also common occurrences in many gymnastic events and at all levels of competitive diving (figure 9.11). In these instances, however, the motions of body parts in rotation (or other motion about the center of gravity) quite often mask the apparent path of the center of gravity. Observation of the trajectory of the center of gravity is also made difficult or impeded by motions

Figure 9.11 Gale Wyckoff, former elite gymnast at USC, starting a parabolic flight pattern as a part of her dismount maneuver from the uneven parallel bars. *Courtesy USC Athletic News Service.*

of the limbs during these events. These limb motions may cause shifts in the relative position of the center of gravity but they do not change the trajectory of the body's center of gravity previously established at takeoff.

In the Fosbury Flop style of high jumping, for example, the center of gravity will lie somewhere between the iliac crests during takeoff, but at the time the pelvic girdle is passing over the bar, the positions of the upper and lower limbs cause the center of gravity to lie somewhere below the body of the performer (fig. 9.12). Although the rules specify that each part of the body must pass over the bar, it is at least theoretically possible for a performer successfully to execute a high jump with his center of gravity never reaching the height of the bar.

In the long jump, the performer must compromise between achieving a high trajectory, which will give him a long time in the air, and maintaining an accelerated horizontal speed, which will carry him as far as possible linearly during this flight time (fig. 9.13). It has been found that the conscious attempt to attain a high trajectory during long jumping has a tendency to decrease speed to such an extent that the total distance of the jump will be lessened. This leads to the technique of running through the takeoff board instead of jumping vertically (fig. 9.14).

Figure 9.12   Shifting
of the center of gravity
during the Fosbury Flop
as executed by Dean
Owens, USC high jumper.
*Courtesy USC Athletic
News Service.*

Figure 9.13   Long
jump flight by Randy
Williams, USC and USA
Olympic team long
jumper. *Courtesy USC
Athletic News Service.*

Relationships between Force and Motion

Figure 9.14 The long jump as performed by Johnny Johnson, former member USA Olympic team. *Courtesy Visual Track and Field Techniques, 292 So. La Cienaga Blvd., Beverly Hills, CA 90211.*

Figure 9.14—*Continued*

Relationships between Force and Motion

Figure 9.14— *Continued*

Landing techniques included in the stereotype of perfect long jump mechanics are designed to allow the center of gravity to drop as low as possible before any body part touches the pit. The performer then moves over the heels without touching the sand in order to obtain the maximum measured distance for the jump. The maintenance of horizontal velocity is important in the landing; because, without falling backward, the body must be carried angularly over the point at which the feet hit. Measurement is made from the takeoff board to the nearest point touched by a body part in the jumping area.

In the triple jump, the maintenance of horizontal velocity is even more important than in the long jump. The horizontal velocity in triple jumping must be carried through a series of motions. Thus, there are three separate trajectories,

Figure 9.15 Steeple-chase trajectory modification as demonstrated by Bill Reilly, former member USA Olympic team. *Courtesy Visual Track and Field Techniques, 292 So. La Cienaga Blvd., Beverly Hills, CA 90211.*

Figure 9.15—*Continued*

Relationships between Force and Motion

and the horizontal velocity must be maintained through all of them in order to obtain maximum distance. For this reason, it is found that the takeoff angle for triple jumpers is generally several degrees lower than for long jumpers. Also, the most accepted form seems to be one in which the "step" portion of the performance covers a smaller distance than either the initial takeoff portion or the final jump. Normally in the execution of the event, the trajectory of the final or "jump" portion of the triple jump has considerably more altitude than in either of the other two parts of the event. The difficulty of the triple jump lies in the fact that the three jumps interrupt the continuity of motion and decrease the average horizontal velocity. Good triple jumpers overcome this biomechanic problem. This is an example in sport where rules of an event predetermine an inefficient series of motions and add to the difficulty of the event by disrupting the sequencing of forces.

One final example related to body motions on a near parabolic trajectory is that of the steeplechaser crossing a race barrier. While these barriers may be hurdled, they are designed in such a way that the performer can place a foot on top, push off, and aid his subsequent motion (fig. 9.15). The movement of the center of mass passing over the steeple will not be a true parabolic trajectory because the athlete uses the steeple as a "stepping stone" at the midpoint of the maneuver. One would expect a plot of the trajectory of the center of gravity to have a parabolic rise followed by an additional rise before falling on the far side of the steeple.

Forces and Counterforces

In a discussion of any motion involved in a sport, exercise, or dance context, the application of various forces and counterforces that occur within the human body is of great importance. As has been noted, almost all body movement must be described in terms of rotation about various joints. Any muscular action causing a motion of one part of the body also causes a compensatory motion elsewhere. This is made more complex by neurologic, reflexive mechanisms which often combine certain types of motions with others at a subconscious level. An example of this is the crossed-extensor reflex where the flexion of one arm tends to cause an automatic extension of the other. This complex force-counterforce reaction is also evident in learned skills such as walking and running, where there is a set of equal and opposite motions occurring in upper and lower limbs. Many sport and dance skills are actually learned complex sets of force-counterforce movements.

Force and counterforce in linear motion have an exact parallel in torque and countertorque of angular motion. For example, during ballistic actions muscles rotate the pelvic region and counterrotate the lumbar-thoracic spine concurrently during the preparatory phase of a throw or kick. This torque-countertorque effect is especially notable in those situations where a ballistic motion leading to generation of maximum force or velocity is of utmost importance. The term *ballistic* is used in biomechanics to designate those movements that take place in a short interval of time having high velocities and/or large forces and torques.

This principle of force and counterforce is also observed in many sport objects where an extreme acceleration in one direction is countered with an equal change of momentum in the opposite direction, as in the recoil of a rifle or shotgun in hunting or sport shooting. The effects of counterforce can often be lessened by use of a large body mass to receive it. For example, if one holds a shotgun during skeet shooting away from the shoulder when firing it, the relatively small mass of the gun will acquire a high velocity, and the impact of the butt of the shotgun at the shoulder will be quite painful. On the other hand, if the gun butt is pressed tightly against the shoulder before firing, the momentum of the "kick" is absorbed by the gun and the body of the hunter. The effects of counterforce, therefore, are much less noticeable (fig. 9.16).

A large number of learned motion techniques are designed to aid in the absorption of excess velocity or momentum. For example, a long jumper in the landing phase neither lands completely extended at the knees, hips, and spinal column, nor does he allow all joints to move into extreme and damaging flexions. Full extension at major joints would involve excessive accelerations, large forces, and potential joint trauma. Upon landing, the performer controls the extent of flexions at the ankles, knees, hips, and spinal column by eccentric (lengthening) contractions of the muscles associated with these areas. This eccentric contraction absorbs the momentum and the energy associated with landing. The same stretched muscle groups contract concentrically immediately after landing to provide the rotation and extension necessary to move the long jumper over the base of support and out of the pit.

A notable application of force and counterforce in ancient sports was the Greek Olympic form of the long jump. The performer came up to the takeoff point holding two hand weights or *halteres*. These were carried high and in front of him. When the jump was begun, the halteres were thrown sharply downward and backward. The force applied to these weights by the performer resulted in a counterforce that carried him farther in the jump. Although this technique is no longer used in track-and-field events, it comes into play in an analogous fashion in competitive swimming, where the basic idea is to shove the water backward in order to provide a forward thrust to the swimmer's body. This interaction will be discussed more fully in chapter 10 in the section concerned with hydrodynamics of water sports.

The absorption of these counterforces is much more difficult when the performer is not on the ground where a base of support can be established against the counterforce. For example, the jump pass in football is no longer considered good technique because the control of the passer's body motions is hampered to the point that distance and accuracy of the pass are lessened. The passer on the ground, on the other hand, utilizes the position of the rear foot to absorb the counterforce induced by the throwing of the ball. Therefore, there is a more effective arm movement, resulting in the probability of a more accurately thrown pass.

In some sports, however, ballistic motions involving forces and counterforces without a base of support are a necessary maneuver. For example, a prime scoring device in water polo is for the offensive player to come upward with his trunk

(a)

(b)

Figure 9.16 Mark
Logan fires a ten-gauge
shotgun to demonstrate
the principle of force-
counterforce: (a) Stance,
(b) Force at firing and,
(c) Counterforce through
displacement of body
mass from the stance
position.

(c)

and shoulders out of the water in full spinal extension. Through a series of torques causing rotation of the lumbar-thoracic spine, he makes a forceful shot toward the goal (figure 9.17). Although the water has a certain viscous resistance, this is minimal and does not provide the necessary base of support for full absorption of the counterforce. The skilled water polo player is able to impart enough velocity to the ball to be effective, but will normally move backward as a result of the arm and shoulder motions. The performer in this maneuver may lose balance; but the fact that he is falling into water during his follow-through rather than onto ground minimizes the results of his disequilibrium.

The use of the foot to provide a stable absorption of the counterforce is essential in the tennis serve. In order to attain an optimal trajectory over the net and into the serving court, the performer attempts to strike the ball as high off the ground as possible. However, if the player actually leaves the ground in attempting to reach the ball before striking it, the body reaction to the swing may sometimes lessen the velocity of the racket head. This decreases speed and has an adverse effect on accuracy. Sometimes a performer is observed leaving the ground during the follow-through, but this has no effect on the speed of the ball.

Torque and countertorque are especially important since basic body movements are rotational in nature. An excellent example of the countertorque is found in the long jump style used by Bob Beamon, the present world record holder. One of the basic problems in long jumping is to keep the legs and feet as high in the air as possible during the latter part of the jump. In order to accomplish this, Beamon used an extreme circumduction of the arms and lumbar-thoracic flexion. The erector spinae musculature plays the major role in the eccentric control of flexion within the lumbar-thoracic region in order to provide this rotation. The countertorque provided by this deep back musculature and gravitational pull facilitates raising the legs through an extreme bilateral diagonal abduction of the hip joints during the latter part of flight. This leg position allows the trajectory to continue to a point where the center of gravity is very close to the ground before any part of the body strikes the pit.

Figure 9.17 Force applied against a minimal resistive counterforce during a water polo shot.

This extreme use of torque and countertorque has a second effect in the landing maneuver that Beamon was able to perform. Just prior to landing, he diagonally adducted the hips. This forces the body upward and drives the heels downward into the pit. The extra lift gained from the heels striking the sand is then used to carry the body over the landing position.

Another example of the use of torque and countertorque is in the balancing mechanism used by pole-vaulters and ski-jumpers while in flight. One of the problems generated by the new heights attained in pole-vaulting is simply landing in the proper position after completing the vault. For purposes of safety, the vaulter attempts to land as flat as possible on the back of the rib cage, with the hips flexed. In order to achieve this position, the vaulter often will perform circumduction of both shoulder joints while in the air. This causes a rotational motion of the arms during the fall. The torque used to cause this rotation of the arms then develops a countertorque, rotating the body backward to achieve the proper landing position.

Likewise, in the sport of ski jumping, one often sees circular or circumduction motion of the arms used to achieve balance in order to maintain the proper position before landing. In this case, the jumper goes into flight in roughly the position in which he desires to land. However, if he has developed a small amount of angular momentum on leaving the ski jump, this tends to rotate him into an unfavorable position which could be disastrous and/or traumatic upon landing. Therefore, he uses circumduction of the arms simply to maintain the initial equilibrium position and keep his feet below him.

One last example of torque and countertorque is the optimal use of spine and shoulder motions during hurdling. As a hurdler approaches the hurdle, the motion of bringing the lead leg up and forward (hip flexion) involves a lateral rotation at the opposite hip just prior to the time the foot of the takeoff leg leaves the ground. This rotation, if not counteracted, could rotate the whole body through the transverse plane. Since this is an undesired motion, a countertorque is provided by a diagonal adduction of the shoulder joint on the side of the takeoff leg. This has the effect of driving that arm forward very rapidly through the high diagonal plane. Also, as the legs come up to pass over the hurdle, a flexion of the lumbar-thoracic spine combined with the shoulder-joint movements causes a rotation of the upper part of the body. This motion has the effect of minimizing the vertical lift of the center of gravity.

Collision   A collision or impact between two bodies is a common occurrence in all types of sport and dance performances. From the point of view of biomechanics, impact may occur between two bodies, a body and sport object, an implement and object, or between a body or object and the ground or floor. As mentioned in the discussion on momentum and impulse, collisions may occur with high forces for very short time periods, or they may be sustained types of interactions with lower average forces occurring for longer periods of time. In either case, although the precise forces exerted at each instant during the collision cannot be analyzed, the change of momentum of each body can be observed to determine the total impulse of the collision.

Because of the difficulty of measuring forces exerted at each instant, the most convenient method of analyzing a collision is in terms of the momenta of the bodies involved. If the total amount of momentum which each body carries into a collision can be measured, the total momentum of the system becomes a known factor. This, in turn, determines the momentum and direction the bodies will move after the collision.

In maximizing the change of momentum in collisions, the importance of the follow-through becomes apparent. Since the impulse of a collision is the product of force multiplied by time, strategies that sustain the time interval of a collision are more effective in changing the motion of surrounding bodies. The sustained block by an offensive football lineman is a typical example of an impact being sustained in order to continue building impulse. The proper technique for performing this maneuver is to make contact with the defensive lineman and continue this contact until such time as a force can no longer be supplied against him. This will often require two to four seconds, or a major portion of the time involved to execute the play. Sustained contact is also necessary in those situations where spin is imparted to an object.

Energy Transfers during Collision

Since a collision always occurs between two bodies or objects moving on a course relative to each other, the initial energy of prime concern is kinetic. During the impact of collision, the change of shape or deformation of the bodies or objects also produces potential energy. Following the collision, as the bodies separate again there is a transformation back to kinetic energy. When considering the human body, sport implements, and objects, the transfer from initial to final kinetic energy is not perfectly efficient, i.e., during the deformation phase a certain amount of work goes into the form of heat or into a permanent deformation of the bodies. The loss of energy into heat is primarily due to internal friction within the bodies rather than surface friction at their point of contact.

If the loss of energy during a collision is negligible or zero, the collision is described as being perfectly elastic. In a perfectly elastic collision, momentum is conserved. Furthermore, kinetic energy remains as kinetic energy. In any other collision, momentum will continue to be conserved, but part of the kinetic energy is transformed into heat. One of the examples nearest to a perfectly elastic collision in sport is the collision between two billiard balls. In this case, almost all of the kinetic energy is conserved as kinetic energy after the collision. In the case of a head-on collision between spinless balls, the conservation of both momentum and kinetic energy during this collision means that the two balls must exchange velocities. Since one of the billiard balls is stationary before the head-on collision, the other ball will be motionless or have slight motion after the collision.

The other extreme in collisions is the type of impact that causes two colliding bodies to remain together. Scientifically, this is called a perfectly inelastic collision. Probably the most common example of a perfectly inelastic collision in sport occurs when a performer or sport object hits the ground and does not bounce. In this case, the body and ground represent the colliding bodies, and in essence they are "sticking together." An important factor in inelastic collisions is the amount of distortion occurring before the colliding bodies achieve the state of cohesion. Since the total relative momentum and relative kinetic energy must be absorbed during the time of collision, a very short period of active collision or small distortion leads to very large forces during the collision. This, for example, is the difference between hitting hard-packed ground and softer ground. A greater amount of distortion means a smaller average force during the time the performer is stopping. Therefore, there is less chance for injury. This principle is used in the design of good wrestling mats. A more extreme example would be the situation of a slalom skier who misses a gate and falls into soft snow. He may travel a very great distance before finally coming to rest; the smaller coefficient of friction within the snow makes this sliding type of fall much less dangerous to the skier than a similar fall of a motorcyclist on an asphalt or dirt track.

Another example of an inelastic collision with reference to horizontal motion takes place during a tackle on a football field. Both the runner and tackler have an initial momentum. As they collide and lock together, due to the arm action of the tackler, both of them move in a path determined by the sum of these momenta.

For example, assume that a 220-pound fullback moving at 25 ft/sec is met head on by a 180-pound defensive back moving at 15 ft/sec:

$$\text{Momentum (fullback)} = \frac{220}{32} \times 25 = 172 \text{ momentum units}$$

$$\text{Momentum (defensive back)} = \frac{180}{32} \times 15 = 84.4 \text{ momentum units}$$

Since the two are moving in opposite directions, the momentum after collision is the difference in their two individual momenta:

$$\text{Momentum (combined)} = 172 - 84.4 = 87.6 \text{ momentum units}$$

Therefore, at the instant of collision and before any momentum is transferred to the ground by other forces, the combination of fullback and tackler has 87.6 momentum units in the direction in which the fullback was originally moving. To find their velocity at this instant—

$$\text{Momentum} = 87.6 = \frac{180 + 220}{32} \times \text{Velocity}$$

$$\text{Velocity} = \frac{87.6 \times 32}{400} = 7 \text{ ft/sec}$$

During the immediate collision, the fullback is slowed from 25 ft/sec to 7 ft/sec, while the defensive back has his velocity reversed from 15 ft/sec to 7ft/sec in the opposite direction. In this situation, the biomechanic advantage is given to the fullback.

In the previous chapter, *coefficient of restitution* was discussed with reference to elastic bodies. Coefficient of restitution is a term utilized in collision-type situations. *This coefficient is the measure of an object's kinetic energy after a* collision, usually with a solid surface, compared to the amount of kinetic energy that it had prior to the collision. This takes into account the amount of kinetic energy lost in the form of heat or deformation of the body.

Earlier it was calculated that the coefficient of restitution of an official handball was approximately 0.64. This means that on each vertical bounce 64 percent of the kinetic energy is retained, and 36 percent is converted into heat. In most sports where this quantity is specified, it is indicated in terms of the height of rebound after a specified drop. The gravitational potential energy before the drop is converted into kinetic energy before the collision. A certain percentage of this energy is maintained during the collision and is again converted into potential energy at the height of rebound.

Tennis rules specify a rebound of 53 to 58 inches for a tennis ball following a drop of 100 inches. This gives a coefficient of restitution of about 0.55, somewhat less than that of a handball. Lacrosse rules specify a rebound of 45 to 49 inches for the lacrosse ball after a drop of 72 inches. This results in a coefficient of restitution of about 0.65, the same as a handball.

The difference in coefficient of restitution becomes quite apparent in multiple-bounce situations. Since a certain percentage of energy is lost with each bounce, a succession of bounces dissipates the energy in a multiplicative fashion. That is, after the first bounce a handball has 64 percent of its original energy; after the second only 64 percent × 64 percent, or 41 percent, and so forth. Figure 9.18 compares a series of bounces of a lacrosse ball and a tennis ball to illustrate this effect.

In sports where multiple bounces are an integral part of rules and strategy, a higher coefficient of restitution is usually needed. Also, it should be pointed out that the coefficient of restitution is often a function of total distortion. A baseball may have a small coefficient of restitution for small forces but a much larger coefficient when struck by a bat or when striking the ground at high velocity.

The actual rebound velocity of a sport object may depend on the condition of both the object and the surface from which it is rebounding. For example, "dead spots" on gymnasia floors are rather common. Dribbling a basketball on such a floor becomes difficult because the ball simply does not bounce back effectively when it strikes a "dead spot." An even more critical situation in regard to basketball courts is the firmness with which backboards are supported. A variation in the "give" of the backboard can determine to a great extent whether

Figure 9.18 Effect of differences in coefficients of restitution.

| | Coeff. | Drop | Height of Bounce (inches) | | | | |
			1st	2nd	3rd	4th	5th
Lacrosse	0.65	100 inches	65	42	27	18	11.5
Tennis	0.55	100 inches	55	30	16.5	9	5

rebounds fall close or away from the basket area. This affects both the positioning and the timing of rebounders. Knowledge of this reflection factor can add to the "home court advantage" in some cases.

The relationship between forces or energies involved in an elastic deformation and the actual distance of body deformation is called compressibility. An object with high compressibility will suffer extensive distortions from relatively moderate forces. On the other hand, an object with low compressibility shows very little "give" in response to extremely large forces. It should be emphasized here that virtually all sport objects are to a certain extent compressible. Probably the least compressible object in sport is the billiard ball.

Because all objects during collision are compressed somewhat and rebound to their original shape, there is a definite time associated with collision. A greater amount of compression results in a longer time of contact; this contributes to a lower average force used to impart a given change of momentum.

It was calculated that the change of momentum of a batted softball was approximately four momentum units. If the ball were new and quite hard, this change might be imparted by the bat in a period as short as $1/100$ second. In this case:

$$\text{Avg. force} \times \text{Time} = \text{Change of momentum}$$

$$\text{Average force} = \frac{4}{1/100} = 400 \text{ pounds}$$

If the ball were used enough to become soft, this time of contact with the bat might be as long as $1/20$ second:

$$\text{Average force} = \frac{4}{1/20} = 80 \text{ pounds}$$

A greater or lesser time of collision may substantially affect a performer's action in striking something with a sport implement, since the amount of contact time between the implement and the sport object determines, to a great extent, such factors as accuracy and amount of spin imparted to the object. This relationship between collision time and the resultant spin from that collision is presented later in this section.

Compressibility is a factor which may vary over a wide range in seemingly identical balls or other sport objects. It should not be assumed by a performer that a series of supposedly identical sport objects will behave the same under all situations. For example, even among a dozen high-grade softballs one may find small but noticeable variations in such factors as compressibility, weight, and surface-roughness characteristics. This lack of manufacturer's quality control has distinct implications for the performances and strategies used during a contest, because each ball may have different interactions and motions during a collision or flight through the air. Highly skilled fast-pitch softball pitchers take full advantage of compressibility variations of softballs.

Most collisions of interest in sport are those where the collision is not "head on." In this type of collision, a sidewise motion is introduced to both objects, due to the angle at which they hit. Since momentum is a vector quantity, it is conserved both in the initial line of motion and in the direction perpendicular to this initial line. This means that both the speed of the objects after collisions and the angles at which they rebound are determined by conservation of momentum and energy.

Glancing Collisions

If a billiard ball moving at 20 ft/sec strikes a second ball in such a way that it moves off at 14.14 ft/sec at an angle of 45° to the path of the first ball, what is the subsequent motion of the first ball? The motion of 14.14 ft/sec at 45° represents components of 10 ft/sec in the original direction of motion and 10 ft/sec laterally perpendicular to this original direction (figure 9.19).

To find the component of motion of the first ball in its original direction, use the conservation of momentum in that direction:

$$\text{Mass}_1 \times 20 = (\text{Mass}_2 \times 10) + (\text{Mass}_1 \times \text{Velocity (parallel)})$$

Since the two masses are the same:

$$\text{Velocity (parallel)} = 20 - 10 = 10 \text{ ft/sec}$$

Since the original motion had no lateral component of momentum, the sum of the lateral components after the collision must be zero:

$$(\text{Mass}_1 \times \text{Velocity (perpendicular)}) - (\text{Mass}_2 \times 10) = \text{Zero}$$

$$\text{Velocity (perpendicular)} = 10 \text{ ft/sec}$$

The first ball would move with a lateral velocity equal and opposite to that of the second ball. The total motion of the first ball would be the sum of these two components, or a speed of 14.14 ft/sec at an angle of 45° to the original direction (figure 9.20).

The problems are more complex if the two bodies do not have the same mass. If a bowling ball moving at 25 ft/sec strikes a pin and gives it a velocity of 28.3 ft/sec at an angle of 45°, how much is the path of the ball deviated and how much does the ball slow down?

Consider an average 3.25-pound pin and a 13-pound ball. A velocity of 28.3 ft/sec at 45° has components of 20 ft/sec parallel and 20 ft/sec perpendicular to the original motion (figure 9.21).

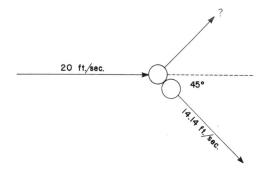

Figure 9.19 Billiard ball collision.

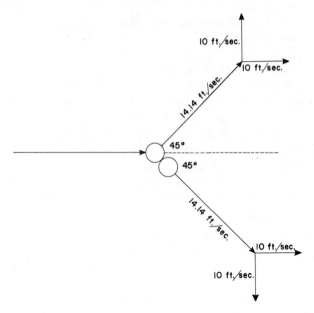

Figure 9.20
Conservation of
momentum components
during billiard ball
collision.

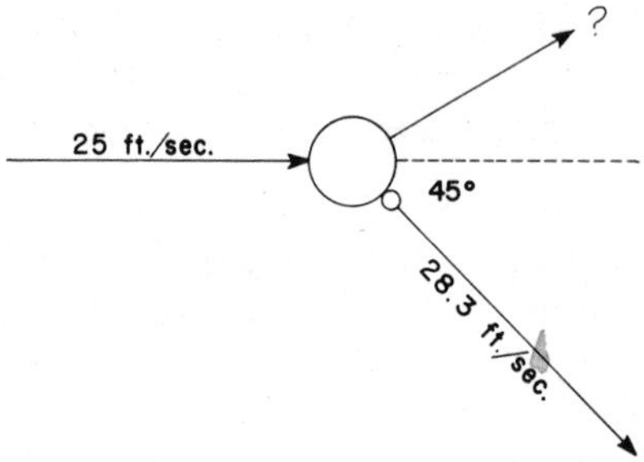

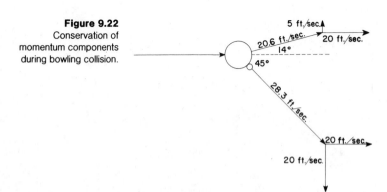

Figure 9.21 Bowling
ball and pin collision.

Figure 9.22
Conservation of
momentum components
during bowling collision.

Relationships between Force and Motion

To find the velocity of the ball parallel to its original direction:

$$\frac{13}{32} \times 25 \text{ ft/sec} = \left(\frac{3.25}{32} \times 20 \text{ ft/sec} \right) + \left(\frac{13}{32} \times \text{Velocity (parallel)} \right)$$

$$\text{Velocity (parallel)} = \frac{(13 \times 25) - (3.25 \times 20)}{13} = 20 \text{ ft/sec}$$

The perpendicular velocity is given by:

$$\left(\frac{13}{32} \times \text{Velocity (perpendicular)} \right) - \left(\frac{3.25}{32} \times 20 \right) = \text{Zero}$$

$$\text{Velocity (perpendicular)} = \frac{3.25 \times 20}{13} = 5 \text{ ft/sec}$$

Combining these two components provides a total motion of the ball of 20.6 ft/sec at an angle of 14° from the original line of motion (fig. 9.22).

It is this deflection of the ball which causes the most effective collision in bowling to be at about a 25° angle with the length of the lane and at a position between the head pin and the number 3 pin for right hand bowlers. At this angle of entrance to the pin pattern, the ball and pins both move in directions to give maximum probability of carrying all pins. A straight ball tends to deflect too much if thrown slowly, and it goes straight through without carrying the back corners if thrown too hard. Thus, the use of the hook into the pocket gives much more leeway for variation with good probability of getting a strike.

The actual results of a glancing collision are highly dependent on a factor known as the impact parameter. *Impact parameter is a measure of the relationship between the direction of motion of colliding bodies and the line joining their centers at the time of collision* (fig. 9.23).

The results of impacts between bodies of equal mass and speed at various impact parameters are:

Zero—pure rebound along the lines of approach

Low—primarily rebound but also with lateral motion

Moderate—primarily lateral motion; each body moves off at near a right angle

High—lateral motion but with each object having a component in the direction of the original motion

Graze—slight change of motion on the part of each body.

For cases where the objects have unequal masses and velocities, where spin is involved, or where surface friction is important, the changes of motion become more complex.

The foul ball in baseball or softball is a good example of a collision between a sport implement and a sport object with a high impact parameter. The contact between the ball and bat occurs at an off-center position as related to their in-

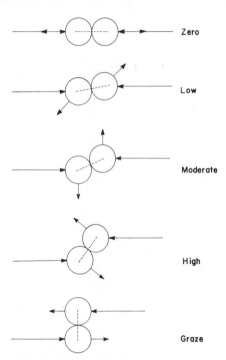

Zero

Low

Moderate

High

Graze

Figure 9.23 Impact parameters.

dividual lines of motion. In this case the curved surfaces of both the bat and the ball add to the effect of the impact parameter. Contact produces a component of motion of the ball perpendicular to its original line of flight. The ball often goes high into the air in foul territory but curves back into fair ground. A foul tip is an example of a grazing impact.

Figure 9.24 illustrates the angles at which softballs leave the bat as a function of the impact parameter for a knuckle ball (spinless). It is assumed that the batter is swinging horizontally, the ball is moving upward at a four- to five-degree angle as it approaches, and there is no appreciable rebound of the bat. The parameter, d, represents the distance in inches below the center of the ball of the line of motion of the center of the bat.

From this it can be observed that in order to successfully hit the ball, the batter must match the center of the bat barrel within the range from ¼ inch above the center of the ball—to hit at least a hard ground ball—to ¾ inch below the center of the ball—to hit anything other than a high fly ball. This range represents the allowable margin of error for hitting.

The glancing block between ice hockey players is another example of an impact parameter. As two players approach one another and bump shoulders, spectators normally do not observe an appreciable slowing of either player. However, the direction of each is deflected to one side of the original line of skating. This is much more exaggerated in ice hockey than it would be in a field sport played on grass, because of the low friction between the performer and the ice surface. Because of this, the ice hockey player moves with the lateral force occurring during the collision in order to maintain equilibrium.

d (inches)	Angle Above Horizontal
0.25	17°
0.50	27°
0.75	38°
1.00	50°
1.25	more than 60°

Figure 9.24 Angles of projection as a result of impact parameters.

A second effect of a glancing collision is the introduction of a spin into the objects involved in the collision. The fact that the direction of the line of centers at the point of collision is not the same as the direction of motion of the body means that the contact force consists of a component perpendicular to the surface, which causes a linear acceleration of the body, plus a component parallel to the surface. The latter is the frictionally related component. The effect of this frictional force is to produce a torque around the center of gravity of the body, resulting in a spinning motion of one or both bodies. This spin factor is also observed in the example of the batted ball. A foul ball has a high spin component; and this must be considered by the player attempting to catch it.

Arm tackling in American football is another example of rotational motion introduced by an off-center collision. Arm tackling is considered a poor technique in American football; in this case, the tackler does not make a head-on collision with the ball carrier but, rather, hits him with one arm and shoulder. The result of this type of collision is often to spin both players; the runner remains on his feet while the tackler loses contact and rotates around him. The momentum of the tackler often causes him to release the runner, and this leaves the ball carrier free to gain extra yardage during the play.

Many times in sport the body segment or limb motions before the collision and/or the position of the sport implement are used purposely to impart an effective spin to the object being hit. As an example, faces of various golf clubs are set at different angles, partially to give lift to the ball during the swing, but also to impart a backspin. Because of the slant of the striking surfaces of the irons, the relative velocity between the club face and the ball at the time of collision includes both a normal force component, which drives the ball, and a frictional component, which imparts backspin to the ball. As will be discussed in the next chapter, this backspin is essential in order to give an aerodynamic lift and greater range to the drive. Designs of club face and ball surfaces provide a high coefficient of friction between these two surfaces during the collision, and thus help to impart an optimal amount of spin to the ball.

In the sport of ping-pong, or table tennis, there is a very special relationship between the angle of the bat and the direction of motion of the arm and hand at the time of contact with the ball. This implement-motion relationship is used to impart a wide variety of spins to the ping-pong ball. In order to produce high forward spin on the ball, a medial rotation of the shoulder joint is used during a forehand shot, thus bringing the bat up and over the ball at the moment of contact. A good backhand shot is made by a diagonal adduction at the shoulder joint, combined with an extension at the elbow. If no lateral rotation of the shoulder is made during this type of shot, the ball will have a backspin. The addition of a lateral rotation in the above motions produces a forward spin during a backhand slam.

Another valuable concept to be considered in collisions between sport implements and sport objects is the *center of percussion* mentioned earlier. This is sometimes called the "sweet spot" on a bat or racket. This concept is immediately concerned with the relationship between the implement and the performer. In the swing of a bat or racket, the implement is acting as an extension of the performer's limbs. This provides extra leverage and has a multiplying relationship to the velocity of the implement. If there is a forced rotation of the implement in the hands of the performer during collision, this leads to less than optimal results. When a ball is struck near the top or bottom of a racket, golf club, or bat, it tends to twist the handle in the grip of the performer. There is also the possibility of a rotation of the implement around a transverse axis of the motion. In this case, the implement tends to either pull out of the grip or "jam" into the hand of the performer. Probably the most common example of this is hitting a baseball either on the end of the barrel or on the handle portion of the bat. In either case, there is a strong motion of the end of the bat that must be counteracted by the performer.

If a collision is made at the center of percussion, there is no tendency of the implement to rotate with reference to the performer. From the standpoint of stability, the tightness of grip is not particularly important when hitting a ball properly at the center of percussion. However, a solid linkage between the implement and the performer is necessary in order to absorb the counterforce of the collision without excessive rebound. This is the reason a solid grip must be maintained through the point of impact (figure 9.25).

Since the very tip of a baseball bat is the part moving at the highest velocity, it might seem that this would be the best place to make contact with the ball. Empirically, however, this is not true. From an analytic point of view, the problem with contact at that point arises from the fact that the bat can rotate about a point along its own length, and this rotation will be in a direction opposite to the rotation of the swing. Impact on the end of the bat tends to rotate it in a direction that will allow the barrel to drop back and the handle of the bat to move forward more rapidly than its center of mass. This causes an extreme force between the handle and the hands—the so-called *sting of the bat.*

Because of the distribution of mass in the outer half of the bat, ball contact made near the center of the bat tends to produce a rotation in the opposite direction. This causes the handle of the bat to jam backwards into the hand. At a point outside the trademark and along the barrel of the bat, a collision does not tend to cause either of these rotations, but simply slows the lateral motion of the bat. This point is called the center of percussion.

Figure 9.25
Importance of solid linkage between batter and bat at point of impact, illustrated by former USC player Fred Lynn. *Courtesy USC Athletic News Service.*

The term "sweet spot" (used as a synonym for the center of percussion) is derived from the positive and subjective kinesthetic feeling experienced by the performer when a ball is struck at this point by the implement. It can be shown that a maximum transfer of momentum from the implement to the object occurs with this kind of collision. Therefore, this is a necessary condition for optimal performance. Often there is a distinct auditory response to a collision at or near the center of percussion. This arises from the fact that a maximum amount of resonant vibrations occurs during a collision at or near the center of percussion. For example, a person with tennis experience, blindfolded and seated near a tennis court, would be able to tell from the sounds of the strokes whether the players were experts or beginners.

Pendular motion is observed in a wide variety of sport actions. As an example, proper delivery in bowling requires a pendulum swing of the ball suspended by the arm from the shoulder joint (anteroposterior plane motions). Many moves in gymnastics from the rings, horizontal bar, and parallel bars are essentially pendular in nature.

In addition, a number of warmup exercises utilize the effect of pendulum swings of the limbs to loosen connective tissue and promote blood flow into muscles. The swinging of a bat in the on deck circle, the preliminary upper limb motions with a discus, and the lower limb pendular motions of dancers and gymnasts are examples of warmup activities. These pendular motions can be performed through any plane of motion.

Effect of Spin on Elastic Collisions

In the last section, the discussion centered around the introduction of spin to a body or a sport object by collisions. What occurs during a collision of an object that is already spinning? In this case, as the surfaces meet, the perpendicular (normal) force is present, as well as frictional interaction along the surface due

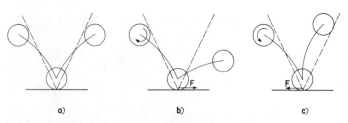

Figure 9.26
Rebounds as a result of
spin ("F" represents the
frictional counterforce
exerted by the surface
on the ball).

to the spin of the object. The tendency to move along the surface results in a frictional component of force which affects the subsequent motion of the object. *It is the frictional component of force that influences the rebound in the direction of the spin during the bounce of a spinning ball* (fig. 9.26). Exceptions to this principle may appear to arise when the component of linear velocity parallel to the surface is larger than the surface velocity due to spin. In this case, the center of mass of the colliding object overrides the point of contact and tends to produce spin rather than react to it.

In figure 9.26, three different possibilities exist for the bounce of a spinning ball from a hard surface. If there is negligible frictional interaction with the surface, the entrance and exit paths of the collision will be symmetric. The straight lines in figure 9.26(a) indicate the angles of incidence and reflection without considering gravity. The actual paths will be parabolic curves because of gravity, as shown by the solid lines.

A topspin against a frictional surface will cause forces in the direction of motion, in addition to the elastic normal force. This causes an increase in the horizontal component of motion and a consequent lower angle of trajectory (figure 9.26b). This is often seen with the topspin service or forehand shot in tennis.

A backspin results in a frictional force opposite to the original direction of motion. This lessens the horizontal component of velocity, causing the ball to bounce more nearly straight up (figure 9.26c). This maneuver is used effectively in the drop shot in tennis or ping-pong and in approaches to the green in golf.

Sidespin causes the line of bounce to deflect. This introduces a new dimension to the path of the ball. In sports such as handball and racquetball, collisions with surfaces located in different planes introduce complex spin interactions into many of the shots. In these sports pinch shots, kill shots, and z-shots derive their effect from the combined interactions of spins derived from multiple collisions.

The effect of spin on rebound appears in many instances during sport activities. One of the most commonly observed is the backspin put on a basketball during a shot. This spin tends to interact with the surface of either the backboard or the rim in such a way as to increase the probability that the ball will go into the basket. When in contact with the backboard, backspin of the ball provides a component of force in the direction of the hoop. On the other hand, any component of spin about a vertical axis tends to cause the ball to move around the rim, and this increases the probability of the ball rotating out of the basket.

The concept of backspin is also used in basketball in the bounce pass, in order to make the ball easier for the receiver to handle. The backspin on the bounce reduces horizontal velocity; therefore, the ball tends to rebound upward in front

Relationships between Force and Motion

Figure 9.27
Neutralizing spin during a
volleyball pass.

of the receiver, allowing easier reception of the pass. Front spin imparted to the basketball during a bounce pass would provide a lower and faster rebound trajectory of the ball, making it more difficult for the player to catch.

Useful examples of problems encountered in handling spin in a sport context are the possibilities that can occur during the "pass maneuver" in power volleyball (figure 9.27). In this case, the defensive player who receives the downward and spinning shot, spike, or serve has the problems of playing the ball back into the air vertically and attempting to remove the spin. This must be accomplished so the second or setup shot can be executed most effectively. Since the dig or pass is made with the surfaces of the forearms or radio-ulnar joint area, the rapidly spinning ball provides tangential forces, tending to pull the ball laterally rather than vertically. These factors, plus the uneven surface of the forearm rebound area, make the pass very difficult.

In considering the total effect of spin, the three outstanding factors are (1) total amount of spin, (2) coefficient of friction between surfaces, and (3) total time of contact between surfaces. For relatively soft sport objects, which would have a long contact time during collision, the effect of spin tends to be greater than for very hard objects with short times of contact.

In the approach shot to the green in golf, the ball is given a high arching trajectory with a large amount of backspin. When the ball strikes the green, backspin causes a frictional counterforce, tending to cancel the horizontal component of velocity. If the green is hard, the time of contact during the first bounce may be as short as a few hundredths of a second. In this case, the total impulse (product of force and time) is insufficient to completely cancel the motion of the ball. Since much of the spin has been lost, subsequent bounces have less effect, and the ball "runs" across the green. If the green is soft, however, the time of contact may be as much as ten times longer. This leads to an impulse large enough to stop or reverse the horizontal velocity of the ball. In addition, a soft green absorbs much of the vertical motion so subsequent bounces on the grass phase out quickly.

When two sport objects such as billiard balls on a billiard table collide, the spin of one of these objects exerts a tangential force that creates an opposite spin to the other object. Therefore, the spin in colliding objects often is transferred from one to the other, but always in the opposing direction. The use of spins acquired during collisions, both with other balls and with the side cushions, is often an advantage for the skilled player in the game of snooker.

| Observable and Nonobservable Quantities | It is necessary to point out that not all physical quantities described in the two preceding chapters are visually observable. Although each has a specific place in an organized description of mechanical motion and interaction, many must be calculated from simpler, more fundamental properties that can be observed either directly or in filmed records. The purpose of this section is to outline those properties which can be observed visually and those which must be calculated. |

Position
The relative positions of two body parts can be observed directly during a performance. With experience it is even possible to make rough quantitative measurements of relatively slow motions noncinematographically. From filmed records, both relative and absolute observations of position can be made, and quantitative measurements can be computed on the basis of film or on projected images or drawings made from film. Accuracy in these quantitative values often depends on such factors as proper scaling, corrections for camera angle, and filming speeds.

Velocity
Relative velocities can be observed directly. By timing a motion over a fixed distance or by measuring the distance covered in a specific time, average velocity over a time interval can be quantitatively determined. From filmed records of a motion, the velocities of various body parts over short intervals can be accurately measured. Ballistic motions require a high filming rate (frames per second) to freeze motion and provide acceptably short intervals. Filmed or videotaped events also allow the breakdown of complex motion into its components.

Acceleration
Acceleration can be detected by direct observation of changing motion. However, any measurement or comparison of acceleration requires the use of film or an electrical accelerometer. From positional measurements on film, both velocity and acceleration can be calculated. This applies to individual components as well as total motion.

Only the *effects of forces* are directly observable during human body motion. From the observed changes of motion and a knowledge of the mass involved, force is calculated by the formula: force = mass × acceleration. Various experimental techniques can be used to measure force by the distortion it causes in elastic objects or implements, but a calibrated instrument is required in order to make direct observation.

Force

The mass of an object or body part is usually computed by weighing or balancing techniques. It is seldom determined by direct observation of motion changes. Mass may also be calculated by multiplying the density of the material involved by the volume of the object or body part. Calculation of the mass of body parts is difficult, because of the need to determine the volume of each irregularly shaped body part. Studies useful in the determination of mass of body parts have been reported in the literature.

Mass

Relative timing of motions can be observed directly. However, this is much more accurate if recorded on film. Precise synchronization of complex motions is best determined by film analysis.

Time Relationships and Synchronization

Pressure can be observed directly by the distortions caused in elastic materials. Accurate monitoring of pressure requires instrumentation or precise measurements of distortion and calibration of elasticity.

Pressure

Frictional coefficients can be observed qualitatively in terms of conditions necessary to produce slippage. Accurate determination requires measurement of force components perpendicular and parallel to the surface of contact.

Frictional Coefficients

Energy is seldom observed directly but is used as a calculated quantity. It is of critical importance because its conservation property can be used in planning transformations from one type of interaction to another. The three most important areas of mechanical energy to be considered in biomechanics are (1) *kinetic energy*—calculated as ½(mass)(velocity)² of a body, (2) *gravitational potential energy*—calculated as the product of the weight of a body times its height above some arbitrary reference level, and (3) *elastic potential energy*—calculated from the force versus distortion characteristics of the body involved. In addition, the concept of energy production from biomechanic reactions is essential to determining the possible mechanical energy production under a given set of conditions by a given set of muscles or muscle groups. *In this context, it is necessary to consider efficiency as the useful work output divided by the total energy input.*

Energy

Momentum can be observed qualitatively in terms of sizes of bodies, body parts, or objects and their specific velocities. Momentum is more often calculated by multiplying the measured or calculated mass of a body part or body by its velocity. Like energy, calculations of momentum are useful in defining changes during transformations from one type of motion to another. Momentum calculations are most commonly involved in collision-type interactions.

Momentum

Impulse	Impulse is almost always a calculated rather than an observed quantity. It is the product of force multiplied by the time of action of the force. It is useful since it represents the transfer of momentum during a collision.
Power	Power is observed only qualitatively, even in cinematographic studies. It is calculated in either of two ways: (1) work done per unit time, or energy transfer divided by time, and (2) force multiplied by velocity.
Elastic Constants	Elastic constants can be observed qualitatively in terms of amounts of distortion present under application of forces, and in terms of noticeable energy losses following collisions or other interactions. These are normally calculated for quantitative information but require specialized equations for each type of distortion. Elastic constants can be calculated in terms of force versus distortion or in terms of energy change versus distortion.
Angular Position, Velocity, and Acceleration	These are observed and calculated in much the same manner as their linear counterparts. It requires more training to make valid judgments of angles and angular changes than to make similar estimates of relative positions. Comparisons of ballistic angular motions usually require the use of film records. Quantitative measurements require careful analysis of such records.
Moment of Inertia	Moment of inertia can be qualitatively observed in terms of general location of various portions of mass with respect to axis of rotation. Quantitative information can be obtained in two ways: (1) by dividing a known or calculated torque by the observed angular acceleration or (2) by summing the products of each part of the mass multiplied by the square of its distance from the axis of rotation, $I = \Sigma(mr^2)$.
Radius of Gyration	Changes of the radius of gyration can be observed qualitatively, but they are normally calculated as the square root of the quotient of the moment of inertia divided by the total mass.
Torque	Torque, like force, is observed only in terms of the changes of angular motion which it produces. It can be calculated in two ways: (1) product of force multiplied by lever arm, or (2) product of moment of inertia multiplied by angular acceleration.
Angular Momentum	Angular momentum can be observed in terms of the relative size of a body and speed of rotation. Observation is essential with reference to the direction of axis of rotation. It is calculated as the product of the moment of inertia times angular velocity.

Alexander, R. McNeill. *Animal Mechanics*. London: Sidgwick and Jackson, 1968.

Dagg, Anne Innis. *Running, Walking and Jumping*. London: Wykeham Publications Ltd., 1977.

Dyson, Geoffrey H. G. *The Mechanics of Athletics*. London: University of London Press, 1970.

Tricker, R. A. R. and Tricker, B. J. K. *The Science of Mechanics*. NY: American Elsevier Publishing Co., Inc., 1967.

Williams, Marian and Lissner, Herbert R. *Biomechanics of Human Motion*. Philadelphia: W. B. Saunders Co., 1962.

Recommended Readings

10
Fluid Mechanics

The interaction of a body or sport object with a fluid has a wide variety of sport applications. The object of many sport maneuvers is to place an object into free flight, where it is affected only by gravity and the forces of interaction with the air. In addition, many rapidly growing sport activities are now concerned with the act of flight itself. These range from flying contests with powered aircraft, sailplaning, and ballooning, to skydiving, hang gliding, and Frisbee games.

In addition, there are numerous sport actions and events associated with water. As variations of the sport of swimming, there are activities such as water polo, synchronized swimming, and underwater diving and swimming, either free or with self-contained underwater breathing apparatus (SCUBA). Performers using a small amount of equipment perform such events as surfing and water-skiing. Also, there are many sport events associated with the use of boats. In the human-powered category, there are sculling, canoeing, and kayaking. Combined interaction of air with water makes possible sailing and yachting. Beyond this, there are the motorized sports ranging from small-craft competition to high-speed hydroplane racing. The latter type of sport also involves the use of forces from air and water.

The term "fluid" as used in biomechanics means any substance with relatively continuous properties that does not maintain a specific shape. As far as this text is concerned, this includes the study of sport object interactions with both air and water, i.e., air and water are both considered to be fluids. From a scientific standpoint, *interactions with air are studied under the title of aerodynamics or the dynamics of compressible fluids. Interactions with water, on the other hand, are the subject matter of hydrodynamics or the dynamics of incompressible fluids.*

Because of the continuous nature of fluids, a new set of quantities must be introduced to describe many of their physical properties. Instead of referring to the total mass of a fluid, it is usually desirable to refer to its *mass per unit volume,* or *mass density.* Likewise, descriptions of weight are usually replaced by the term *weight density. Weight density is defined as the weight per unit volume.* This is normally expressed in units such as pounds per cubic foot. The ratio of the mass or weight density of a material to the density of water is designated as the *specific gravity* of the material.

Many of the other quantities that have been discussed in the preceding chapters are also applicable to fluid mechanics. However, instead of using total forces exerted by or within fluids, it is more common to indicate the force per unit area or *pressure.* In energy considerations, the energy per unit volume or energy per unit mass is often quite useful. These and similar quantities can always be converted to the total properties considered earlier by multiplying by the total volume of fluid involved.

While both air and water are technically considered fluids, there are differences in their properties that become critical when considering their contribution to sport action. One of the principal differences is density. Although the average density of air at sea level is 0.075 pounds per cubic foot, the average density of fresh water is 62.4 pounds per cubic foot. This difference, combined with water's resistance to compression, results in notable contrasts in a performer's ability to generate or use forces and counterforces when interacting with water as compared to air.

Fluid flow is defined as the continuous deformation of a mass of fluid, and can be a result either of a movement of the material from one point to another, or of the movement of a body or object through a fluid medium. In either case, the particles of the fluid have a motion relative to the solid objects involved. It is the relation of this motion to the various physical quantities describing the system that is significant in the study of fluid dynamics.

If the flow is smooth and unchanging, each fluid particle will follow well-defined and predictable paths. These will be straight lines or smooth curves, and are known as *streamlines*. The kinetic energy involved in such flow is efficiently used, with little wasted in the form of heat. The velocities along different streamlines are not always the same but may vary smoothly from one part of the fluid to another. For example, in smooth flow in a pipe, the velocity at the center is greater than at the walls, and there is a smooth transition in the regions in between. As a result, the fluid can be pictured as sheets or *lamina* of material sliding over one another. This model gives the name of *laminar flow* to this type of fluid motion (fig. 10.1).

Most observed flow of real fluids involves another element of motion, in addition to this ideal laminar flow. Some particles are moving in other directions, creating eddies and whirlpools and eventually dissipating most of their energy in the form of heat. This motion is not as predictable as the laminar motion and detracts from the efficient movement between the fluid and its surroundings. Such nonuniform, unpredictable, and inefficient motion in a fluid is known as *turbulence* or *turbulent flow* (fig. 10.2).

These same terms are used when we are primarily concerned with the motion of a body through a fluid. In laminar or streamline flow, the particles separate and move in lines around the body, then close together behind it in such a way that the particles of fluid have little motion after the passage of the body. In the case of turbulent flow, the passage of the body creates excess motion in the fluid

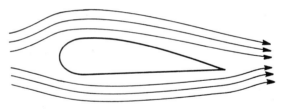

Figure 10.1 Laminar flow.

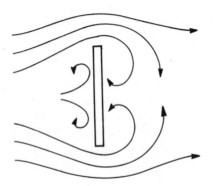

Figure 10.2
Turbulent flow.

particles; this leaves a wake of disturbed fluid along the path of motion. This turbulent wake contains kinetic energy taken from the body by the fluid and indicates a force-counterforce situation between the object and the fluid. Laminar flow causes a minimum loss of energy to the moving object, while the existence of turbulence indicates the performance of work against opposing forces.

Another important concept in the description of fluid flow is that of *viscosity* or *fluid friction. Viscosity is the internal friction of a fluid.* Since the various layers of fluid are moving with different velocities, they are, in effect, sliding over one another and creating frictional forces because of this sliding motion. The amount of energy lost to this viscous friction varies widely from one fluid to another. In most gases the viscosity is relatively low, although it accounts for an appreciable energy loss within a moving gas. In liquids the amount of viscosity varies over an extremely wide range, from the alcohols, which have very low viscosity, to materials like syrups with high viscosity where the major energy loss originates in this interlayer conversion of mechanical energy into heat. Air and water tend to be intermediate in viscosity.

Besides the internal friction attributable to viscosity, there is also friction between the surface of contact of a liquid or gas and an object immersed in it. This is due to the fact that the object in its motion tends to move parts of the liquid; therefore, forces are exerted on these liquid particles. The liquid then exerts a counterforce on the object. It should be emphasized that this is not a force due to a pressure difference because of the motion of the object through the fluid or a difference in the density of the fluid in front of the object. It is a frictional force which the fluid exerts as it slides along the surface.

Buoyancy   An interaction taking place whenever a body is immersed in a fluid is the phenomenon known as *buoyancy*. In any fluid, pressures exist at all points sufficient to support the weight of fluid above them. If part of the fluid is displaced by an object, the pressures remain the same and exert an upward force on the submerged body. This upward force is known as *buoyant force*. The magnitude of the buoyant force is just enough to support the displaced fluid against gravity, the quantity which was earlier designated as the weight of the fluid. *The buoyant force acting on an object immersed in a fluid is always equal to the weight of* *the fluid displaced by the object.* The existence of this upward force means that less force is then required to support the object against gravity. It appears that the object within the fluid weighs less.

Because of the low density of air, the buoyant effect of this fluid on performers and most sport objects is negligible. Since an average performer displaces about 2½ cubic feet of air, the buoyant effect can be calculated as follows:

$$\text{Buoyant force} = \text{Weight of displaced air}$$

$$= \text{Density of air} \times \text{Volume displaced}$$

$$= 0.075 \text{ lb/cu. ft.} \times 2.5 \text{ cu. ft.}$$

$$= 0.188 \text{ lb} = 3 \text{ ounces}$$

This force is negligible in comparison to the other forces affecting human performance. However, as will be seen below, the much higher density of water causes buoyancy to be an important factor in most water sports.

Sport ballooning is one example of an area where the buoyant force of air is quite necessary. The balloons are filled with hydrogen, helium, or hot air. These substances all have densities lower than the atmosphere in which they are immersed. Therefore, the displaced air weighs more than the gas-filled balloon, and this causes a net upward force that can be used to lift a payload. But even in these circumstances, the densities are so small that several thousand cubic feet are necessary for passenger balloons.

In contrast to air, the buoyant forces exerted on objects in water have magnitudes often comparable to the weights of the objects. Objects whose specific gravity is greater than that of water will appear to weigh less when submerged by an amount equal to the weight of the water displaced. Those objects with specific gravities less than that of water will experience buoyant forces greater than their weight and will tend to float to the surface.

Figure 10.3 represents the actions of a stone, a fish, and a block of wood of equal volumes when suspended in water. Since each displaces the same amount of water, the buoyant forces on each are equal. The weight of the stone is greater than this buoyant force, so it sinks to the bottom, where contact forces provide the extra necessary support. The buoyant force just matches the weight of the fish, so it remains suspended. The buoyant force is greater than the weight of the wood; this forces it to move upward until it displaces less than its volume of water. The buoyant force then is decreased until it matches the weight of the wood and the wood floats in equilibrium.

Since the human body contains a high percentage of water and has a specific gravity very near 1.0, buoyant and gravitational forces very nearly cancel each other for performers in water. Adipose tissue, or fat, is less dense than water, while muscle tissue is slightly more dense. Some people having a high muscle-to-fat ratio (lean body mass) will sink when relaxed in fresh water, but most people will float nearly submerged at the surface. The nonfloating condition most often occurs in young athletic males, often the type who participate in compet-

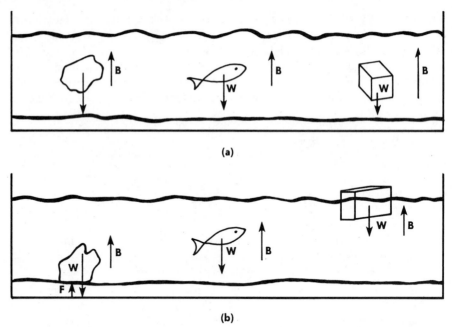

Figure 10.3 Effects of buoyancy on objects of different densities.

(a)

(b)

itive swimming. Long distance swimmers, in contrast, tend to have larger amounts of body fat. This not only increases buoyancy but tends to reduce the dissipation of body heat into the surrounding water. These factors are reasons why women do well in events such as swimming the English Channel and other long distance swims, since women have more adipose tissue than men.

Because the upward buoyant force and downward gravitational force effectively cancel each other for performers in water, no action is required to support oneself against gravity. The performer can move as though weightless. This condition allows the slow maneuvers and rotations observed in the sport of synchronized swimming. In competitive swimming, this weightlessness facilitates the use of strokes and kicks which are designed to generate horizontal forces, rather than the vertical forces which would be necessary to support the body.

Since ocean water has a higher density than fresh water, due to the dissolved salt content of the former, buoyant forces in the ocean are about three percent greater than on an equivalent body in fresh water. For this reason, a swimmer or floater in ocean water tends to ride with a larger percentage of the body above the surface than the same performer in fresh water. This is one reason that major long-distance swimming records have been established in salt water rather than in fresh water. In these long-distance swimming contests a female swimmer would theoretically have an advantage over a male swimmer.

Since the average density of the human body at maximum inhalation is about 0.98 that of water, the submersion of 98 percent of the body will displace a weight of fresh water sufficient to counter the weight of the person. Under these conditions the person can float in a relaxed state with 2 percent of the body above water. In sea water, the relative density would be $0.98/1.03 = 0.95$. This means that equilibrium would require only 95 percent submersion, leaving about 5 percent of the body above the surface. It should be emphasized that the relative percentages of muscle, bone, and adipose tissue cause the average density to vary several percent. In ocean water few relaxed people are able to sink even in a fully exhaled condition.

The density of the water in the Great Salt Lake is greater than 1.15 relative to fresh water. A person can float relaxed in that lake with nearly one-fifth of the body above the surface. When upright, this allows the head and shoulders to remain out of the water; when on the back, the head, hands, and feet can all be above the surface at the same time.

An increase of the density of water also causes craft such as sculls, kayaks, and canoes to have a smaller portion of their volume underwater during a race. This reduces the total drag on the craft and allows a slightly higher speed to be maintained for the same thrust effort.

The most important application of fluid dynamics to sport actions is in the description of performers and objects moving through fluids. In these situations, forces and counterforces between the fluid and the body influence the path and speed of movement. Because of the continuous nature of fluids, these interactions are often complex and difficult to analyze; but a series of terms has been used to classify the interacting forces.

Movement through Fluids

The most common terms found in the literature describing movement through fluids are *drag, thrust,* and *lift.* These refer to the components of force which are at work, and are separated by their directional relation to the velocity vector of the moving body (fig. 10.4). Drag refers to those components of motion opposite to the velocity. *Drag* always tends to oppose the relative motion between the body and the fluid. *Thrust* refers to those components of force in the direction of the velocity vector, i.e., those forces that tend to cause or maintain the motion. The term *lift* refers to the components perpendicular to the velocity vector. These neither increase nor decrease the speed, but they affect the direction of motion. Lift is most commonly observed vertically upward, but it may occur in a horizontal direction, or even downward under certain circumstances.

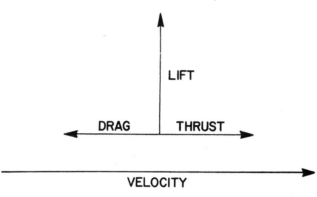

Figure 10.4 Vector relationship between thrust, lift, and drag forces on a body moving through a fluid. *Courtesy of Mike Coulson.*

Drag   Any time an object is moving through the air, there will be a drag force tending to oppose the motion. This force is due almost entirely to the turbulence created by the motion. The energy lost from the body goes into kinetic energy of the air in the path of motion.

The amount of drag a body will experience in moving through the air is a complex function of its shape, surface, and velocity with respect to the air. Because of the complexity of this function for most situations, the relationship is expressed simply in terms of a *drag coefficient*. Equations for drag force are written in the following form.

$$\text{Drag force} = (\text{Constant})\ (\text{Area})\ (\text{Drag coefficient})\ (\text{Velocity})^2$$

Constant is determined by the density of air and the system of units used. *Area* represents the cross-sectional area in the direction of motion. *Drag coefficient* is determined by the shape of the body, its surface roughness, and the range of its airspeed.

As examples of calculation of the drag forces on three distinct and different sport objects, consider the following:

1. When considering the cross-sectional area of a sport object in square inches and its velocity in ft/sec, the constant for this equation equals approximately 0.00001. For a sphere moving at nominal velocity, the drag coefficient equals approximately 0.7. Therefore, the drag force on the shot during the shot put would be—

$$\text{Drag force} = (0.00001)\ (19.6\ \text{sq. in.})\ (0.7)\ (40\ \text{ft/sec})^2$$

$$= 0.22\ \text{pounds} = 3.5\ \text{ounces}$$

A 3.5-ounce force acting on a 16-pound shot will have no noticeable influence on the trajectory. Even a strong head wind, which increases the velocity of the shot relative to the air, should show little effect on the total distance.

2. The same calculation for a table-tennis-ball slam shot is the following:

$$\text{Drag force} = (0.00001)\ (1.2\ \text{sq. in.})\ (0.7)\ (60\ \text{ft/sec})^2$$

$$= 0.03\ \text{pound} = 0.5\ \text{ounce}$$

This one-half-ounce drag force is five times the one-tenth-ounce weight of the ball. Therefore, the velocity of the slam is very quickly reduced to only a few feet per second where the drag force becomes small with respect to the weight of the ball.

3. A golf drive actually leaves the tee at such a high velocity that the drag coefficient is only about 0.2. The drag force can then be calculated—

$$\text{Drag force} = (0.00001)\ (2.06\ \text{sq. in.})\ (0.2)\ (200\ \text{ft/sec})^2$$

$$= 0.16\ \text{pound} = 2.6\ \text{ounces}$$

This is still appreciably greater than the 1.62-ounce weight of the ball, so the ball slows rapidly. As the ball slows down through a certain critical velocity, the drag coefficient increases rapidly to the 0.7 value. At this part of the flight, the drag force actually increases as the object slows down. This produces a marked change in the appearance of the flight. Precise calculations of drag forces on the golf drive are complicated still more by the fact that the spin of the ball causes changes in the effective drag coefficient.

For the various speeds with which most sport objects and performers move, the actual drag force tends to vary as the square of the velocity. This means that if velocity is doubled, force increases by a factor of four.

There has been considerable literature relating to the effect of wind on drag forces affecting the human body during sport events. For example, in all sprint and jumping events in track, there is a maximum allowable wind used in establishing new records. This concept implies that the wind will blow the runner along the track or assist the jumper through the motion. As noted earlier, a good sprinter can move nearly 30 mph, so a 10 mph tail wind would not actually push him. However, there is a beneficial element in such a wind, because the effect of the wind reduces drag forces. Therefore, performance time would be aided.

As an example of the effect of wind, consider a bicyclist riding at 40 ft/sec in a wind of 20 ft/sec. It is assumed that the cyclist presents an effective cross-sectional area of 500 square inches and has a drag coefficient of 0.6. The resistive force can be calculated from the equation above, and the power output needed to maintain velocity is the product of the resistive force times the velocity.

In still air, the cyclist has a relative velocity of 40 ft/sec.

$$\text{Force} = 0.00001 \times 500 \times 0.6 \times (40)^2$$

$$= 4.8 \text{ pounds}$$

$$\text{Power} = 4.8 \times 40 = 192 \frac{\text{ft.lb}}{\text{sec}} = .35 \text{ Horsepower}$$

With a tail wind of 20 ft/sec, the velocity relative to the air is reduced to 20 ft/sec.

$$\text{Force} = 0.00001 \times 500 \times 0.6 \times (20)^2$$

$$= 1.2 \text{ pounds}$$

$$\text{Power} = 1.2 \times 40 = 48 \frac{\text{ft.lb}}{\text{sec}} = 0.09 \text{ Horsepower}$$

But, with a head wind of 20 ft/sec, the relative velocity with the air is increased to 60 ft/sec.

$$\text{Force} = 0.00001 \times 500 \times 0.6 \times (60)^2$$

$$= 10.8 \text{ pounds}$$

$$\text{Power} = 10.8 \times 40 = 432 \frac{\text{ft.lb}}{\text{sec}} = .79 \text{ Horsepower}$$

While a 20 ft/sec tail wind does not push the cyclist at this speed, it reduces the resistive force and the power requirements to small factors. On the other hand, a head wind increases the power requirements so much that this speed could be maintained for only a short time. In races involving a circular course, the energy and time required to travel against the wind greatly outweigh the gains made when moving with the wind.

This example also shows the importance of position in performances involving high relative airspeeds. A cyclist riding without the standard crouch pedaling position not only presents a much greater cross-sectional area but also has a higher drag coefficient. The resistive load is increased and the quality of the performance is decreased. Other sports involving this streamlined position are ski racing and speed skating. Such activities as sprinting and long jumping involve the performer in very poor positions from the standpoint of streamlining, and these events are adversely affected by head winds.

When considering forces acting on bodies falling through air, there is a determining concept known as *limiting velocity*. Since the drag force of air is proportional to the square of the velocity, a falling body will accelerate until it reaches a velocity at which the drag force is equal to the weight of the body. At that point, the net force acting downward on the body is zero, since the gravitational force is being exactly balanced by the upward force of the air. *This critical velocity beyond which acceleration does not occur is known as limiting velocity.* Limiting velocity is dependent on shape, size, and total weight of the falling body.

The purpose of using a parachute is to achieve a configuration with such a high drag coefficient that its limiting velocity is only about 20 mph. This is the approximate velocity with which most sport parachutists strike the ground on landing. The limiting velocity for a human body during free fall from high altitudes in a tumbling condition is calculated as follows:

$$\text{Drag force} = (0.00001)\,(\text{Area})\,(\text{Drag coefficient})\,(\text{Velocity})^2$$

If it is assumed that the tumbling body approximates a sphere with a cross-sectional area of 500 square inches and that at the limiting velocity the drag force would balance the 200-pound weight of a man and equipment:

$$200 = (0.00001)\,(500)\,(0.7)\,(\text{Velocity})^2$$

$$\text{Velocity} = 239 \text{ ft/sec} = 163 \text{ mph}$$

This is approximately the limiting velocity observed in skydiving, but it will vary with the size and weight of the sky-diver, the wind conditions, and other factors.

If the performer assumes a position with upper and lower limbs extended and abducted, the effective cross-sectional area is increased to approximately 1,200 square inches, and the shape factor increases the drag coefficient to about 1.0. Then:

$$200 = (0.00001)\,(1,200)\,(1.0)\,(\text{Velocity})^2$$

$$\text{Limiting velocity} = 129 \text{ ft/sec} = 88 \text{ mph}$$

In order to slow the limiting velocity to 20 mph for landing purposes—

$$200 = (0.00001) \text{ (Area) } (1.0) \text{ } (29.3 \text{ ft/sec})^2$$

$$\text{Area} = 23,250 \text{ square inches} = 161 \text{ square feet}$$

The parachute would need an effective interacting area of about 160 square feet, which would require a radius of about 7 feet. Actual design of the parachute changes the relation between the true radius and the effective area.

Skydiving groups utilize this concept in performing various maneuvers and patterns. A diver above the group can move into a flexed position at the lumbar-thoracic spine, hip, and knee joints. This will cause the diver to fall fast enough to rejoin the group. By abduction of the shoulder joints, hyperextension of the lumbar spine, abduction and extension of the hip joints, and extension of the knee joints, it is possible to control horizontal movements to achieve various configurations with the other divers. The latter position of the body increases the drag coefficient.

In sport applications other than skydiving, sport objects seldom fall far enough for limiting velocity to become a significant element. Limiting velocity, for example, is not a factor in three-meter diving. However, in a large number of applications, a sport object will be projected with a velocity greater than its limiting velocity for air near the earth's surface. Under these conditions, the body will gradually slow down, approaching the limiting velocity by deceleration.

A body or object moving through water experiences much greater drag forces than those involved in motion through air. The fact that the water adheres to many surfaces increases the frictional component of drag while the greater density of water leads to high turbulent energy losses. Limiting velocities in water are often only a few feet per second, even under rather large steady driving forces.

Frictional drag occurs because of the actual slipping motion between the body moving through water and the water itself. In this frictional component, the performer tends to move water in the direction of motion. Therefore, the counterforce exerted by water on the performer is opposing his motion. To minimize this type of drag force, many performers use various oils on the skin or shave hair from the torso, arms, and legs. Theoretically, this may be effective to reduce drag and assist the performer psychologically, but laminar drag is only a minor factor in the resistance of water to moving bodies.

The major source of resistance to a body moving through water lies in the turbulent drag. Higher density means that moving water possesses more kinetic energy so performers moving through it are doing more work and meeting more opposing force than they would while moving through air. The design of the human body does not allow efficient operation within a water medium. All motions involved in swimming actually result in rather large amounts of energy being dissipated in the water, due to turbulent eddying motions following the competitive swimmer.

In order to minimize the drag effect of water on the performer, it is necessary that the longitudinal axis of the swimmer be maintained as nearly as possible in the direction of motion. Spinal rotation to one side or slanting the body in the water presents a much greater surface in the direction of motion; therefore, much greater turbulent action follows the swimmer. Because of the requirement to keep the cervical spine hyperextended and the body at an oblique angle when playing water polo, the swimming in this sport is far less efficient than that in competitive swimming. The water polo performer has a far more extensive output of energy in moving the same distance through the water than a competitive swimmer uses when executing the front-crawl stroke over the same distance.

An especially ponderable source of viscous drag in motion through water develops at the surface between water and air. Because of the high surface tension of water, forces exerted parallel to a surface tend to move water into wave formations which move away, carrying energy with them. This wave effect formed by moving through water is responsible for the wake seen behind swimmers, boats, and other objects moving at the surface. The energy loss at this point is so drastic that frictional force tends to develop as the cube of the velocity rather than the square, as is common in most fluid motion. In order to double the velocity of an object moving along the surface of water with waves, eight times as much energy must be expended. For this reason, swimming motions producing the least amount of waves at the surface will be most efficient in moving a body through water.

In swimming in a competitive pool, the wave motion at the surface introduces other problems. A competitive pool is normally overflowing beyond the gutter level to reduce the reflected wave motion; but considerable choppiness at the surface remains during a race. The leading performers tend to swim in much smoother water than those following by a body length, or more, behind. Therefore, in swimming events there is a biomechanic as well as a psychologic advantage in staying ahead of the competition. The wave problem becomes especially significant for trailing swimmers at the end of the pool where flip turns are being made. The leading swimmer, in making the flip turn, creates a wave of water which washes back directly into the face of those who are swimming behind the leader. This wave force increases their drag coefficients and makes the maintenance of stroke mechanics more difficult.

In sports in which a high velocity can be attained with respect to water, the technique known as *planing* tends to reduce these surface losses. In the sport of water-skiing, for example, before a skier is able to "get up on the skis," he or she is literally plowing through water and is exerting a large drag on the towing boat. However, once a standing position is attained, only a small part of the surface area of the ski is in actual contact with the top of the water. At that time, the drag coefficient is greatly reduced. The actual surface area in contact with the water during planing may be quite small. As an example, the trick skis used for high maneuverability are rather short in length, and the ultimate trick in skiing is to eliminate the skis altogether and plane simply on the soles of the feet. In order to perform this maneuver successfully, however, the velocity of the tow boat and the skier behind it must be exceptionally high. In performing turning maneuvers during water-skiing, it is necessary to interact strongly with the water so the resultant drag will supply centripetal force necessary for curved motion (fig. 10.5).

Figure 10.5
Increasing drag to
provide radial
acceleration. Note the
lean of the skier to
provide equilibrium
between weight and
centripetal force.

As described earlier, *thrust is the component of interaction that results in a force*   **Thrust**
in the same direction as the velocity vector of the object. Under most conditions,
this force is purposely generated by the performer as a counterforce to actions
which push the fluid backward. By causing a change in momentum of the fluid
opposite to the direction of motion, the performer receives aid in accelerating or
maintaining the motion desired (fig. 10.6).

Because of the very low density of the air, it is virtually impossible through
human motions to obtain enough interaction to generate a noticeable thrust.
Human beings have no way of interacting with a large volume of air fast enough
so that its change of momentum per unit time will help in sport actions. Motorized machines utilize high-speed propellers to overcome this handicap, and recent
pedal-powered flight systems have been designed which allow a measure of controlled flight by human motions.

Thrust is an important interaction in water, however, since the forces obtainable in the direction of motion can cause appreciable changes of motion of
the body. Although the actual mass of water moved is usually only a small fraction of the mass of the human body, the change of momentum per unit time is
larger than the frictional and viscous resistant forces, and this can be used in
sport actions. The term *propulsive force* is used by many swimming coaches as
a synonym for thrust.

There is a close relationship between the interactions of thrust and drag in
many sport actions and maneuvers in water. If a part of a body or body segment
is moved backward under conditions leading to a high drag coefficient, while the

remainder of the system is positioned to have low drag, the result will be a net forward motion of the body. In swimming strokes, the hands and arms are positioned for large drag, while the body lies level in the water to minimize the resistance to its motion. A motion such as cervical hyperextension during a front-crawl stroke causes a slanting of the body in the water. This greatly increases the associated drag and lessens the effectiveness of the stroke.

As a numerical example, consider the case of a hand being brought through the water in the front-crawl stroke. The drag force would be calculated as follows:

Drag force = (Constant) (Area) (Drag coefficient) (Velocity)^2

Constant for density of water = 0.014
Area of an average hand in the anatomic position is equal to approximately 35 square inches
Drag coefficient for the extended hand is equal to about 1
Maximum velocity with respect to water is equal to about 7 ft/sec

Drag force = (0.014) (35) (1) (7)^2 = 24 pounds

This would seem to represent the maximum force a swimmer could achieve during a stroke. The average force is much smaller because of the small fraction of time during which the hand is moved at maximum velocity. The obtainable force is also highly dependent on the size of the hand and its position. Cupping the hand not only reduces the effective area but also reduces turbulence and may drop the drag coefficient by as much as half. It is estimated in some studies that in order to maintain a body speed of five to seven feet per second in moving through the water, a swimmer must maintain an average thrust of about ten pounds. This average force would not be enough to support the weight of the body against gravity, but that function is supplied by the buoyant forces of the water.

Much of the same concept of exerting thrust is found in various sports where implements are used for propulsion through the water. In the sport of crew, for example, the basic strategy is to plant the oars in the water so the pulling motion of the rowers forces the shell forward. The limited amount of motion of the oars requires great precision in the timing, stroke length, and stroke power in order to maintain a straight course (figure 10.7). The design of the oar and shell is such that the oar has a large drag coefficient, while the shell has a small drag coefficient. As a consequence, when the oar is planted in the water, its linear motion is minimal, but the motion of the shell across the top of the water is maximal in terms of stroke force.

The same general motion with paddles is used in canoeing and kayaking, but with a much wider variety of strokes. Again, the basic idea is to plant the paddle with high drag coefficient and then exert a force to cause as much forward motion of the boat as possible. Since the strokes must be applied off-center to the boat, they actually constitute a torque, and any single stroke has a tendency to turn the canoe or kayak off the line of motion. Therefore, it becomes imperative to apply a balancing torque from each side or to use a feathering of the stroke in such a way as to counteract the generated torque. In kayaking with the double-bladed paddle, the lower body motion of the performer is often used to exert torques and actually turn or maneuver the craft.

In motorized craft, thrust is obtained by propelling smaller amounts of water backward at higher velocities. In this case, interaction of water with the propeller exerts a force backward on the water. This action results in a counterforce for-

Figure 10.7 The USC women's crew illustrating the timing and coordination necessary to combine maximum thrust with minimum drag. *Courtesy USC Athletic News Service.*

ward on the boat. This same principle can be used with air-driven boats, but it requires motion of very large volumes of air. Extremely large propellers are utilized on these boats, as compared to water-driven craft. One of the newer concepts used in powerboat design involves taking water into tubes and thrusting it backward from the boat at extremely high velocities. This provides a water jet action for the forward propulsion.

In the various aquatic activities involving swimming, the techniques and maneuvers used by performers are those designed to be most efficient in providing forward thrust. The arm movement provides a high drag coefficient, while the trunk is used to minimize drag. This results in the body being moved linearly. A large amount of water is being displaced, but with a relatively low velocity. The shoulder-arm motions of the front crawl, as an example, are designed to "plant" the hand and forearm. The diagonal adduction, seen in the pull of the arm through water, subsequently moves the body over the "plant" position. In this context, the body is analogous to a racing shell, and there is similarity between the arm action of the swimmer and the oar action by the oarsman. Carrying this analogy one step farther, there is a definite parallel between the oarlock and the enarthrodial or ball-and-socket arrangement of the shoulder joint; both are multiaxial in nature.

An essential function of arm motion while swimming is to maintain body orientation and equilibrium during the execution of the stroke (fig. 10.8). For efficient motion through the water, rotation of the body about its longitudinal axis must be minimized. This type of motion tends to increase drag and represents wasted energy. Equilibrium functions and thrust provided by arm and leg motions are commonly called "stroke mechanics" by swimming coaches.

Leg motions in swimming strokes contribute to equilibrium but are primarily designed to provide extra thrust. Both the flutter and whip kicks are intended to generate a scissor-type action that forces water along the legs and back past the

Figure 10.8 Arm action of Tom McBreen, USC swimmer, helps maintain body orientation and equilibrium. *Courtesy USC Athletic News Service.*

Fluid Mechanics

feet (figs. 10.9 and 10.10). In both kicks, the most efficient action comes with the knees and ankles dynamically stabilized. The major internal force is provided by muscular actions occurring at the hip joints. *Dynamic stability implies a co-operative contraction of all muscles associated with the joint without maximal fixation of the joint involved.* This arrangement allows a limited amount of motion to occur within the joint while it remains stable. This condition is referred to in the literature as the Lomac Paradox (Logan and McKinney, 1970).

It is actually the laminar flow along the lower limb, beginning near the pelvic girdle and extending to the feet, which causes the resulting water motion and counterforce against the body during the flutter kick. The fact that a transverse motion of water is induced and turbulence results from the kick means that this is not a maximally efficient method of achieving thrust. By anatomic configuration and physiologic mechanisms, man is not adapted to move efficiently through water.

The motion by which man most nearly imitates fish and aquatic mammals is the dolphin kick used with the butterfly swimming stroke (fig. 10.11). In this

Figure 10.9 Flutter kick for propulsive force during the back crawl.

Figure 10.10 Whip kick action to produce thrust.

Figure 10.11 Dolphin kick during the butterfly stroke.

technique, the primary propulsive movement is a lumbar-thoracic flexion-extension-hyperextension. The hips, knees, and ankles remain dynamically stabilized. This provides a sinuous motion of the lower half of the body, which traps water and propels it backward by laminar flow. This is the same motion many fish use in swimming through a different plane of motion. Unfortunately, the musculature and skeletal structure associated with this motion in man is not developed to the point that it can be continued for long periods of time.

Lift   In addition to the forces of drag and thrust, it is possible for an object to interact with the air in such a way that a force perpendicular to the direction of motion will be achieved. *This component of force perpendicular to the direction of motion is called lift.* Although substantial lift forces are possible in water, there are few sport actions in which hydrodynamic lift is used. This is partly because, with the exception of motorized boat sports, velocities in water tend to be rather low. In air, the term *flight* is commonly used to denote those motions where lift is significant.

In the absence of lift, a sport object will follow a path through the air nearly corresponding to a parabolic trajectory. However, when an object is flying in the presence of lift forces, the trajectory will be greatly different from this parabola. Due to their weight, the shot and hammer do not fly, but follow parabolic trajectories. A discus, javelin, baseball, or golf ball can develop lift forces from its interaction with air, and experience varying flight patterns. No specific name is given to these flight trajectories, since they are highly variable and depend on such parameters as flight speed, amount of spin, and orientation of the object with respect to the direction of motion.

There are three different characteristics which produce an interaction with air to develop lift: (1) an airfoil shape, (2) a slanted orientation (angle of attack) with respect to the motion of the body, and (3) a rotation of the body. The effect in all three cases is to exert a force causing air to move downward. Since the object itself is exerting a downward force on air, it follows from the action-reaction principle that air is exerting an upward force on the object. In many cases, the analysis of lift is easier to understand by considering the effect of the object on the air.

In the case of an airfoil shape, the lift is caused by what is scientifically known as the *Bernoulli Effect*. A form of the Bernoulli equation is:

Pressure at point 1 — Pressure at point 2 =

½ (Density) [(Velocity at point 2)² — (Velocity at point 1)²]

The Bernoulli Effect states that the transverse pressure exerted by air becomes lower as the velocity of air becomes higher. Since air flowing over an aircraft wing is moving faster than the air flowing under the wing due to its longer path, the pressure exerted on top of the wing is less than that on the bottom. Therefore, there is a net lift on the wing. Air passing over a wing tends to acquire velocity downward, and the reaction from this change exerts a force upward on the wing. *Within the realm of sport, there are few cases where a purely airfoil lift critically affects a sport object moving or flying through air.*

If an object moving through air is oriented in such a way that the front portion is slightly higher than the rear portion, the net effect will be to exert a downward force on the air. Since the object is forcing air downward, the counterforce of the air exerts an upward force on the object. This is the aerodynamic situation observed in flights of the javelin, discus, arrows, and thrown or punted footballs. It should be noted that in all of these cases, the objects are given a spin around their longitudinal axes at the time of release in order to provide stability to maintain proper flight orientation with respect to air.

In the sport of ski jumping, the form used by modern performers provides lift by a combination of an airfoil described by the body and the slant effect of the skis. The body position over the skis actually presents a near airfoil shape to the air through which the ski-jumper is moving. In addition, the basic slant of the skis, and the body, provide an additional lift.

The third type of interaction with air resulting in lift is rotation of the moving object (figure 10.12).

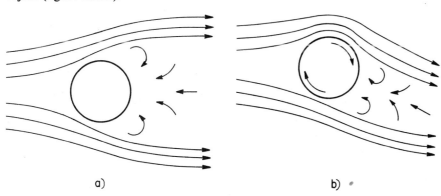

Figure 10.12 Lift as a function of spin: (a) flow without spin and (b) flow with spin.

a)

b)

In a common example, air behind a ball is forced in the direction of motion of the back of the ball. This means that the counterforce is then perpendicular to the direction of motion of the ball. In the case of a backspin, the force on the ball will tend to be upward. The trailing air will be moving in the direction of motion of the back portion of the ball. Force on the ball will be in the same direction as motion of the front portion of the ball. The actual value of the force is a complicated relationship between forward velocity, angular velocity, and frictional interaction between the air and the ball. This relationship is expressed as the *lift coefficient*.

The numerical calculation of lift is similar to the drag equation:

Lift force = (Constant) (Effective area) (Lift coefficient) (Velocity)^2

Lift force depends to some extent on the square of the velocity. The lift coefficient is exceedingly difficult to calculate; it depends on such factors as shape, spin, surface roughness, and pattern of airflow around the object in flight. Exact calculations for examples from sport are beyond the scope of this text. In such actions as the flight of a discus or javelin, the lift force is appreciable, but not equal to the weight of the object.

Objects subject to lift have a much greater range than would be given by a parabolic trajectory. In the case of a golf drive or batted baseball, the lift force may actually exceed the weight of the object during the early part of the flight. This is shown by the fact that the initial trajectory of these sport objects often exhibit a curvature upward. As the object rises, however, its velocity decreases, and the lift force becomes smaller than the weight. As a result, the final portion of the trajectory curves downward and usually approaches the parabolic pattern.

Almost all sport objects have a certain amount of flight, but it is only when the lift-to-weight ratio becomes relatively large that the effect of flying becomes apparent. As examples, a baseball hit with a small amount of spin would have a very low lift-to-weight ratio. Therefore, it would follow a parabolic trajectory and not meet the criteria of flight. A ball hit with extreme reverse spin, however, has a higher lift-to-weight ratio; consequently, it is actually projected into flight. The latter is usually the case when a baseball player hits a "tape measure" home run.

Lift and drag forces impacting on sport objects moving through air depend on the density of air as well as on the parameters of the sport object. For air with high density, large values of drag and lift result. Low densities of air produce correspondingly lower lift and drag forces. At high altitudes one would expect that trajectories which rely very little on lift would be increased in range. Drag forces would not counter the motion of the object to the same degree. However, trajectories which depend on lift tend to be diminished at high altitude. The available amount of lift force is decreased with the decrease in density of air. Quite often the deadening effect of a high humidity is erroneously attributed to an increase in drag forces on the objects. Actually, high-humidity air is of lower density than dry air at the same pressure and temperature. In this case, the available lift is smaller than it would be on a dry day; therefore, the objects tend not to fly so well.

The direction of lift force is always considered to be perpendicular to the line of flight of the object. This means that if the path of the object is highly slanted upward, there are lift force components upward and backward with respect to the horizontal motion of the object.

The effect of the lift can hold the object in the air and decrease its forward horizontal motion. Since the speed is also being decreased by a gain in altitude, a condition can be reached where the velocity drops to a very low value. When this happens, lift essentially disappears. This is the situation known as *stalling*. In this case, the trajectory has a sharp upward motion, followed by a virtual stoppage of the object in air. It then falls with very little horizontal component of motion. This is seen at times when a sport object such as the javelin is released at an angle too high with respect to the ground. One sport skill which utilizes a form of stalling is the football punt. In order to gain a maximum interval of time for downfield coverage by the punting team, the punter will attempt to "hang the ball in the air." This is accomplished by contacting the ball with the dorsum of the foot while the ankle is plantar-flexed (fig. 10.13). Flight angles and spin are both determined at the point of contact. In this maneuver the punter kicks the ball through a high trajectory with an extreme slant, causing stalling at the peak of the trajectory. The fact that velocity decreases almost to zero during the upper portion of flight gives the football a maximum amount of time in the air. This "hang time" is used for punt coverage by members of the kicking team.

Figure 10.13 Punt trajectory is partially determined by the foot angle at impact. The punt follow-through phase is demonstrated by Dave Boulware, former USC punter. *Courtesy USC Athletic News Service*—Robert Parker.

If the spin of a round ball moving through air is not a perfect backspin, the direction of the lift force will not be entirely vertical. For example, a spin about a vertical axis causes a lift horizontal to the left or right. This results in a curved path of the ball during flight. This is the basic spin used by baseball and softball pitchers who throw curves.

In order to impart this horizontal spin, the right-handed baseball pitcher throwing through the high diagonal plane utilizes a combination of lateral rotation of the shoulder joint, elbow extension, supination and ulnar flexion, together with an off-center finger pressure as the ball leaves the middle finger. For the right-handed pitcher, this imparts a clockwise spin, causing the ball to curve away from a right-handed batter. For the left-handed pitcher, the opposite direction of spin imparted at release results in the opposite motion of the curve. *The curvature of the ball is in the same direction as the motion imparted to the front of the ball by the spin.* A complicated relationship between velocity of the ball and the rate of spin accounts for the change-of-pace, slider, and slow curve thrown by many pitchers.

A fastball pitcher with an extreme backspin will sometimes gain enough lift to cause an upward curvature of the trajectory as the ball reaches the plate. This desired trajectory modification is the so-called "hop." In softball, it is quite easy to give an extreme forward spin to a ball, resulting in the pitch known as the "downer" or "drop." In all these maneuvers, the proper relationship between the orientation of the seams of the ball and the axis of spin provides an optimal effect both in stabilizing the ball and in maximizing interaction with air.

Golf balls are purposely designed to have large lift coefficients during flight. The strong backspin provided during a drive causes a lift much larger than the weight of the ball during the early stages of flight. This leads to the characteristic flight pattern of a well-hit drive. However, if the ball is given a horizontal component of spin, the forces of interaction will combine with the long flight path to produce a significant deviation from the intended direction of flight. These are the "hooks" and "slices" sometimes experienced by golfers. The techniques of a perfect golf swing are designed to give the desired velocities and spin to the ball without introducing these unwanted components of spin.

A body in flight, due to its lack of contact with the ground and frictional forces, is not in a position where stability can be maintained easily. A small imbalance between forces either to air or internal body motions can cause drastic changes in either the direction of flight or the orientation of the body. Unbalanced pressures due to the motion of air cause torques about the center of mass and may also change flight direction. Examples of this lack of stability are observed in the responses of sport objects to sudden gusts of wind or turbulence during a long flight. A discus or javelin quite often will oscillate when a wind gust hits. Throwing a football accurately on an open field in gusty wind is difficult. This is especially true if the object itself does not have considerable spin to maintain a strong angular momentum. Probably the best-known example of erratic motions for a nonspinning object is the behavior of the knuckle ball in baseball and softball. Very slight changes in pressure, either from the characteristic of air into which the nonspinning ball is moving or from turbulence forming behind the ball, can cause sudden deviations from the path of motion. Similar types of flight deviations may be developed by the application of saliva or Vaseline to a ball.

Because of the size and construction of the human body, and the low density of air, the lift forces from air are incapable of greatly changing the motion of a performer who is not in contact with the ground. An exception to this rule is the case of skydiving where the relative velocity between the performer and the air is three to five times greater than that experienced in any other sport event. This means that forces acting on a sky-diver are ten to thirty times greater than those working against an athlete diving from a ten-meter platform. Movements of the platform diver are not designed to interact with air in order to elicit changes in position or velocity. However, the diver will conserve angular momentum to provide specific body orientations and rotations. The center of mass of the platform diver will follow a parabolic trajectory predetermined by the forces exerted against the platform by the diver.

Combined Water and Air Interaction

Certain sport applications, such as sailing, wind surfing, and surfing, involve an interaction with both air and water. The motion of the performer and craft is the result of a large number of forces partly interacting to assist, but also to counteract one another. In sailing, for example, the primary force used to drive the boat is the force of the moving air or wind. The setting of the sail is designed to select and utilize a component of this force moving in or near the direction the sailor wishes to travel. The angle of the keel and the rudder is set to take advantage of the combination of wind force on the sail and counterforce in the water to provide the desired motion.

Figure 10.14 indicates a typical situation in small-craft sailing. With the boat pointed at an angle 30° into the wind, the sails are close-hauled to give a force of about 180 pounds in a direction 15° forward of a line drawn amidships. This force will then have components of 46.6 pounds in the direction of motion and 174 pounds in a sideways direction. Because of the design of the hull and keel, the drag coefficient is high for such sidewise motion, and the 174-pound force causes a drift in that direction of only 0.75 ft/sec. In contrast, the design of the boat is such that the 46.6-pound component will not be balanced by drag until the boat achieves a velocity of 8.33 ft/sec (5 knots) in the forward direction. These two components of velocity combine to give a total motion of 5 knots in a direction only 5° away from the direction the boat is pointed. As a result, the boat achieves a speed of 5 knots in a direction 25° into the wind.

It should be emphasized that this is only one set of typical numerical components, and that the actual motion depends on a great number of factors such as design, wind velocity, and water conditions. The angles at which the sails can be set are limited to those that do not cause stalling. These, in turn, restrict the most efficient angles at which the boat can be pointed. Moreover, wave conditions and angle of tilt of the boat cause great variation in the drag coefficients and in the final result of the wind force. The expert sailor will learn to match the known performance characteristics of the craft with the existing weather and sea conditions in order to achieve optimal results with minimal risk of capsizing or stalling.

A phenomenon of interest in sailing is the difference between sailing downwind and sailing crosswind. In sailing downwind, the maximum possible velocity of the boat is slightly below the velocity of wind. As the craft approaches wind velocity, the difference in velocity between boat and air becomes so slight that the force is balanced out by the drag of the boat moving through water. On the other hand, when sailing crosswind it is often possible for the boat to reach velocities greater than that of the prevailing wind. This is due to the fact that as the boat moves crosswind, the downwind component of the velocity with respect to the sail remains a constant regardless of how fast the boat is moving.

The ultimate in the use of the aforementioned concept in sport lies in the techniques used in iceboating where a wind of 10–15 mph is capable of propelling the iceboat at speeds near 100 mph. The iceboat actually has an advantage over a sailboat because forward frictional forces in iceboating are almost zero, regardless of the velocity, and sidewise drag or frictional forces are nearly infinite, i.e., there is very little resistance to forward motion, and it is almost impossible to slide the iceboat sideways. In the case of a boat moving through water, however, there are drag forces which increase as the cube of the forward velocity. The fact that the keel does not exert a perfect drag means that there is a sidewise or leeward drift downwind against the action of the keel.

If the craft in figure 10.14 were an iceboat instead of a sailboat, the 174-pound component of force would be countered by a sidewise force of the runners against the ice, with no component of motion in that direction. Also, the forward

Figure 10.14 Force and velocity components in the sailing example.

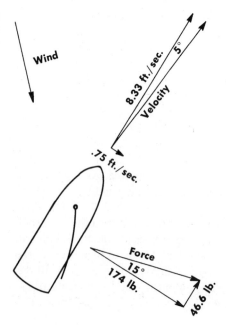

Fluid Mechanics

velocity might reach 75 ft/sec before the 46.6-pound component was overcome by the drag caused by the air. The iceboat would be moving at 75 ft/sec (51 miles per hour) in the same direction it was pointed. However, wind conditions and stability limit the possible sail settings, and the condition of the ice causes great variation in possible performance alternatives.

Another problem in relation to forces affecting a sailboat arises from the fact that the sail force is impacting above the surface of water, while the keel force is exerted below the surface. This means that the boat is always acting under the influence of a torque about its longitudinal axis. Improper setting of sails, especially on small centerboard craft, can very quickly allow this torque to tip the boat. While racing small craft, the person on board, in order to allow the use of a large component force from the wind without tipping the boat, at times will move out to the side as far as possible to provide a gravitational countertorque (fig. 10.15).

Force due to air on a sail is a combination of a high-pressure region behind the sail caused by an actual deflection of wind, the so-called "jet effect," and a low-pressure area created in front of the sail by the passage of wind around a shape resembling an airplane wing. This is another example of the Bernoulli Effect. The total surface area and shape factors of the sail are used to give maximum impetus in the desired direction, with minimum resistive effects in the various forms of drag. The decisive effect of the Bernoulli portion of the force on the sail is illustrated by the behavior when a sail stalls. Tilting a sail at too small an angle with respect to wind tends to lose the wing shape. The sailcloth then forms a series of ripples. At this point, the effective force on the sail drops nearly to zero. Maneuvers must be made to refill the sail with air before effective motion can be resumed. The possibility of stalling is greatly enhanced if the wind is gusty rather than steady. Therefore, with the same average wind speed, a steady wind drives a boat much more efficiently than a gusty one.

Figure 10.15
Providing a gravitational countertorque during sailing.

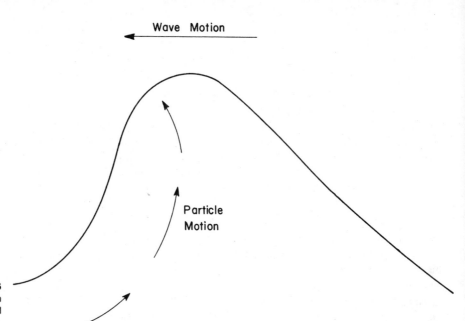

Wave Motion

Particle
Motion

Figure 10.16
Observed wave motion
compared with actual
motion of water particles.

In the sport of surfing, the force affecting the motion arises from an inter-action of gravitational and water forces. The wave on which the surfer is riding consists of water moving toward the surfboard. While it appears that the wave is rushing toward the shore, the exchange of water composing the wave means that the water particles are moving upward (figure 10.16). The surfer, therefore, is actually sliding downhill on a water surface which is moving upward with enough velocity to maintain the total distance above the normal water level. Skilled performers maintain this balance by various tactics such as sliding diagonally along the wave, lifting or dropping the nose of the surfboard, or by maneuvers involving a shift of body weight on the board.

Stabilization with respect to direction is usually provided by the drag of a very small keel located near the tail of the surfboard. One good illustration of the upward velocity of the water on which the surfer is riding is a surfer's "wipe-out." After the performer leaves the board, the board is normally thrown high in the air by the upward action of the water.

As the surfboard is moving at a relatively high velocity with respect to air, there are drag forces between the surfer's body and air. Arm and upper-body motions are designed to counteract this force as well as to maintain equilibrium during the various maneuvers on top of the board.

The velocity of a surfer with respect to water can be increased by moving the center of gravity forward on the board. This causes it to tilt more nearly parallel with the surface of the wave. The surfer then gains speed and literally "moves downhill." The surfer moves toward the shoreline by taking a diagonal path along the front surface of the breaking wave (moving "in the tube!"). Near the end of the ride, there is a shift of weight farther back on the board, causing the nose of the board to rise. This increases the drag coefficient and causes the surfer to move up and over the back of the breaking wave (fig. 10.17).

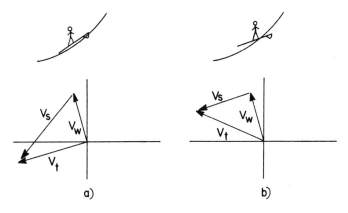

Figure 10.17
Comparison of technique for moving downward (a) or upward (b) on a wave.

Counsilman, James E. *The Science of Swimming*. Englewood Cliffs, NJ: Prentice-Hall, 1968.

Hertel, Heinrich. *Structure-Form-Movement*. NY: Reinhold Publishing Corp., 1966.

Marchaj, Czeslaw A. *Aero-Hydrodynamics of Sailing*. New York: Dodd, Mead, 1980.

National Committee for Fluid Mechanics Films. *Illustrated Experiments in Fluid Mechanics*. Cambridge, Mass: The MIT Press, 1972.

Proctor, Ian. *Sailing Strategy: Wind and Current*. London: Adland Coles, Ltd., 1977.

Recommended Readings

11

Principles of Biomechanics

In the preceding chapters of this section on the physics of sport, a basic mathematic-physics approach is utilized to present the principles of biomechanics. There are students who do not readily relate to concepts when numbers are used as the primary symbols to present ideas. In fact, numbers have a tendency to place some students in a state of "symbol shock." When a person is in this condition, communication of ideas is exceedingly difficult. In such cases, other symbols are often substituted to transmit or communicate ideas.

The purpose of this chapter is to restate many of the principles discussed in chapters 8 through 10, but the principles will be expressed with words in order to communicate the basic principles of biomechanics to the reader. The principles are stated briefly, and each is preceded by a short "recall phrase" which should assist the reader in remembering the principle, once it has been comprehended beyond the rote memory level. This comprehension is greatly enhanced by a thorough understanding of material already presented in chapters 8 through 10. *Readers wishing a more detailed explanation of any term should consult the index for references to the pages of expanded discussion.* This chapter also serves as a glossary of terms.

A physical educator must understand these biomechanic principles in order to conduct noncinematographic and basic cinematographic analyses. These are the most common forms of biomechanic analysis. The following are the principles most frequently confronted in the areas of exercise, sport, and dance. These are the principles which performers ordinarily utilize, whether effectively or ineffectively, while performing in athletics. Therefore, numerous examples from athletics are presented to clarify the implications for performances when these principles of biomechanics are utilized.

Velocity
Specific Velocity

Most sport actions involve the optimization of velocity of a specific body segment, sport instrument, or sport object.

This statement implies that for any given sport action there is a particular component of the motion that needs to be optimized. Any extraneous motion which detracts from the desired effect of this specific movement represents a lost or impeding effort on the part of the athlete. One of the characteristics which distinguishes the beginner from the championship-class performer is the appearance of extraneous motions that do not contribute to attaining the desired velocity of the object or limb. It is imperative that the professional physical educator should recognize which are the important velocities in a given event or performance. Thus he or she can aid the performer in optimizing these velocities. For example, in the javelin throw the specific velocity of the throwing limb is the critical factor determining the distance which the javelin will ultimately travel. All motions, starting with the toes and ending with the fingers, must be carried out in such a way as to maximize throwing-limb velocity. Any motion tending to diminish this velocity also reduces the ultimate horizontal distance covered by the javelin. *The velocity of an object, implement, or body segment is the result of several velocities measured relative to one another.*

In almost all sport motions, the final motion of a limb or of a sport implement or object is the result of the summation of several of these interrelated velocities.

Summation of Velocities

In many sport actions, a whole body motion is added to the velocity of a body segment relative to the whole body in order to produce the final specific velocity desired. As an example, in the basketball jump shot the velocity of the performer going into the air is the result of the summation of an ankle plantar flexion, knee extension, and hip extension (fig. 11.1). The maximum height obtained is directly

Figure 11.1
Summation of lower limb joint velocities to achieve maximum height for the jump shot by Bruce Clark, USC forward. *Courtesy USC Athletic News Service.*

dependent on the summation of these three velocities. The jump shooter who fails to utilize the best possible summation of these velocities will not be able consistently to clear the defensive player with his shot.

Sequential Velocity

The optimizing of a specific velocity is achieved by the proper application of the time at which each component velocity reaches its greatest contribution in sequence to the final performance.

The optimal results often will result from the application of joint movements critically timed to the nearest thousandth of a second. These motions may occur simultaneously or sequentially, and more often occur in an overlapping type of sequence. For example, in the *long* jump shot in basketball, the motion of the arm and shoulder, and especially the forceful extension of the elbow, must be initiated at the proper moment of the flight upward in order to have the ball released at or slightly after the peak of the jump (fig. 11.1). This will lead to optimizing the final key movement of the wrist joint (flexion) at the highest possible position and with maximum force against the ball. Any deviation in this sequence of motion will lead to a loss in accuracy, height of release, or velocity in this final wrist action.

Angular Velocity

Because of the configuration of the human body, almost all sport actions are the result of angular velocities about joints by limbs or body segments.

Even in such apparently linear motions as running or sprinting, the primary actions are angular velocities about the ankles, knees, and hips. As performers begin to recognize the combination of flexions, extensions, and rotations at the joints, they can better correlate the desired results with the kinesthetic sense of motion. Frequently the maximizing of specific velocity is actually dependent upon maximal angular velocity or rotation of body segments at joints. As an example, the discus thrower tries to attain a maximum angular velocity at the shoulder in terms of diagonal adduction at that joint (fig. 11.2).

Radius of Movement

Linear velocities normally result from an angular motion multiplied by the radius of movement or distance from the axis of rotation.

Continuing with the above discus example, the maximum range of the discus requires a maximum linear velocity at the time of release. This linear velocity results from angular velocity at the shoulder (mentioned previously), multiplied by the radius of movement which is maximized by the full extension at the elbow and wrist joints. This maximum radius of movement, combined with maximum angular velocity, provides an optimal linear velocity to the discus and assures the greatest potential range of flight.

Acceleration

Positive Acceleration

An increase in speed of an object or body segment results from an acceleration in the same direction as the velocity.

Negative Acceleration

A decrease in speed of an object or body part results from an acceleration in the opposite direction from the velocity.

Principles of Biomechanics

Figure 11.2 The point of maximum angular velocity of the throwing limb of Ludvik Danek, Czechoslovakian discus thrower. *Courtesy Visual Track and Field Techniques, 292 So. La Cienaga Blvd., Beverly Hills, CA 90211.*

A change of direction of motion results from an acceleration perpendicular to the direction of velocity.

Change of Direction

The ability to distinguish the relationship between direction of acceleration and its effect on total motion is most important in applied biomechanics. These differences are exemplified in the movement of a basketball guard working against the various players on the defensive team. In the fast break he or she may increase speed very rapidly. At times the guard may stop or slow down quite suddenly to allow a defensive player to pass by, and at other times may suddenly change direction in order to throw an opposing player off balance.

At times the unwanted existence of a negative acceleration may lead to a diminishing of athletic performance. For example, a long jumper who either adjusts his/her stride near the end of the run, or who plants the foot very strongly in leaving the board, tends to negatively accelerate with respect to horizontal motion. This results in traveling forward more slowly through the trajectory. The result of this negative acceleration is decreased total distance for the long jump.

It should be emphasized that the use of acceleration to decrease or modify motion is often fully as important as is the maximizing of speed. In the case of a pulling guard preparing to run interference in American football, the initial acceleration will be utilized to provide maximum velocity for the athlete to get into a position where the play is developing. After reaching this position, however, it is often desirable that the guard slow down in order to be in the most effective position to block the defensive player approaching the ball carrier. Maximum velocity at this point will often carry the guard past the position where the block should be made.

Sequential Acceleration

Most movements of body parts or sport implements are the result of a timed sequence of individual accelerations.

The importance of combining a properly timed sequence of velocities in order to obtain a maximized specific velocity has been emphasized. Likewise, in the development of the stereotype of perfect mechanics for many sport actions, the proper sequencing of change of velocity or acceleration is highly important. Again, this involves the starting or stopping of accelerations relative to one another, within a few thousandths of a second. An example of such a sequential acceleration occurs in the sport of baseball when players try to stretch a single into a double. As the runner approaches first base, it is necessary to decrease velocity with a negative acceleration in order to minimize the radius of curvature. Once the turn is made, velocity must be increased toward second base. It has been found that the optimal sequencing of these accelerations can be performed best if the player attempts to hit the corner of the bag with the left foot on rounding first base. An example of an improper sequence occurs when players attempt to jump to reach the first-base bag on a close play. The very act of planting the foot to jump decreases horizontal velocity. This means that the actual time necessary to reach the base is longer than if the athlete had simply run across the bag (fig. 11.3).

Figure 11.3 The result of proper sequencing of accelerations while running to first base— Rod Towe, SMS baseball player. *Courtesy SMS Public Information Office.*

Most sport actions are influenced by constant gravitational acceleration.

Since most human body actions take place at or near the surface of the earth, the downward attraction of gravity is a constant factor that must always be considered. Weight is the component associated with this acceleration. The motion of any body in flight will follow a trajectory determined primarily by gravitational acceleration. Muscular forces used in all jumping events are designed to counteract this acceleration of gravity, in order to maintain optimal times or distances in the air. In interaction with fluids, buoyancy and/or lift forces in flight are necessary to counteract gravitational acceleration. Knowledge of gravitational accelerations can lead to optimal angles of projection in such events as the discus, javelin, and shot put. On the other hand, too much concern for overcoming gravitational acceleration can be detrimental to a performance. For example, a long jumper may sacrifice considerable horizontal velocity in a firm planting of the foot in order to obtain maximum lift at takeoff from the board. This negative horizontal acceleration results in a loss of total linear distance for the jump.

Every acceleration is associated with an unbalanced external force acting on the object. Acceleration is in the same direction as the force and is proportional to the force.

This principle relates the observed phenomena of acceleration and velocity to the kinesthetically felt phenomena of forces. In the analysis of sport skills, it is necessary to be able to identify, describe, and communicate the movements as observed, either noncinematographically or cinematographically, in terms of forces kinesthetically felt by the performer. In limb actions requiring considerable ranges of motion, the performer is unable to determine visually limb positions, velocities, and accelerations. Therefore, the performer is dependent on the sensory inputs developed by the proprioceptors located in muscles, tendons, and joints. For example, in the butterfly stroke in swimming the concurrent circumduction of both shoulder joints carries the arms well past the limit of visual contact, so one phase of training in this event must be the development of the proper kinesthetic feeling which accompanies the optimal range of motion of the arms.

One reason for the extreme importance of this principle is that the direction of applied force is a performer-controlled variable of the action. *In order to produce optimal motion in the performance, force must be applied in the intended direction.* To return to the example of the butterfly stroke, circumductions at both shoulder joints must be concurrent and symmetrical in order to achieve the best motion through the water, i.e., the forces must be applied with the same strength in the same direction with both arms at the same time. Failure to do this leads to extraneous body motion and to a variable path in the water. This results in a longer time for the total distance covered during the race and to greater drag forces on the body of the swimmer.

The total effect of force on motion of a body is the product of the magnitude of force and the time during which it operates.

This principle has been discussed earlier as the term *impulse,* in connection with changes of momentum. It is essential to producing maximum motion or changes of motion. It is not only important to generate optimal forces; it is also necessary to hold them for a maximum interval of time.

This principle is closely associated with the stretch reflex and the muscle length-force ratio, which states that the force capability of a muscle increases as it is stretched to greater length. There is, therefore, a strong enhancement of the force-time effect developed by the increased range of motion in a given joint. This increased range of motion not only leads to the greater internal force being generated by the muscle groups most involved, but it also allows this force to be exerted over a greater time span. For example, during a baseball pitcher's preparatory diagonal abduction at the shoulder joint, the diagonal adductors that will be used immediately following diagonal abduction are placed on great stretch if the pitcher has gone through a complete range of motion for diagonal abduction. This increase in range of motion for diagonal abduction means that the pitcher will have more time for the subsequent diagonal adduction prior to release of the ball. This is where there is a critical enhancement relationship between the muscle length-force ratio principle and the force-time principle of biomechanics. Anything less than a full range of motion will lead to a subsequent decrease in the release velocity of the ball. This interrelationship between the force-time product of biomechanics and the stretch principle from anatomic kinesiology is of utmost importance in all ballistic actions.

Force-
Counterforce

Whenever an object or body segment exerts a force on another object or body segment, the latter reacts with a force equal in magnitude and opposite in direction.

This principle applies in the case of a performer exerting forces against an instrument or object or against another performer, or where one part of the body is being used to exert forces against another segment. This application of force and counterforce is a basic ingredient of combative sports like boxing, wrestling, or judo. In wrestling, for example, when one performer exerts a substantial force against his opponent, he must be prepared, either through his body position and balance or his contact with the mat, to receive and nullify the effects of the equal and opposite counterforce exerted on him (fig. 11.4).

In weight lifting, the force which the lifter is exerting against the weight in order to raise it above the head produces a counterforce against muscles of his or her body. This counterforce must be controlled while moving the weights to the floor in order to maintain equilibrium. In an unsuccessful lift, often the lifter will attempt to push the weights forward. This results in a counterforce which moves the body backward, and sometimes results in a fall or backward roll by the athlete.

Internal Forces

In the action of one body segment against another, the effects of both force and counterforce will be apparent in motions of the body.

The principle of internal forces is of utmost importance in determining the relative positions and velocities of body parts during athletic performances. Every

Figure 11.4 Force-counterforce by wrestlers during a match. *Courtesy SMS Public Information Office.*

muscle has two attachments to enable it to pull (contract) on body segments. Therefore, in biomechanics one must consider muscular contractions from both force and counterforce points of view. These should be carefully observed when dealing with a body in flight, since it is impossible to negate effects of counterforces by contact with the ground. As an example of the use of forces and counterforces during flight, consider the competitive diver who is capable of transferring from a twisting motion to a tumbling motion by making the desired arm movements. In this case, forces used to move the arms also provide the desired change of motion of the body by their counterforces.

The principle of internal forces is also critical when external forces are being used to help propel the body. As an example, during running the actual external force used to propel the runner is the counterforce of the ground against the feet. However, a complex and sophisticated set of internal forces are used by the experienced runner in order to make the most efficient use of this counterforce by eliminating extraneous body motions and controlling equilibrium. In essence, the skill of running includes the utilization of internal forces to allow the most efficient use of external forces through an optimal series of alternate equilibrium losses and gains. This timing is especially critical in distance races, because muscles require relaxation time between the contractions necessary for ballistic motions of lower limbs.

Concentric-Eccentric Forces

In the human body, muscular forces can be used both to cause motion (concentric contraction) and to control motion or absorb external forces (eccentric contraction).

Concentric or shortening contraction of a muscle group is the *causative force* for all body motions against gravity or resistance. This action is used to produce greater motion or to add kinetic energy to the moving body. Eccentric or lengthening contractions, on the other hand, provide the *controlling force* for a body moving with gravity or receiving other external forces. These are used for a decrease of total motion or an absorption of kinetic energy. Eccentric control of a joint in motion is used both to control range of motion and to limit speed of motion. In this sense, eccentric control is a built-in safety mechanism of the body. As an example, in the sit-up exercise involving lumbar flexion, the abdominal muscles are the muscles most involved. They contract concentrically during the first 90° of the exercise. As the trunk is lowered slowly during lumbar extension with gravity, the abdominal muscles are contracting eccentrically in order to regulate both the extent of motion and the speed. While in the position of 90° of lumbar flexion, the force of gravity by itself is capable of causing a return to the supine position through lumbar extension. This action without the control of eccentric contraction by the abdominals could cause injury in some cases.

Summation of Forces

The effect of more than one force acting on a body can be found by the summing of forces, taking into account the direction of each.

This vector or directional addition of forces is apparent in situations where both a gravitational acceleration and another desired acceleration are present. The need for both vertical and horizontal components of force results in a forward leaning of the body for a runner who is accelerating, and a leaning to the side for a runner changing direction of motion. This principle is also seen in cycling (fig. 11.5), auto racing, bobsledding, and lugeing. In such activities, the performer has the option of remaining low on a banked turn with decreased velocity or maintaining a high speed and going to the upper parts of the bank. While the performer moving higher on the bank travels somewhat farther, the higher average velocity often leads to a minimum time around the curve. In these examples, the combination of the force of gravity downward with the force of the earth or ice upward and inward produces a net result of an inward centripetal force, giving the desired acceleration for change of direction.

Figure 11.5 Control of centripetal force on an inclined surface during cycling by Mark Logan.

Principles of Biomechanics

The term *summation of forces* is also applied to the sequencing of internal forces in order to provide an optimal motion for the performer or athlete. Whenever this sequence is interrupted, even for a period of a few microseconds, it may lead to a less efficient total motion for the athlete. This will produce diminished results for the performance, particularly in ballistic skills.

Equilibrium of a body or object is obtained when each force acting on the body is balanced by an equal but oppositely directed force or force component.

Equilibrium

The condition of equilibrium in a body is a situation in which motion of the body remains unchanged, i.e., a stationary body remains stationary or a moving body continues moving at the same velocity in a straight line. The condition of zero acceleration implies a zero net force (fig. 11.6). This does not mean that there is no force acting on the body, but rather, that each force is being balanced by another equal but oppositely directed force, as illustrated by a gymnast in static equilibrium on parallel bars. Another example of this situation is the sailing of a boat crosswind at a constant velocity. In this case, the cross component of wind is balanced by an equal but opposite force of water against the keel of the boat. Meanwhile, the forward component of wind is balancing the drag of water. This maintains a constant linear velocity for the boat. It is necessary for the sailor to maintain the proper sail setting in terms of both angle and tension in order to keep these forces in balance and to continue moving in a straight line down the course.

Figure 11.6 Lefty Adams, a noted bear wrestler, demonstrates static equilibrium in his "casual stance" designed specifically to give the bear a false sense of security before Lefty summates his forces.

Pressure *Since it is pressure and not total force which often causes pain or tissue damage, the application of force should be spread over as wide an area of the body or body part as possible.*

Pressure is a term used to denote force per unit area. High pressure can result from an extremely large force or from a small or moderate force concentrated on a small area of the body. It is therefore a useful rule to spread forces as widely as possible in order to prevent tissue damage. A classic example of this principle in sport lies in the falling techniques used in judo. The judoka lands on the mat with as much body surface area as possible to dissipate the force. A properly executed baseball slide utilizes the same principle (fig. 11.7). This is also the reason that a small rubber tip is placed on the foil in fencing. Although small in surface area, this increases the area of the foil and prevents actual stabbing or penetration. This is a significant principle considered by manufacturers of athletic equipment. For example, baseball gloves, football pads, and other items of this type need to be designed to dissipate forces over as large an area as possible. Changes in design of the catcher's mitt, as one example, have provided a greater surface area over which the force of the pitch is absorbed.

Weight and Mass *Weight is the downward force of gravity acting on a body. Mass represents the total amount of matter in a body and its inertial resistance to changes of motion in any direction.*

There is often considerable confusion and uncertainty regarding use of the terms *weight* and *mass. The term "weight" actually refers to a force, the force*

Figure 11.7
Dissipation of force to minimize pressure during a slide by Bill Helfrecht, SMS all-American baseball player. *Courtesy SMS Public Information Office.*

Principles of Biomechanics

with which the earth attracts the body (pounds). The term "mass" is used in evaluating the amount of force necessary to cause a change of motion of a body, i.e., the larger the mass, the larger the amount of force necessary to cause the same change. Although these terms differ greatly in meaning, the distinction is not exceptionally important in sport actions, since these normally take place at the surface of the earth where weight and mass will always be proportional, i.e., an object that weighs more than another object will always have a greater mass. However, as previously discussed, it is necessary when making numerical calculations to express both weight and mass in their proper units. In gravitational fields other than those found at the surface of the earth, the mass-weight relationship may change drastically.

External forces alter the location of the center of gravity of a body, while internal forces alter the motion of various body parts with reference to that center of gravity.

Center of Gravity

The center of gravity of a body is a theoretical point at which the weight of the body could be located to produce the observed biomechanic effects during motion and balance. For the human body in the anatomic position, the center of gravity is located near the umbilical region. However, as the limbs are moved through various planes at shoulder and hip joints or during spinal column motions, the center of gravity may move outside the region occupied by body segments. This is the situation occurring during leaping, diving, and many gymnastic movements.

In the analysis of motion of a performer it is necessary to consider the motion of the center of gravity as well as the motion of various body segments with reference to the center of gravity of the body. For example, the experienced hurdler uses hip, shoulder, and lumbar motions to keep the center of gravity as low as possible during the flight phase to minimize vertical motion. In contrast, the inexperienced hurdler who "jumps" the hurdles experiences a large amount of vertical motion of the center of gravity. This necessarily requires a slower average horizontal velocity, and demands a longer time to complete the hurdle race. During events where a performer is in the air, the center of gravity moves through the gravitational parabolic trajectory, while the internal forces (muscle contractions) control orientation and rotation about the center of gravity.

The frictional force between two surfaces is always in a direction parallel to the plane of contact of the surfaces. Its magnitude depends on the materials involved, the smoothness of the surfaces, and the force pressing the surfaces together.

Frictional Forces

One of the decisive forces affecting sport activities is frictional force. Through the action of friction on the surface of the ground all running and jumping events acquire their horizontal velocity. Friction is also the main factor to consider when analyzing grips on sport implements. Negative aspects of frictional losses tend to be emphasized. Consequently, there are many examples in sport where frictional force needs to be maximized to achieve an optimal performance. Since the available force of friction depends on the force pressing the surfaces together, there is an optimal angle at which a force should be exerted against a surface in order to obtain the greatest component of thrust parallel to that surface. If, for example, a runner plantar-flexes slightly and pushes against the ground with too shallow an angle, the component of force holding the foot to the ground is not great enough to allow a large value of friction. The foot will slip, and summated motion will be interrupted. On the other hand, if a runner plantar-flexes too much during the downward-thrust motion of the lower limb, then the athlete does not develop a large enough horizontal component of force to give the needed acceleration.

In any sport situation where a spin is to be imparted to a sport object, there must be friction at work between the object and the hands of the performer (fig. 11.8). Resin and other compounds are often used to furnish a higher coefficient of friction and thus prevent slippage during the execution of numerous throwing and striking skills.

The principles involving friction have special meaning for people who design equipment and facilities for use in sport. Both the material used and the surface condition can be varied in order to increase or decrease available frictional forces. As one example, the frictional coefficient is especially important to the owner or manager of a bowling facility. In the first place, the material for surfacing the lanes has to be one that provides a relatively low coefficient of friction between the ball and the floor, but provides a high coefficient of friction between the bowler's shoes and the floor. Secondly, it is extremely important to experienced bowlers that the lane be consistent both from one side to the other within a single lane, and as one moves from one lane to another. If one lane tends to have a high coefficient of friction while another has a low coefficient, the bowler cannot adjust

Figure 11.8
Frictional force is an important factor in gripping a sport object. This is demonstrated by Pat Haden, former USC quarterback, as he grips the football while looking for a pass receiver downfield. *Courtesy USC Athletic News Service.*

Principles of Biomechanics

his lift on the ball in order to give it the proper curve for obtaining strikes. It might appear to the spectator that the bowler is being inconsistent, while the inconsistency actually lies in the differences of the frictional coefficients of the lanes. Learning on such lanes would also be difficult and inefficient.

A change in the direction of motion of a moving body requires a force directed inward toward the center of curvature of the path.

Since a change of direction of a moving body represents an acceleration, the force associated with this acceleration depends on the mass of the body and on the velocity with which it is moving. *This inward force is known as centripetal force and increases as the mass of the body increases and as the square of the velocity.* A body moving twice as fast requires four times as much centripetal force. *The equal and opposite reaction on the object or body providing centripetal force is known as centrifugal force.* These forces may become decisive where a ballistic rotary motion of a sport implement is involved such as in tennis, baseball, jai-alai, hammer and discus throwing, and golf. It should be recognized that these forces are present, so the performer can counter them in order to maintain body stability during rotary ballistic motions. In this context, foot placement is of extreme importance in providing the countering force against the ground to maintain body stability.

Centripetal-
Centrifugal Forces

The rotational effect of a force is directly proportional to its distance from the axis of rotation.

It should be recognized that forces not only provide motion of the center of gravity of a body, but also cause rotations of a body about some axis. Muscle attachments at major joints of the human body are arranged to provide increased range or velocity of motion, rather than to multiply force exerted by muscles. The purpose of many sport implements is to increase this leverage still further. For example, the velocity with which a person can strike a handball can be increased by the addition of a racquet as used in the games of squash and racquetball.

Torque
Lever Arm

Inertia with respect to rotation of a body depends on the total mass involved and the average distance of this mass from the axis of rotation.

In order to provide a rapid rotation of a body or implement, it should be held with the limbs or portions of the implement as close to the axis of rotation as possible. This allows the torque generated either by the feet against the ground or by action of internal forces within the body to produce maximum acceleration and velocity of rotation. As an example, the experienced discus thrower keeps the hand holding the discus located very near the lumbar spine as the rate of body spin in the circle is increased. After the body has attained maximum rotational velocity, the discus is moved away from the body. If the discus thrower were to keep the arms and discus away from the longitudinal axis of the body at all times, a smaller rotational velocity of the body would result, producing a smaller linear velocity at the time of release of the discus. The end result would be a shorter distance of flight.

Moment of Inertia

Torque-Countertorque *Whenever one body exerts a torque on another body or body part, it receives an equal and opposite countertorque.*

This principle applies to rotational motion in the same way that the force-counterforce principle applies to translational or linear motion. A good example of this lies in the kinesiologic construct known as the Serape Effect discussed earlier. The torque exerted on the rib cage in order to provide rotation of the upper body results in a countertorque being exerted on the pelvic girdle. If the pelvic girdle is moving in the direction of this ballistic motion, the angular velocity which it is able to give to the rib cage is greater. This ensures a higher velocity for the ultimate ballistic motion. If the pelvic girdle is stationary at the beginning of the ballistic motion, it will recoil and will provide a smaller ultimate velocity to upper portions of the body.

Equilibrium or Balance *Rotational equilibrium of a body requires that each torque acting on the body be effectively balanced by an equal and opposite countertorque.*

This principle deals with the effect of externally applied torques. The concept of balance in sport actions corresponds to the maintenance of a rotational equilibrium. Normally, this means an equilibrium with zero rotational velocity. This requires that any torque exerted on the body, often due to the effect of gravity acting on an extended limb, must be counteracted by an equal and opposite torque against the base of support. *In the maintenance of equilibrium, it is essential that the center of mass of the body be kept above the area known as the base of support.* The width of a stance is very significant in establishing a broad or substantial base of support and allowing the effect of forces at the feet to have larger lever arms in exerting torque to recover or maintain equilibrium. In some sport actions, this width of stance is not allowed due to either the rules, skill execution, or the equipment used. For example, for a gymnast performing on the balance beam, the limited width of the beam at times negates the possibility of establishing a wide base of support. In this case, internal muscular forces must be used to establish and maintain equilibrium. Loss of equilibrium is normally overcome only by some type of compensating body movement and/or eccentric muscle contraction.

It should be noted that even when equilibrium is lost, there is a time period of about one second before the rotational motion is sufficient to cause a complete fall. During this period of disequilibrium a skilled athlete often is able to adjust limb orientations to complete the attempted sport action and recover equilibrium. For example, a skilled basketball player will frequently make a leap off balance to make a shot. He will use follow-through motions from the shot and other compensatory limb actions to regain equilibrium without falling (fig. 11.9). On the other hand, the unskilled athlete does not possess this ability to recover, and an extreme disequilibrium will almost always result in a fall. For this reason, the athlete of this type constantly strives to maintain a state of near-equilibrium. When limited to conservative actions, a performer often misses the opportunity to complete a successful sport maneuver. Athletes who are conservative and inhibited in their motions rarely become highly skilled.

Figure 11.9
Controlled motion in a state of disequilibrium by Paul Westphal, former USC basketball guard. *Courtesy USC Athletic News Service.*

The transfer of energy from one body or body segment to another is accomplished by means of a force acting through a distance.

While the most familiar method of describing motion is in terms of forces and accelerations, this description can also be expressed in terms of energy transfers. The transfer of energy from an athlete to a sport object or implement, for example, is accomplished by the exertion of a force through a distance of motion of that object or implement. Maximum transfer of energy requires the most efficient use of force-producing motions in the body and the maximum range of motion at the point of contact with the instrument or object. This maximizing of the range of force in a throwing skill is exemplified in the body and limb actions observed during a javelin throw. If the thrower does not go through an extreme range of body motions before release of the javelin, the result will be lower release velocity and a shorter distance for the throw.

The change of speed of an object requires a force acting through a distance, since kinetic energy is dependent on the square of speed.

To increase or decrease the speed of a body segment, implement, or object, work must be performed or absorbed in the process. As indicated in earlier chapters, the term used for the energy associated with motion is *kinetic energy*. This depends on the mass of the object and its speed. One area where applications of this principle are of vital importance is in the stopping of objects in motion. Because of their motion, such bodies possess a large amount of energy. In bringing the object to rest the performer must absorb this energy. As an example, consider the dismount move at the end of a gymnastic routine. The performer coming off the final movement from the parallel bars, uneven parallel bars, horizontal bar, or rings, has a considerable amount of kinetic energy. As the gymnast comes in contact with the mat, he or she must absorb kinetic energy and control body motion by eccentric contraction of the muscle groups within the lower limbs associated with major joints. The extent to which the performer does this without extraneous movements is an important consideration in judging the performance.

Potential Energy

When work is performed while distorting the shape of an object or lifting it against gravity, the effort or work is stored in the object as potential energy.

The term *potential energy* pertains to an object's ability to perform work due either to its position above some base surface or to an elastic distortion of the object itself. As the object or body returns to its base level, or as the distortion disappears, this energy is then converted into some other form. As an example, the sport of competitive diving is highly dependent on the storing of potential energy in the board during the final approach step on the board itself. If the diver fails to store an optimal amount of energy at this point, it cannot be returned in the form of kinetic energy as he or she leaves the springboard. Therefore, the diver will have less total time in the air in which to complete the maneuvers and make a successful entry into the water. Manufacturers of diving boards try to design them so the competitive diver can store a maximum amount of potential energy more easily, and the adjustable fulcrum allows each diver to adapt the leverage to his or her own style and weight.

Energy Transformation

Almost all human body motions represent a series of transformations from one form of energy to another; and the stereotype of perfect mechanics is usually that one in which energy transformations are carried out most smoothly and efficiently.

In the transfer of energy from one form to another it is necessary to do this with greatest efficiency, since lower efficiency means a loss of total energy by the athlete during performance of the event. In the case of ballistic and antigravity events, which are controlled by the final energy available to the body or object, this loss of energy necessarily means a less satisfactory performance. The smoothness of transfer is stressed in order to prevent the dissipation of energy into extraneous motions. One characteristic observable in a beginner who is still learning a sport skill is the presence of extraneous motions. As a result, the beginner may actually expend more energy during the performance of a particular athletic movement than the skilled performer, but with lesser results.

In the human body, the best transfer of energy occurs from large muscle groups in the direction of smaller muscle groups. The stereotype of perfect mechanics for most athletic performances involves an initial motion with the large muscle groups, followed by sequences of motion controlled by smaller muscle groups within the body. This proper sequence of motion is often referred to as the "kinetic chain" associated with performance.

The rate at which energy can be supplied to muscle fibers regulates the possible muscular performances over extended periods of time, and this rate depends upon the physiologic efficiency of the individual.

Energy Expenditure

During an athletic performance involving reciprocal movements, a given set of muscles or muscle groups are not constantly undergoing contraction. Following an expenditure of energy during contraction, there is a reduction of intensity of contraction within the muscle, but motion continues due to momentum. As an example, this alternate contraction-reduction process is repeated during each limb motion in a running action. If the average energy being expended by a muscle group is greater than the rate at which it can be replenished by the oxygen consumption and other factors within the body, the condition known as fatigue will set in. In such activities as middle- or long-distance running, the rate at which muscles can be replenished with energy determines the average velocity with which the race can be run. The training effect tends to increase the rate of delivery of oxygen (energy) to the muscle group. This increases the average velocity and permits a given distance to be run in a shorter time. Thus, the possible energy expenditure is increased by means of the training effect. Conversely, any motion involving excess tension or excess muscular contraction uses energy inefficiently, and thereby decreases the total amount of energy available for the running action. Coaches often refer to this excess tension in terms of the runner being "tight." Indicators are a clenched jaw, elevated shoulder girdle, and excessive flexions in the hands and at the elbow. This extra energy expenditure necessarily means that the average velocity during a race will be lower. Therefore, total time necessary for a given distance will be greater.

Power
Force-Velocity Relationship

Maximum force applied during a ballistic motion may be limited by the rate at which power can be developed within a muscle group.

Scientifically, *power* is defined as work performed per unit time, or the product of force times velocity. Since the availability of power is a limiting factor in many human body performances, it follows that at times maximum force can be administered only at the sacrifice of a maximum velocity of motion. This limitation can often be overcome by the use of a greater range of motion, thus providing a higher limb velocity in many throwing or other ballistic motions as discussed earlier under the force-distance relationship.

Length of Application

Short-interval ballistic actions deliver high power, while long-interval continued actions deliver large amounts of energy.

The power that can be delivered by a muscle group is highly dependent on the time the action must be continued. In situations requiring a very short but intense effort, the quantity delivered is *power*. An example of this would be the situation discussed earlier in weight lifting where an athlete was capable of exerting up to seven horsepower during the snatch lift. The time interval involved in actions of this type would be measured in seconds or fractions of a second.

Other examples of these power-delivering moves are the explosive charge of a football lineman from a stance at the beginning of a play, the jump of a rebounder in basketball, or the starting motions in either swimming or track (fig. 11.10). In most of these applications, the need for high power comes from the necessity to overcome inertia.

When an action requires delivery of larger amounts of energy, there is need for a *less-intense continuous action of longer duration.* For example, in an event such as the rope climb, which was an NCAA gymnastic event many years ago, the total energy necessary was more than could be delivered in a single burst. Therefore, the event required a sustained effort in order to reach the top of the rope. A 160-pound man climbing the 20-foot rope in a time of three seconds represents an average power output of about two horsepower. During greatly extended periods of time such as long-distance running, the average athlete can only perform at about one-half horsepower. The total amount of energy expended during such a run is hundreds of times greater than in the previous examples.

Linear Momentum

Momentum

The total amount of movement carried into a situation by a body or object depends on the mass of the body or object and on its velocity.

The term *momentum* or *quantity of movement* is used to describe the product of the mass of a body or object times its velocity. As discussed earlier in the concept of force, external forces provide changes of momentum of a body. In order to obtain the same inertial results in a sport or athletic performance, a smaller body requires a higher velocity than a larger body. In the impact sports such as American football or ice hockey, a larger, more massive athlete can achieve the same type of blocking results while moving at a lower velocity than a smaller or less massive one.

Principles of Biomechanics

It should be noted that the term *inertia* refers to a resistance to change of motion, i.e., a stationary body can have inertia against being set in motion in any given direction. Or, a body already in motion has inertia against being stopped or diverted in its motion. The stances commonly observed in various sports are designed to allow the performer to overcome static inertia to begin a sport motion. Within the stances, the muscles are placed on stretch in order quickly to apply the large forces necessary to overcome inertia and place the body in motion.

In any two-body interaction, the change of momentum of the individual bodies are equal but opposite in direction.

This principle is simply another facet of the force-counterforce principle discussed earlier. Since the result of an external force is to cause a change of momentum of a body, the result of the counterforce will be to provide an equal and opposite change of momentum. The concept that bodies of large mass will experience smaller changes of velocity in any given interaction than will bodies of smaller mass becomes important. Whenever there is a collision or impact between two performers in a sport, the smaller one will suffer the greater change in velocity.

A moving body tends to maintain its same state of motion except when acted on by an external force.

This conservation of linear momentum is most apparent when an object is not in contact with the ground to enable the external forces to be applied as readily. A body or object in air will maintain the same horizontal momentum and will gain a downward vertical momentum due to gravity unless it can strongly interact with another body or object or with air. For example, a ball after leaving the foot of the kicker will follow its prescribed trajectory through the air, involving conservation of this momentum unless it is touched by an opposing player. Such action would provide an external force and deflect its horizontal momentum into a new direction.

Many of the techniques of footwork are designed to enable the performer to obtain external forces from the ground in order to provide necessary changes of momentum for actions. Experienced athletes will often make a certain action or performance "look easy" because they have learned the necessary footwork to adjust their linear momentum to perform the skill. Beginning athletes, without this ability, often seem to be fighting against their own momentum or to be in a state of disequilibrium. This is one reason for assigning a high priority to footwork fundamentals during the instructional process of most sport skills.

Most efficient sport movements should be viewed in terms of a linear motion of the center of gravity of the body combined with angular rotations of the various body parts about this center of gravity (fig. 11.11).

All motions involving positions of various segments of the total body can be viewed in terms of angular motion about the center of gravity. These motions

Figure 11.11
Combination of linear
motion and angular
rotations by a hitter, Dan
Cook, SMS baseball
player. *Courtesy SMS
Public Information Office.*

are controlled primarily by the internal body forces described earlier in this chapter. A determining aspect in the analysis or observation of human body movement is the orientation of various body parts with respect to the axis of rotation.

An example of linear and angular motions of body segments combined to produce an optimal result occurs during the throwing action through the high diagonal plane of the shoulder, as observed in an outfielder's throw in baseball or the quarterback's pass in football. The linear velocity in this case is being supplied primarily by a diagonal adduction at the shoulder joint. A critical addition to this motion is the medial rotation of the shoulder joint just prior to release of the ball. This action not only adds linear velocity, but also helps impart the decisive spin to the ball. In some unskilled performers, this rotation occurs as a part of the follow-through *after release,* rather than while the hand is still holding the ball so the force transfer can be made. In this case the final motion of the throwing limb may look the same, but the rotational component of motion has not been transferred to produce motion of the ball, since it occurred following release. In the noncinematographic analysis of ballistic movements, it is very difficult to observe this release. In the analysis of ballistic movements, observation of this critical timing of shoulder medial rotation during throwing and striking skills must be done cinematographically.

Action-Reaction   *Any time one body or object changes the angular momentum of another body or object, it receives in turn an equal but opposite reactive change.*

This concept of action and reaction has been discussed in connection with the principles involving forces, torques, centrifugal and centripetal forces, as well as linear momentum. Because of its use by the body to maintain an equilibrium position in many sport actions, this factor is exceedingly important with respect to angular momentum. This principle of biomechanics is fundamental to many of the reflexive movements observed in human performance.

A rotating body tends to maintain the same quantity and direction of angular momentum unless it is acted on by an external torque.

Conservation of Angular Momentum

Conservation with respect to angular momentum has certain connotations which were not present in linear momentum. Since angular momentum depends not only on the total mass of the body, but also on its extension with respect to the axis of rotation, the actual rotational velocity of a body may be adjusted through the conservation of angular momentum. If the body mass is brought closer to the axis of rotation by a series of flexions, the total moment of inertia will be decreased, since the average radius from the axis to body segments is less. The principle of conservation of angular momentum then specifies that the angular velocity will increase. On the other hand, if the body is extended so the mass is located farther from the axis of rotation, conservation of angular momentum will cause a decrease in rotational velocity. As an example of this effect, during a tennis serve or overhand smash involving diagonal abduction and diagonal adduction (rotary action) of the racket limb through the high diagonal plane, a flexion of the elbow decreases the average radius and increases the velocity of the racket at impact and of the tennis ball during flight (figure 11.12). A similar effect occurs in the case of knee flexion used in sprinting. Sprinters utilize extreme knee flexions to allow the lower limbs to be brought through an anteroposterior plane of motion with a higher rotational velocity at the hip.

Effects of conservation of angular momentum are often apparent in whole body motions as well as limb motions. For example, a diver leaving the board with a "twisting action" has a much higher angular velocity during the "twist phase" while rotating about the longitudinal axis of the body. The degree of difficulty of a dive is judged on the basis of the ability to generate and conserve angular momentum. For example, a dive in a full layout position involving rotation(s) of the total body is much more difficult than the same dive performed in a pike or tuck position, because of the larger total angular momentum required to make a set number of rotations during the time of flight.

The magnitude of forces involved in an impact varies inversely with the amount of time of contact between the two colliding bodies.

Collisions
Impact Time

When two bodies collide, producing a change of momentum for both, the product of force and time is called the impulse of collision (fig. 11.13). Whether it is a short- or long-time action this impulse will vary over only a small range for most collisions. Therefore, for very short-time actions, the total forces involved must reach much higher levels of magnitude. To minimize the possibilities of injury during a collision either with another body or with the ground, impact time should be as long as possible to keep forces at a minimum. As an example,

Figure 11.12
Conserving angular
momentum within the
racket limb prior to
impact increases limb
velocity. *Courtesy USC
Athletic News Service.*

football players are taught to land on the ground using a shoulder roll, in which case their motion is dissipated over a longer time than if they were to simply land on the acromion process of the shoulder. The latter impact would probably result in a dislocation of the acromioclavicular joint and/or sternoclavicular joint.

Rebound Angles *For an object bounding off a surface with no spin, the angle of rebound will be symmetrical to the angle of incidence.*

During a collision from a surface where spin is not involved, the component of momentum parallel to the surface will remain approximately constant, and only the component perpendicular to the surface will be reversed. An example of this type of collision is the throwing of the ball against the backboard in basketball on a driving layup. Since spin is not involved at this point, the angle at which the ball rebounds toward the hoop will be symmetric to the angle at which the player projects the ball toward the backboard. When possible, the approach angle of the drive for this shot should be about 45° toward the basket.

Figure 11.13
Impulse of collision
between two wrestlers.
*Courtesy SMS Public
Information Office.*

If the object bouncing from the surface has a spin, frictional forces must be considered because they modify the motion of the object parallel to the surface. This is the case in most rebounds of tennis balls from the racket surface (figure 11.14). Another example would be the reverse layup shot in basketball, in which the player passes under the basket, shoots the ball, and uses spin to bring it back into the basket. The control of a collision involving spin presents new factors of difficulty. Collisions with a nonspinning object should be used, when possible, in preference to those which involve spin.

Two bodies colliding are said to rebound elastically if they separate and maintain the same total kinetic energy which was carried into the collision. **Elastic-Inelastic**

A collision is said to be perfectly inelastic if the colliding bodies adhere to one another.

Actually, most sport collisions are between perfectly elastic and inelastic in their nature.

Coefficient of restitution is a measure of the total amount of kinetic energy derived from a collision, divided by the amount of kinetic energy carried into the collision. **Coefficient of Restitution**

The collision of a number-one wood with a golf ball during a drive off the tee serves as a good example of this principle. The swing and the elasticity in the shaft of the wood are used to impart a maximum kinetic energy to the club head before collision. During collision, this kinetic energy is transferred into a deforming potential energy of the club head and ball. As the ball rebounds from

Figure 11.14
Rebound angle of the tennis ball from the racket face during a backhand drive by George Taylor, USC tennis player. *Courtesy USC Athletic News Service.*

Figure 11.15 Energy transformation during a field hockey drive.

the collision, this potential energy is retransformed into kinetic energy, primarily in the ball. The coefficient of restitution in this case is a measure of kinetic energy of the ball after the collision, compared to the kinetic energy which the club carries into the collision. A less efficient energy transformation occurs when a field hockey ball is struck during a drive. Here the coefficient of restitution is determined by the materials and weight of the sport implement, the stick, and the sport object (figure 11.15).

It is the purpose of equipment manufacturers to make balls or surfaces with as high a coefficient of restitution as possible in order to effect maximum energy transfers during the sport action. This must be accomplished within the rules set forth for the various sports.

The subsequent motion of two bodies after a collision depends to a large extent on the angle between their relative velocity and a line joining their centers. **Impact Parameter**

The concept of impact parameter involves the use of such terms as *head-on collisions* and *glancing collisions*. In the case of a head-on collision where the line joining centers of two bodies is in the same direction as their relative velocity, there is a maximum transfer of energy from one body to another. In a case where the collision is slightly off center, there is less transfer of energy and more variation in the directions of motion of the bodies following the collision. In the case of maximum impact parameter, a grazing collision, little energy is transferred from one body to another, and only slight changes in the original direction of motion are noted.

In many sport applications such as a hit in baseball, a slap shot in hockey, or a drive in field hockey, optimal conditions occur with the smallest possible impact parameter, i.e., maximum velocity will be imparted to the ball or puck by a head-on collision. A slight deviation from the head-on collision tends to give the sport object a high velocity, but in the wrong direction. An extensive deviation or a glancing collision tends to prove ineffective in causing a change of motion of the object.

There are situations where the ability to deliver a precise but glancing type of collision becomes decisive. For example, some batters in baseball become masters at protecting the plate by fouling off pitches until they get the pitch they want or work the pitcher for a walk. In these glancing-type collisions where the impact parameter is not zero, a large amount of spin will normally be introduced into the motion of the sport object. This is also commonly observed in good punting techniques in both soccer and American football. As the ball is released from the hand before the kick, it has no spin, and the proper kicking technique imparts an optimal stabilizing spin.

When two bodies, one or both of them rotating, suffer a collision, the effect of spin is to transfer rotation from one to the other and provide a frictional force which causes a rebound in a nonsymmetrical elastic direction. **Effect of Spin**

As noted earlier, when a nonspinning object strikes a surface, the angle of reflection tends to be equal to the angle of inflection. This is because the horizontal velocity or horizontal component of momentum is maintained in such a collision. The effect of spin is to provide a frictional force with the surface, which either adds to the horizontal component of motion or subtracts from it. For example, a tennis ball striking the court surface with backspin tends to lose its horizontal velocity and to rise vertically. A tennis ball with a large forward spin tends to take a very low, fast rebound as the frictional force due to spin adds to the horizontal component of velocity. Often such a collision will result in a change of spin, causing a change in subsequent rebounds.

For example, in the sport of handball, a lateral spin on the ball driven into a side wall will either add to its forward velocity or subtract from it, depending on whether the spin can be considered forward or backward with respect to the wall. However, in the case where the spin diminishes the speed of the ball, this spin is generally reversed by contact with the side wall so the subsequent contact with the front wall results in a different type of rebound than would be expected from spin that the performer originally placed on the ball.

In court sports where spin is imparted to a ball before a rebound from a wall or playing surface, a skilled performer needs to understand the original trajectory and direction of motion of the ball, and also must be able to "read spin" before and after each rebound in order to predict successfully the motion of the ball in future rebounds. The ability to "read spin" both in direction and magnitude also becomes important when considering that spin affects the motion of the ball in the air, determining whether it will rise, drop, or curve. This ability to "read spin" is one of the most useful skills of the experienced and well-coached athlete, especially in sports such as baseball, softball, and basketball.

Aerodynamics

Drag

The resistance of air to the motion of an object passing through it depends on the shape of the object and on the relative velocity between the object and air.

The term *drag* is used for that component of force directly resisting the motion of an object or body through a fluid. This force becomes more significant as the velocity is increased, since for commonly used shapes and velocities the force varies as the square of velocity. For a given velocity, force depends on the amount of streamlining caused by the shape and the cross-sectional area in the direction of motion. The human body fully flexed at all major joints meets with a much smaller drag force at a given velocity than when standing erect with limbs extended. For example, the long jumper who maintains equilibrium during the flight phase while abducting both shoulder joints may be sacrificing distance due to an increase of the drag coefficient which is contributed by this position.

It should be noted here that the counterforce to drag—the force of the object on the air—produces a forward motion of air behind the moving object or body. This forward-moving volume of air, sometimes called the slipstream, can be beneficial to a performer who can use it to decrease his own drag resistance. For example, in the sport of cycling, riders tend to maintain a minimum distance behind the leader in order to utilize the slipstream of air created by the drag resistance. This principle is particularly noted in the teamwork used in team-pursuit cycling as one cyclist will take the lead for a short period of time and then drop back behind his or her teammates. Each person spends a certain part of the race time countering a maximum drag and creating a slipstream, allowing other members of the team to "rest."

When objects move through air, it is essential that they have a minimum amount of sidewise rotational motion, otherwise known as "yaw," during the flight. **Yaw**

This question of orientational stability during flight is exceedingly important, since an object that begins with this problem usually will not counteract it during flight. The effect of yaw is to interact with air in such a way as to increase the drag coefficient by losing the natural streamlining which comes from the aerodynamic design of an object such as a football or arrow. An example of this problem is the archer who places too much pressure on the nock of the arrow, with the index and middle fingers at the time of release. This causes a yaw as the arrow clears the bow. The effect of this motion decreases horizontal velocity by an extreme amount and causes the arrow to fall far below the intended target.

When a body or object interacts with air in such a way that the air is forced downward, counterforce from the air exerts a force upward, thus helping the body overcome gravity. **Lift**

The concept of lift or flight often seems to have a mysterious connotation involving various shapes, spins, or other aspects of motion. The basic consideration in obtaining lift from a motion through the air is to achieve interaction in such a way as to push the air downward. This may be done by an aerodynamic shape, or it may be accomplished by the angle of attack as, for example, with the flight of the discus or javelin.

One of the commonly observed techniques of lift in sport is that due to the spin of an object. In a well-hit golf ball or baseball or a well-kicked soccer ball, the backward spin of the ball as it travels through the air brings air behind it downward. The resulting force upward then causes the ball to fly much farther than it would in a simple parabolic trajectory. This same effect is also responsible for the curve, rise, and drop balls thrown by a softball pitcher.

A modern example of the use of lift is in the fast-growing sport of hang gliding. This is also seen in association with water-skiing, where a performer will use a large kite and the proper angle of attack in order to give a lift greater than the weight of the body and kite. This causes the glider to fly at the end of the towrope. The towing boat is providing the energy to maintain a wind velocity, and the performer can adjust the lift or interaction with air by the angle at which he or she holds the kite.

A body interacting with air will not follow a simple parabolic trajectory. **Trajectory Patterns in Flight**

Only by understanding the effect which air is having on an object can a performer be prepared to predict the angle of approach or distance that such a body will traverse before coming back to earth.

One of the most valuable techniques for a performer is the ability to judge the motion of flying objects. For the performer attempting to obtain maximum distance with the javelin or discus in a field event, the ability to read the effects of wind on a given day and to adjust the throwing style and angles accordingly, becomes of utmost importance. Failure to compensate for head winds, tail winds, or crosswinds will result in a less than optimal performance. In other events, for example the shot put, the interaction possible with air is so small compared with other forces involved that it may be disregarded, and the trajectory will follow a classical parabolic pattern.

Imparting spin to a sport object in order to affect its flight pattern or trajectory is an integral aspect of many sport skills. In sports ranging from tennis, handball, and table tennis to the large-field sports such as baseball and soccer, the actual flight of the object during a given contest will often be drastically affected by the spins involved. The flight pattern will deviate from the normal parabolic trajectory in the direction of the spin. A ball with backspin will tend to be lifted and carry farther and stay in the air longer than a ball with forward spin. In the latter case, the front part of the ball is moving downward. For example, baseball coaches should teach all players at all positions to throw the ball by gripping across seams and imparting a backspin at release. (Pitchers, of course, use a variety of grips.) This grip with respect to the seams insures a maximum interaction with air, and the backspin gives maximum lift to overcome gravity as long as possible. It is also essential that the performer be able to "read" the spin of an incoming sport object in order to interact with it in an optimal manner. For example, a tennis player should be able to detect the spin of all serves and ground strokes coming toward him or her in order to determine not only what their flight trajectories will be and where they will bounce, but also the effect the spin will have on the bounce. If the tennis player fails to observe this spin or observes it too slowly, he or she will not be able to get the proper foot and body positions to hit a well-placed return shot.

In considering the range that will be obtained by an object in flight, the angle at which it is projected from the ground is of extreme importance. In the case of an object which has no lift, this angle should be 45° for an object released near the ground. Release of the object significantly higher than the ground, as in the shot put—or deviation of the trajectory due to spin of the object, as in a golf drive—requires that the angle of projection be less than 45° in order to obtain maximum distance. In other cases, such as the long jump, the angle is determined, not by a theoretically optimal flight pattern, but rather by the limitations of the human anatomic structure, which will allow speed to be maintained only for relatively low angles of takeoff. Still another example occurs in the sport of football, where at times the purpose of a punt is not to attain a maximum range, but rather to assure a maximum time in the air to allow coverage. In this case, an extremely high angle of projection is desirable.

Flight Stability
An object not in contact with the ground is susceptible to various types of instability of motion, especially with respect to orientation.

Since an object in flight is capable of interacting only with the air, stability during the initial phase of flight is especially important. An object launched into the air with yaw or a tendency to turn in an unwanted direction has little capability of changing or stopping this motion while in flight. Stability during flight is aided, at times, by the design of the sport object. Examples are the raised seams on various balls and the fletching on arrows. At other times, stability is obtained through the conservation of the direction of angular momentum by causing the object to spin. This is the case in the flight of the discus, rifle bullet, and javelin. An object without rotation will have a much more erratic flight pattern, especially if affected by a side wind or other nonsymmetrical force. In order to make optimal use of this principle, a spinning object must be placed into the air without any yaw. It is for this reason that the technique of release of a sport object is such a decisive phase of many sport skills.

For a performer in the air, stability is often maintained by using internal forces. Muscle contractions cause adjustments of limb positions, and slight deviations from stability of the body can be controlled. In extreme cases where disequilibrium is apt to occur, countertorque to the rotational tendency is achieved by circumduction of one or both arms. This technique can be used either to absorb excess rotation or to create rotation needed, as in the case of the pole-vaulter during the falling phase after passing over the bar.

In those situations where the performer desires to enter the air with an angular momentum in the form of a "twisting" or tumbling motion, as in diving or rebound tumbling, it is necessary that these motions be initiated while still in contact with the board or trampoline. Once in the air, any "twist" of one portion of the body is counteracted by an equal but opposing "twist" of some other portion of the body. For example, if a performer enters the air with no spin and attempts to change his or her position by a forceful diagonal adduction of the right shoulder, the body will turn to the right around the longitudinal axis. However, when the performer has reached the limit of his or her range of motion, both the adduction of the arm and the rotation of the body will cease. The performer can change position but cannot introduce a total spinning motion. These extraneous actions of the limb will not change the flight pattern or trajectory.

Hydrodynamics
Drag

Due to the density and viscosity of water, objects passing through it are retarded greatly by drag force.

Drag force is that component of force directly opposite the direction of motion. In water, the forces are from 10 to 100 times greater than those encountered in air at low velocities. For all activities involving motion through water, streamlining of shape and orientation of the object with reference to direction of motion become extremely important. For example, in the case of competitive swimming strokes it is a necessity to minimize, as much as possible, lateral plane movements such as shoulder and hip abductions. These add to the cross-section presented to the water in the direction of motion, and thus decrease the streamlining while greatly increasing drag. Any motion causing the body to lie obliquely rather than horizontally in the water has the same effect.

The design of kayaks, sculls, or sailboats becomes of vital concern in minimizing the drag forces which will be encountered while moving through water.

Literally millions of dollars are expended for engineering and designing hulls of yachts to be used in the sport of yacht racing.

In any situation where there is turbulence in the water, the drag problem is greatly compounded. Since turbulence represents water moving in many different directions, there is no possibility of always meeting the water in the most streamlined position. In the sport of competitive whitewater kayaking, the problem is made still greater by the fact that the kayaker must at times pass through the gates with, against, or across the flow of the stream. The stream itself shows turbulence and eddies due to its various obstructions.

Buoyancy   *Objects in water are buoyed upward in such a way that the effect of gravity is partially or completely balanced.*

Motions necessary in water sports are somewhat different from those in sports performed on land, because the buoyant effect of water offsets the need for antigravity muscular actions. There are exceptions to this rule. For example, the water polo shot is definitely an antigravity motion due to the vertical force component required of the athlete. However, in most swimming events, and in such sports as kayaking or canoeing, the primary forces necessary are horizontal. While competitive swimmers may not be completely buoyant because they have a high percentage of lean body mass, the total vertical force which they must use to remain at the surface of the water is at most only a few pounds. Their execution of stroke mechanics is developed to the point that this lack of buoyancy is negated.

Thrust   *The density of water allows major thrust forces to be developed through standard limb motions associated with swimming.*

In contrast with human body motions in air, the motion of limbs through water interacts with a mass of water representing a large fraction of the mass of the human body. Therefore, force exerted against these masses can be large enough in magnitude so the counterforce of water provides noticeable acceleration to the body. These forces constitute the thrust obtainable in water.

In a sport such as alligator wrestling, the thrust developed by the human opponent through the use of limb motions is minimal while trying to control and overpower his opponent in deep water (figure 11.16).

The stroke technique in competitive swimming is designed to obtain a maximum amount of thrust while keeping the body in such a position as to create minimum drag. In this way, the highest average velocity can be maintained through the water. For this reason, any motion involved in swimming which either decreases the thrust interaction with the water or increases drag will lead to less than optimal results. For example, in the kicking process for the front crawl and back crawl, a kick which brings the foot out of the water represents a loss of thrust during the part of the motion while the foot is above the surface. On the other hand, a kick that goes too deep in the water increases the drag, thereby decreasing the ultimate speed. It is also desirable that in considering the total time of a stroke, time in which thrust is developed should be as large a fraction as possible, when compared to the time the swimmer is simply gliding through water. This glide portion is an almost unnoticeable part of the stroke in competitive swimming, but it may become a decisive phase of such activities as recreational swimming or lifesaving. In a lifesaving situation, the conservation of energy is more important than ultimate speed. Therefore, use of the side and breast strokes is indicated to maximize the glide phase.

Figure 11.16 Red Carson, a football center who is also an alligator wrestler, minimizes drag in the water and maximizes the forces known as thrust as he accelerates prior to hitting the "gator" in the posterior-cervical region. According to Carson, "It would be easier to apply greater force if the 'gator' were a middle guard moving toward me on land."

4

Theory to Practice: Techniques of Sport Motion Analysis

12

Noncinematographic and Basic Cinematographic Analyses

Cinematography, as delimited for use in this book, is the study of sport performance through the use of motion-picture film or videotape records.

The human eye is incapable of stopping ballistic motion of performers. In order to have a permanent record of a given performance and to study the components of athletic or sport events, the teacher-coach should use film. Film helps to objectify the otherwise subjective elements of viewing, while preserving performances in a relatively permanent state.

The teacher-coach must utilize noncinematographic techniques daily when attempting to improve performances of students or student-athletes. The physical educator must use a professional, disciplined system to observe performances of students when film is not available. If the physical educator does not have a systematic approach to analysis without film, very little objectivity or precision will be obtained; therefore, the quality of the subsequent teaching will be considerably diminished. "Seeing in depth" without the use of film is essential to teaching-coaching, because this analytic technique is used most frequently in physical education. *The physical educator must train himself or herself to understand what is actually taking place as opposed to what one thinks should be occurring in the performance of the student or student-athlete.*

The procedure for basic cinematographic analysis is essentially the same as that which is utilized in noncinematographic analysis. The basic difference lies in the fact that the analysis is made from film instead of from observation of the performer on the field or elsewhere. In this context, any type of motion picture can be utilized for analytic purposes. This includes the game film frequently taken at football games, basketball games, track-and-field meets, and other events. Game films are often overlooked, in spite of their value in terms of biomechanic analysis. Although game films do not meet the critical criteria of biomechanic research film, they can be utilized extensively for biomechanic observations and formulating teaching-coaching suggestions based on sound scientific and technical principles.

Analysis Guidelines
The procedures and suggestions listed below are recommended for undertaking noncinematographic and basic cinematographic analyses when the physical educator wishes to identify and define motion problems, and provide the student with positive teaching-coaching suggestions to improve the overall performance of the skill:

1. *Stereotype of Perfect Mechanics:* The technical aspects or techniques of the skill must be fully understood by the physical educator before any analyses

or teaching can be undertaken. This technical knowledge serves as a reference or base point for subsequent motion adjustments. It does not imply that all performers can or should be expected to duplicate the perfect mechanics model for each skill. *Teaching physical skills is an art based on science;* therefore, physical educators must be cognizant of individual differences of their students, as well as functional variations of motion mechanics for any skill. If these factors are understood, students should be given logical teaching suggestions adjusted for individual differences; these instructions will assist them in achieving a reasonable amount of success in the skill. The *pursuit of perfection* is very challenging in and of itself.

The utilization of technical and scientific knowledge during motion analyses helps the physical educator bridge the gap between theory and practice. Good theory, technical and scientific, is always functional when applied. The starting point for technical theory is what we term the stereotype of perfect mechanics, with its variations for any skill in any sport. The analytic process combines the scientific theory from biomechanics and anatomic kinesiology with the technical theory. Therefore, teaching-coaching suggestions developed from the skill analysis should have logical and rational bases. Theory becomes practice, and performance improves under the direction of a knowledgeable and professional physical educator who can communicate what he or she knows to students. Therein lies one of the challenging and artistic elements of teaching.

2. *Skill Phases:* Virtually all sport skills can be divided into logical, timed sequences. This should be done prior to undertaking an analysis, and based on the stereotype of perfect mechanics. The usual skill phases are: (1) stance, (2) preparation phase, (3) movement or motion phase, (4) follow-through phase, and (5) recovery phase.

The various phases noted are not common to all skills; therefore, adjustments regarding timed sequence phasing of any skill must be made accordingly. As an example, a forehand ground stroke in tennis could logically be divided into the five phases noted above. There is a stance, preparatory motion in terms of movement to the ball and the "backswing" of the racket, a motion phase to the point of impact, a follow-through, and recovery of body position in anticipation of a return shot. However, a javelin thrower does not have a stance. The preparatory phase is a complex run with intricate footwork as the throwing limb is diagonally abducted prior to the throw. The motion phase is complex and ballistic. The follow-through is necessary to dissipate force and provide equilibrium to remain behind the "scratch line." No recovery is essential, because the javelin is not going to be thrown back toward the performer! (An analysis example of javelin throwing is provided in this chapter.)

Skill phasing is unique to the skill being analyzed. The language used for naming skill phases should be descriptive of the skill under study. For example, preparatory motion for a springboard dive could be called "the hurdle phase." The motion phase could be termed "the flight phase." Those phase terms can enhance teaching by better communication of ideas to students while discussing their performances through the phases during the viewing of film or videotape.

3. *Biomechanic Observations:* A major objective of any skill analysis should be quantitative and qualitative determination of the effectiveness or ineffectiveness of motion. This should be monitored during the analysis by asking and answering relevant questions based on biomechanic principles. These principles were summarized in Chapter 11; and each principle had a recall phrase such as "sequential velocity," "summation of force," or "torque-countertorque." The principles and terms discussed in Chapter 11 should be kept in mind, and they should be referred to when the performer is either effective or ineffective in observing a specific biomechanic principle. The extent of the motion problems during the skill, along with the causes, should be explained biomechanically in terms of the effect on the outcome of the performance. The cause and effect relationships of joint motion problems noted during analyses should be discussed in reference to the use or misuse of biomechanic principles.

4. *Anatomic Observations:* Total skill analysis must include the intricate interrelationship between biomechanic and anatomic observations. Which observation is made first is irrelevant; what does matter is that the physical educator analyzes the students' motion problems by synthesizing knowledge from both areas. This synthesis of scientific knowledge objectifies the cause of the motion problem(s), and thus can lead to the functional improvement of skill, because the analyzer has *defined precisely why the performance was not effective.*

There must be a thorough, disciplined approach to joint motion and body segment analyses. This is necessary to qualify biomechanic observations. As an example, if there appears to be a "force-time relationship" problem by a performer during a ballistic skill, the analyzer must look for the cause by making critical observations of the extent and direction of motion within the major joints and body segments. The limited ranges of motion within joints may not provide the performer with the necessary time to develop the "specific velocity" needed to reach the skill objective. To solve the problem, the physical educator would have to undertake range of motion changes in the body segments involved in order to make the force-time relationship more efficient within the execution of the skill. Because of their cause and effect involvement determining effective and ineffective motion, it is imperative to explain and integrate human motion in both biomechanic and anatomic terms during noncinematographic and basic cinematographic analyses. Generally, the cause of biomechanic ineffectiveness is found by making thorough anatomic observations of a performer.

5. *Multiple Observations of Performances:* Throughout the analytic process, there must be multiple observations of each joint and body segment motion during each skill phase. The extent, direction, and velocity of these moving body parts must be observed independently and in relationship to their timed interaction with each other during the execution of the skill. In the course of noncinematographic and basic cinematographic analyses, these motions are simply qualified by the analyzer. However, in the intermediate level analyses, these observations of moving body segments are further objectified by using mathematic procedures. These are discussed in Chapters 13 and 14.

The observations of a skill during analysis should be made from several angles. During noncinematographic analysis it may be possible to observe a performer several times from lateral, posterior, and anterior perspectives. When filming, one should try to obtain a variety of views at angles critical to the motion planes. That is very difficult when filming some team sports.

Multiple observations from several angles are essential to determine accurately the segment motion within planes. If, for example, a runner is only viewed from a lateral perspective it may appear that all body segments are moving through anteroposterior planes. An anterior observation of the same performance would reveal the fact that there are also several transverse, diagonal, and lateral plane motions. (See figure 12.10).

It is recommended that the observations of moving joints and body segments follow a set procedure for all skills. *The following sequence for making multiple observations of moving segments during each skill phase is one disciplined procedure. Several observations are made of each segment during each skill phase:*

1. Pelvic girdle—lumbar-thoracic spine (center of mass)
2. Base of support
3. Head—cervical spine
4. Total upper limbs
   a. shoulder girdle
   b. shoulder joint—humerus
   c. elbow joint
   d. radio-ulnar joint—hand
5. Total lower limbs
   a. hip joint—femur
   b. knee joint
   c. tibia-fibula
   d. ankle—foot

This *segmental analysis* procedure is used in providing several examples of motion analyses later in this chapter.

6. *Definition of Performance Problems:* Following the motion analyses and preceding teaching-coaching, the performance problems must be clearly defined. When these are evaluated technically with the stereotype of perfect mechanics as a reference, and scientifically from biomechanic and anatomic perspectives, it frequently becomes apparent that all of the motion problems observed are not equal in terms of their negative contributions to skill execution. As a result, the physical educator is obligated to rank the defined performance problems in terms of their severity. The most serious motion problem would be given top priority, while the least serious problem would be ranked last. The rationale for these rankings would be founded in the synthesis of technical and scientific subject matter applicable to the skill observed.

7. *Teaching-Coaching: Quality teaching-coaching always follows motion analyses where the performance problems have been succinctly defined and placed in proper perspective.* This eliminates guessing about what one believes to be wrong when a performer does not execute skills consistently. Professional

physical educators and coaches are obligated and paid *not* to make teaching suggestions based on guesswork or superficial considerations of their students' performances.

The focus of attention should be on the most serious motion problem identified during noncinematographic and/or basic cinematographic analyses. This problem is a *negative factor* for the performer. This fact must be a prime concern for the physical educator. However, *when communicating with the student, the quality teacher-coach does not emphasize the negative.* The specific and precise teaching-coaching suggestions should be *positive* in nature. It is this positive suggestion which is stressed during teaching to eliminate a negative motion problem. Moreover, the teacher-coach with a positive perspective does not attempt to eliminate more than one motion problem at a time. Once improvement has been made with respect to the most serious problem, the same positive approach is made regarding the second motion problem.

| Noncinemato-graphic Analysis | The first rule for observing performers without the aid of film is to watch carefully the same skill executed a number of times. |

The first rule for observing performers without the aid of film is to watch carefully the same skill executed a number of times. Too often, extensive teaching or coaching suggestions are made after the coach has observed the performer executing a skill only once. It is entirely possible that a coach may not see a consistent or major performance problem while observing the performance once or a few times only. A coaching suggestion in this context may only serve to compound the performance problem; therefore, the performer should be observed many times prior to making any type of teaching suggestion. This situation is analogous to a physician making a diagnosis and prescribing medication. A physician usually does not prescribe medication until he has completed a total examination of the patient's symptoms. To prescribe medication on the basis of one symptom described by the patient would be poor professional practice. The same thing can be said of physical educators who make performance recommendations based on minimal observations of performers.

To continue with the analogy—like the physician who makes numerous analyses of the patient's various systems, the physical educator must have a systematic approach to viewing performers. One way of doing this is through focussing attention on specific moving body segments or joints since it is very difficult to visualize cause-and-effect relationships when viewing the overall performance. In other words, it is unrealistic to attempt to generalize about specific motion of a given body segment when the observation includes the total performance only. The sport fan looks at the total performance, while the trained physical educator looks at body segments moving in relation to each other. Moreover, the observation of various moving body segments must be made in a systematic manner as discussed earlier in this chapter.

To improve performances, the physical educator must ask and answer questions based on biomechanic principles and techniques for skill execution, according to the accepted stereotype of perfect mechanics for the skill. These biomechanic principles are presented in chapters 8 through 11. The type of questions asked by the physical educator depends on the individual skill under anal-

Noncinematographic and Basic Cinematographic Analyses

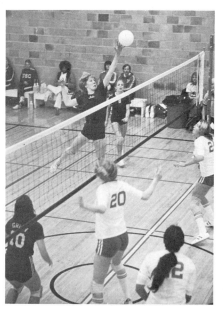

Figure 12.1 A coach must rely upon a disciplined noncinematographic analysis technique if critical adjustments of individual and team performances are to be made. *Courtesy USC Athletic News Service.*

ysis. For example, a coach analyzing a discus thrower would ask entirely different questions than would a dance teacher making an analysis of a dancer's leap. This is one reason why there is no one "best" procedure for undertaking an analysis without the use of film. Each person must develop a system for controlling observations of the performances under his or her direction. A *recommended procedure* for undertaking noncinematographic analyses is presented in this chapter. This method provides the teacher-coach with focal points to consider while the performers are executing skills in classes, practice situations, or athletic contests (fig. 12.1).

In order to assist the reader in acquiring a greater comprehension of analytic techniques, several examples of noncinematographic and basic cinematographic analyses are presented in this chapter. Each analysis should be read slowly, with frequent reference to the figure. Four noncinematographic analyses are presented, including a sprint start, shot put, high jump, and hurdling. Four examples of basic cinematographic analyses are presented, including the hurdles, high jump, javelin throw, and kicking. In order to clearly demonstrate differences between the three levels of biomechanic analyses, the hurdles analysis will be continued through the intermediate level in chapters 13 and 14. This will enable the reader to compare and contrast the differences in the three analytic techniques. In addition, an intermediate cinematographic analysis of the discus throw is presented in chapters 13 and 14.

Although many of the analysis examples are from the sport of track and field, they include basic motion components common to a wide variety of performance areas:

1. **Track Start:** linear motion of the total body mass from a stationary stance. This is also seen in such activities as baserunning, movement from football stances by wide receivers, goal play in soccer, and start-stop motions in tennis and basketball.

2. **High jump and hurdling:** linear motion converted to vertical displacement in varying degrees due to the performance objective differences between the two skills. This general type of movement is seen in any performance where the center of mass must be raised to a given height to achieve the performance objective. For example, dance leaps, volleyball spiking and blocking, basketball rebounding, gymnastic floor exercise skills, and jumping events in equitation, all require that linear motion be converted to vertical displacement.

3. **Shotput and javelin:** angular and ballistic skills through high diagonal planes at the shoulder joint. Throwing patterns of this type are an integral part of any sport in which sport objects must be projected into the air with or without the use of a sport implement.

4. **Kicking:** integration of linear and angular motions involving a ballistic motion of the lower limb through a diagonal plane. Most ballistic motions of the lower limbs will involve diagonal and/or anteroposterior plane motions.

5. **Discus:** angular, ballistic motion through a low diagonal plane at the shoulder joint. Patterns through low diagonal planes and an anteroposterior plane are common in the sports of softball, bowling, golf, badminton, and many others.

Sprint Start   Figure 12.2 includes a sequence of the sprint start. The following is an example of how a noncinematographic analysis would incorporate *segmental observations*. The analysis should begin with repeated observations of the performance until the observer has a general feeling for the various execution parameters. By doing this, the coach should gain some insight into the potential general outcome of the skill, as well as the overall timing or interrelated aspects of the body segments of the performer in motion. Also, this gives the physical educator an opportunity to compare the observed performance with the stereotype of perfect mechanics for the sprint start.

In the example of the sprint start, the preparatory phase begins in frame three and ends in frame four. Frame four also indicates his stance in this case. The movement phase begins in frame five and ends in frame thirteen. Frames fourteen through twenty show the follow-through for the sprint start. At that point, the athlete is in the acceleration stage of the sprint itself, and has successfully executed the sprint start. To point out the observation difficulties from a noncinematographic standpoint, it must be kept in mind that the coach would be watching frames three through twenty in a time interval of one-half second. This fact is a strong argument for the use of film in coaching.

The **pelvic area and rib cage segments** of the body are useful to observe because they are relatively slow-movement areas and are pertinent orientation points for analytic purposes. The performer should be observed from more than one angle. In the example of the sprint start, the coach should observe the athlete from posterior, anterior, and lateral positions.

Questions the track-and-field coach might ask and attempt to answer while observing the **pelvic girdle and rib cage** of the athlete during the sprint start are: How high is the center of mass above the base of support throughout the entire skill? Are there any extraneous rotations of the spinal column and pelvic girdle

Figure 12.2 The sprint start of Tom Jones, USA Olympic team. Preparatory phase: frames 3–4. Movement phase: frames 5–13. Follow-through: frames 14–20. *Courtesy Visual Track and Field Techniques, 292 So. La Cienaga Blvd., Beverly Hills, CA 90211.*

Figure 12.2—*Continued*

Noncinematographic and Basic Cinematographic Analyses

Figure 12.2—*Continued*

through the transverse plane? Is there an undesired hyperextension of the lumbar spine during the sprint start? Does the path of the pelvic girdle deviate from a straight line at any time during the skill? In the illustration shown, the performer exemplifies ideal pelvic, spinal column, and rib cage position in frames twelve and thirteen. From frame five, or the start of the movement phase, the lumbar spine moves from flexion to extension in frame thirteen. This is accomplished with no extraneous transverse rotations. There are, however, slight lateral rotations of the pelvic girdle when the athlete is in a unilateral weight-bearing position. This is accomplished without excessive movement while the pelvic girdle, lumbar spine, and thoracic spine remain dynamically stabilized. If the pelvic girdle were not dynamically stable at this time, force executed through the blocks would not be effective in driving the center of mass linearly.

The **base of support** should be observed from lateral and posterior positions. The posterior view, which is not shown in the illustration, is advantageous from the standpoint of seeing undesired transverse and diagonal plane movements of the lower limbs. During the preparatory phase of the sprint start, the athlete is in a four-point stance. Are the hands and right foot bearing most of the weight? Is the block placement adequate to provide optimal force and range of motion within the ankle, knee, and hip joints at the start of the movement phase? During the recovery phase, are the changing bases of support following movement through anteroposterior planes of motion? Is there a prolonged period of time when both feet are off the ground before the recovery process begins? Are the feet too wide apart or too close together throughout the performance? The answers to these types of questions relative to this performance help determine the degree of efficiency with which the athlete transfers internal forces from muscular contractions into accelerating forces for the center of mass. If the coach observes considerable extraneous motion within the base of support during the sprint start, the athlete will be less effective in transferring these forces.

Another important body segment to observe during any athletic performance is the **head and cervical spine.** The most important question the coach should ask regarding the positions of the head and cervical spine during the sprint start is: How soon does the cervical-spine movement occur to allow the athlete to look down the track toward the finish line? Any hyperextension of the cervical spine will cause a corresponding reaction in the lumbar spine. These two major movements have a tendency to extend the trunk and greatly increase the resistive drag force caused by air. Cervical hyperextension also tends to raise the body to a point that the horizontal force component is diminished during the acceleration phase. Both factors contribute to a decreased velocity. In this example, the cervical spine and head are stabilized in an extended position in all three phases of the skill. This is a highly desirable technique. The athlete adjusts the eyes only to look down the track.

Arms and hands are vital segments to observe specifically during noncinematographic analyses. During the sprint start, the arms must provide a powerful counterforce from the end of the movement phase to the start of the recovery phase. In order to be in this position, the arms must be moved to a position as high as possible during the power phase of the sprint start, while the left foot in the example is still pushing against the blocks. The diagonal movements of both upper limbs are made in such a way as to provide an optimal state of equilibrium. It must be remembered that during the sprint start the sprinter is in a dynamic state of disequilibrium. The sprinter is protected from falling only because he is undergoing extreme acceleration. Movement of the arms through the diagonal plane during the movement phase is extremely important as a means of maintaining equilibrium. Movement through the diagonal plane provides a greater moment of inertia than movements at the shoulder joints through anteroposterior planes. Movements of the hands and fingers are of interest. Are the fingers flexed or extended? If the sprinter has a flexed hand or fist, this is an indication of extraneous internal tensions that will ultimately inhibit or decrease linear velocity. In the example, actions of the arms are excellent.

Legs of the athlete must also be observed critically during noncinematographic analysis. Specifically, the motions of the ankles, knees, and hip joints must be viewed to determine their overall contributions to the skill. Are the flexions in these joints adequate to provide an optimal range of motion and subsequent force? Are there any transverse rotations at the knee and hip joints that might inhibit performance? Are the extensor movements performed throughout their complete range of motion prior to the time the sprinter leaves the blocks? Are the knee flexions adequate to provide optimal angular velocity through conservation of angular momentum? These questions would be answered affirmatively related to the sprinter in the example.

The culmination of noncinematographic analysis of a sprint start should be in the recovery phase of the runner as he or she starts into the second aspect of this skill: the actual sprint. Many of the same questions asked above, and others, are relevant during the subsequent phases of the sprint through the finish line.

Shot Put
From a noncinematographic standpoint, it must be remembered that the shot-put sequence shown in figure 12.3 takes approximately two seconds. Frames one through eleven show the preparatory phase of the shot put. During this preparatory phase, the performer is concerned with basic body equilibrium and placing muscles on stretch to provide the subsequent internal force necessary during the critical movement phase. The movement phase occurs between frames twelve and thirty-two. The important follow-through phase occurs from frame thirty-three through frame thirty-nine.

The greatest velocity is attained by the performer during the movement phase. From a noncinematographic standpoint, it is very difficult to perceive all basic movements during the movement phase, because the performer executes this phase in approximately one second. Therefore, it is essential that the physical educator watch the important movements within the various large body segments several times and from at least three angles.

From the standpoint of biomechanics, it is imperative that the shot-putter release the shot at its maximum velocity, optimal altitude, and at a release angle of slightly less than 45°. Whatever would interfere with any of these biomechanic objectives should be observed by the coach and ultimately eliminated by adjusting joint motions for the athlete.

During the preparatory phase, the **pelvic girdle and lumbar spine** are being moved with gravity and maintained over the supporting leg, keeping the athlete in a state of equilibrium. In frames seven through twelve, the center of mass is being lowered eccentrically; this places the knee extensors of the right leg on stretch for forceful concentric contraction at a critical point during the movement phase. This lower center of mass is accomplished by flexion with gravity of the right hip and knee joints. Is the lowering of the center of mass consistent with placing the critical muscles of the knee and hip extensors on stretch? The pelvis remains dynamically stabilized from frame twelve until frame twenty. The coach watching the shot-putter's pelvic girdle and rib cage should determine whether there is a torque-countertorque effect during the final weight shift of the

Figure 12.3 The shot put as performed by Randy Matson, USA world record holder. Preparatory phase: frames 1–11. Movement phase: frames 12–32. Follow-through: frames 33–39. Courtesy *Visual Track and Field Techniques, 292 So. La Cienaga Blvd., Beverly Hills, CA 90211.*

Theory of Practice: Techniques of Sport Motion Analysis   329

Figure 12.3—*Continued*

Figure 12.3— *Continued*

Noncinematographic and Basic Cinematographic Analyses

preparation phase. In other words, does the pelvis rotate to the left—left transverse pelvic girdle rotation—while the lumbar-thoracic spine rotates right at this point of the performance? The phase movement motion of the lumbar spine and rib cage would be left rotation through the transverse plane. The major torque would be supplied by the previous rotation of the pelvic girdle, which would have placed the critical muscles on stretch. In the example, the torque-countertorque between these areas is not great enough to place critical muscles on stretch. From frame twenty-five through frame thirty-nine, the lumbar-thoracic spine is simply continuing to rotate to the left through the transverse plane as the athlete releases the shot, follows through, and recovers, maintaining his equilibrium within the ring. During these rotations of the pelvis and lumbar-thoracic spine, there is a sequential elevation of the center of mass through the time the shot is released. In frame thirty-two, as the shot is released, it should be noted that the center of mass is at its highest point and is lowered slightly thereafter. This athlete needs to be given drills and exercises to increase these pelvic and spinal rotations.

In putting the shot, the sequence of movement for the **base of support** shifts from (1) a complete unilateral weight-bearing position, (2) to a bilateral weight-bearing position, and (3) returns to a unilateral weight-bearing position. The velocity across the ring is of extreme importance in terms of ultimate velocity and distance obtained by the shot. During the preparatory phase, does the initial unilateral base of support shift in either direction? During this phase, all adjustments of weight and other body segments should be centered above the unilateral weight-bearing right foot. It should be noted that the base of support does not move until the performer has completed the preparatory phase and starts into the movement phase. At this time, the athlete is starting his movement and showing a strong linear acceleration. The left foot has no supportive function until the performer has moved his body laterally across the ring. Is the shifting of weight from the base of support synchronized with the timing of the torque-countertorque effect seen in the pelvic girdle and rib cage? Is the left foot placed in a position advantageous for torque of the pelvic girdle and lumbar-thoracic spine? Does the actual placement of the left foot in the ring interrupt linear motion or the summation of motions? Are plantar flexions observed in both ankle joints immediately prior to release? If not, the summation of motion will be weakened. Are the motions of the base of support adequate to make maximum use of the diameter of the ring? What is the status of the base of support at the time the shot is released? Does the center of mass come back over the unilateral base of support during the follow-through and recovery phase? This is absolutely essential to maintain functional rotation for dynamic equilibrium.

The **head and cervical spine** of the shot-putter should remain in a relatively stable and extended position throughout the performance. A coach viewing this performance from the lateral view should see a cervical hyperextension—frame twenty-seven—synchronized with the weight shift to the left leg and the forceful start of left lumbar-thoracic rotations. This movement of the cervical spine allows the shot to follow its linear path uninterrupted. Excessive cervical rotations would cause undesired compensatory rotations elsewhere in the spinal column.

The coach should observe the performer posteriorly several times while viewing **arm and hand actions.** One noncinematographic observation should be of the path the shot is actually following as the performer moves linearly and rotationally across the ring through the point of release. There should be no lateral movement or vertical undulations of the shot as the performer goes through the basic motions of putting the shot. Such extraneous movements would contribute to excessive disequilibrium and other compensatory movements on the part of the athlete. When the skill is observed from the lateral view, the left arm should be viewed to gauge its contributions to maintaining equilibrium during the preparatory phase and early part of the movement phase. During the latter part of the movement phase, what is the contribution of the left arm to the torque processes taking place in the lumbar-thoracic spine? Does the movement of the left arm "lead the spinal column rotations"? Does the action of the left arm contribute to the flexor-extensor reflex mechanism and equilibrium during release and ultimate follow-through? If not, the athlete most likely will "scratch." Is the shot held on the distal end of the arm lever? If not, the decrease in the length of the lever will negate the ultimate concentric contractions and movement capacities of the wrist-finger flexors at release.

The coach should specifically watch for finger abduction and wrist flexion by the athlete immediately after the shot is released. If the wrist remains in hyperextension or extension, this will indicate that a maximum velocity of the putting limb has not been attained at the time the shot was released. Maintaining the shot near the ends of the fingers at the time of release provides three to four more inches of altitude which would result in roughly an equal increase in horizontal distance. In regard to the arm holding the shot, does it maintain its relative position throughout the preparation and movement phases until the transverse rotations of the lumbar-thoracic spine have virtually been completed? If the shot is moved away from the position relatively close to the neck, this will increase the length of the lever and cause a decrease in rotational velocity, which, of course, is not desired at this phase of the put. Does the right shoulder become medially rotated and diagonally adducted immediately prior to the forceful elbow extension that precedes release? What is the action of the right arm during release and follow-through?

Actions of the **knees and hips** are critical in shot-putting. Are the right knee and hip joints being flexed with gravity to optimal angles for placing critical muscles on stretch? This should be observed during the latter part of the preparatory phase. If these muscles are not placed on stretch, the subsequent kinetic chain of events will not attain an optimal level of force. During forceful knee extension as the performer is moving linearly across the ring, do the left knee and hip move into complete extension? At the time of spinal column rotation, are both knees flexed with gravity and in a position ultimately to exert forceful concentric contractions for knee extension? If the knees are not flexed at this point, they will be unable to contribute to the rotary movements and vertical lift necessary in putting the shot. At the time of release, is the left knee totally extended? If not, this may contribute to an inadequate angle or altitude of release that potentially would lead to a diminished horizontal distance.

The first two examples of the noncinematographic analysis technique included a basic linear motion and a motion involving a short linear acceleration combined with rotational accelerations of the body mass. The following example of a high jumper using the Fosbury Flop style of jumping was selected because it involves a very strong vertical force component. The kinetic energy of the high jumper is transformed into potential energy through work accomplished against the gravitational field. The concept of segmental analyses in a noncinematographic sense is essential in looking at the high jumper, because it is the movement of body segments in a sequential framework over the bar that comprises the major portion of the skill aspect of high jumping.

In the high-jumping sequence shown in figure 12.4, frames one through fifteen constitute the preparatory phase for the high jump, and frames sixteen through thirty-eight are the movement phase of this skill. The follow-through aspect of high jumping is relatively unimportant because, as shown in frames thirty-nine and forty, after clearing the bar the body is in free fall and the only concern for the athlete and coach is that in landing in the high-jump pit the athlete is not incapacitated.

Figure 12.4 The Fosbury Flop as performed by Kestutis Sapka, USSR. Preparatory phase: frames 1–15. Movement phase: frames 16–38. Follow-through: frames 39–40. *Courtesy Visual Track and Field Techniques, 292 So. La Cienaga Blvd., Beverly Hills, CA 90211.*

Theory of Practice: Techniques of Sport Motion Analysis

Figure 12.4—*Continued*

Noncinematographic and Basic Cinematographic Analyses

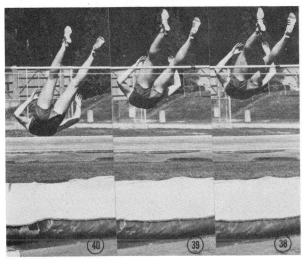

In viewing the **pelvic girdle and lumbar-thoracic spine** during the preparatory phase, the coach needs to observe the nature of horizontal motion of the center of mass. This should be a relatively smooth linear acceleration without excessive deceleration prior to final placement of the left foot at the time of vertical lift. As the performer nears the final stride, the center of mass will be lowered to make internal bodily adjustments for the final vertical motion of takeoff. If lowering of the center of mass is exaggerated, this will negate the potential internal force component of the subsequent contraction of muscles most involved, in addition to causing diminished linear momentum. Both of these factors would have a negative influence on the height subsequently attained in the jump. It should be emphasized that some high jumpers do raise the center of mass significantly at frames thirteen and fourteen. If this is done, it starts the vertical lift transformation and uses some of the force generated from the right foot exerting pressure back against the ground. In our example, this is not a significant aspect of this jumper's skill.

In observing the pelvic girdle and lumbar-thoracic spine during the movement phase, the coach should be aware of the transverse rotations of the lumbar-thoracic spine immediately following the moment when the athlete becomes airborne. Is there a left transverse rotation of the lumbar-thoracic spine? Does the pelvic girdle remain dynamically stabilized during this transverse rotation? These rotary aspects of this skill move the shoulders into a position where they lie parallel to the high-jump bar. As the athlete is moving vertically, is there a pronounced hyperextension of the lumbar spine? This internal force component caused by the spine extensors elicits a counterforce in the thoracic spine and ultimately lifts the pelvic girdle. In order to observe this phenomenon from a noncinematographic standpoint, the coach should be positioned at the end of the high-jump bar opposite the takeoff point of the athlete. When the athlete is directly above the bar in a position of lumbar hyperextension, is there a return to lumbar extension? If so, this initiates an action-reaction between the lumbar spine and both hip joints. It is imperative that both hip joints be slightly flexed to facilitate bar clearance, prior to the follow-through aspect of the jump. Frequently jumpers will fail at this point in the jump, because they maintain the lumbar spine in hyperextension throughout the remainder of the jump. During the follow-through phase of the high jump, most jumpers will move into a flexed position of the lumbar-thoracic spine in order to present a broad surface to the landing pit. This is good technique because it dissipates a large percentage of force over as wide a surface area as possible instead of concentrating it on a small body area where injury might occur.

The **base of support** observations should focus primarily around the latter aspect of the preparatory phase and the initial portion of the movement phase or lift-off. The critical leg to watch in this example is the left leg. At the conclusion of the preparatory phase, is the base of support in the last stride adequate to place the left foot at an appropriate angle for optimal plantar flexion in the ankle immediately prior to vertical lift-off? If the final stride prior to lift-off is too long, the effect will be a negation of horizontal momentum. On the other hand, if the final stride is too short, there will not be an optimal range of motion to transfer horizontal momentum into vertical momentum. Obviously, during the airborne portion of the movement phase there is no base of support. Upon landing, as much of the back as possible should strike the surface to dissipate the force.

There should be no excessive movements of the **head and cervical spine** throughout any phase of high jumping. During the preparatory phase, are the head and cervical spine maintained in a relatively stable and extended position? During the movement phase, does the cervical spine rotate or laterally flex? There is a very human tendency during the Fosbury Flop for an athlete to want to "look where he's going." This usually takes the form of cervical rotation to the right or lateral flexion to the right as in the case of our example. If cervical rotation occurs, this is usually at the expense of a counterrotation elsewhere in the spinal column. This has a tendency to upset the horizontal position needed to clear the bar. In terms of biomechanics, a track-and-field coach might well teach the athlete performing the Fosbury Flop to hyperextend the cervical spine at the same time the lumbar spine is hyperextending during the flight aspects of the movement phase. The combined hyperextensions of the lumbar and cervical spinal regions would have a positive effect in the counterforce that would be evident in the "lifting" of the pelvic region. This is a commonly observed series of spinal column movements in such activities as diving, rebound tumbling, and tumbling. Therefore, it might be good teaching practice to ensure that a high jumper who executes the Fosbury Flop has considerable training in the area of gymnastics.

Motions of the arms during noncinematographic analysis of high jumping should center around two important aspects of the performance: (1) What do the arms contribute to vertical lift? and (2) What do the arms contribute in terms of body equilibrium during the airborne phase? As noted in our example, both arms of the athlete are moved from extreme hyperextension into a forceful flexion at the time of vertical lift. This is an absolute necessity to add to the vertical momentum already being generated elsewhere in the body. During the movement phase, actions of both arms contribute significantly to spinal column rotations and then stabilize for a brief period of time to contribute to body equilibrium as the performer is moving diagonally across the bar. It must be remembered that motion of the high jumper includes velocity components both along and across the bar.

Noncinematographic observations of the legs should focus on specific joints within both legs during the three phases of the performance. During the preparatory phase following the initial run, are the series of flexions at hip and knee joints occurring with gravity at angles that will not interrupt the summation of forces and cause a negative acceleration? Will the flexions in the left knee and hip joints at the end of the preparatory phase during the stride provide an optimal range of motion and a vertical lift? At the end of the preparatory phase and start of the movement phase, is the velocity of the limb during the movement from right hip extension to right hip flexion great enough to contribute significantly to vertical lift? Is the range of motion great enough for right hip flexion at this point to add to vertical lift? Is the right hip extension-flexion pattern at the time of lift-off synchronized with flexion movements in the shoulder joints? Does the knee go through a complete range of motion for knee extension at the time of vertical lift? During the airborne phase, does the right hip joint move into an extended position? The fact that hip extension occurs at this time contributes to rotations observed concurrently within the spinal column, and it also places the legs in a symmetrical position for the final portion of movement over the bar. As the pelvic girdle or center of mass crosses over the bar, do the hip joints move into flexion and do the knee joints move into extension? If these movements do not occur, the performer is in danger of knocking the bar from the stands by hitting the bar with his heels or some part of the legs.

Hurdling
The example of the hurdler shown in figure 12.5 differs from previous illustrations because in this case the athlete is trying to obtain an optimal vertical lift component while maintaining a relatively large horizontal velocity. This combination of factors makes hurdling a skill more complex than the track start, shot put, or high jump skills.

The preparatory phase for hurdling is the movement of the athlete in running between the hurdles, which does not involve any projection into the air. The movement phase for hurdling can arbitrarily be divided into two aspects: (1) lift-off and (2) flight pattern or trajectory. The recovery occurs when the performer touches the track after clearing the hurdle and leads effectively into the subsequent running pattern in moving toward the next hurdle. In our example, the preparatory phase is from frame one through frame four. The lift-off phase is from frame five through frame eleven, and the flight phase is from frame twelve through frame twenty-eight. Frames twenty-nine through thirty-two show the recovery aspect of the skill. Hurdling, unlike our preceding three examples, requires considerable effort on the part of the performer to control counteracting ballistic movements of the upper and lower limbs. Such movements would interrupt the hurdler's summation of forces and decrease horizontal velocity.

Observations of the **pelvic girdle and lumbar-thoracic spine** are quite important in hurdling. During noncinematographic observations of the pelvic region, the coach should see a relatively smooth transition from horizontal motion into a low parabolic trajectory of the pelvic girdle as the athlete clears the hurdle. If there is a sudden vertical lift of the pelvic region during the lift-off and flight phases, this indicates that horizontal motion is being converted into undesired vertical motion. This would decrease the horizontal velocity and increase the amount of time to traverse the 110 meters or 120 yards.

During the preparatory, movement, and recovery phases, are there any rotational components through the transverse plane of the lumbar-thoracic spine during the airborne or flight phase? Is the upper body, lumbar-thoracic spine, flexed? If this is not done, there will be a greater surface area of the body presented to the air, which will create a greater resistance or drag and decrease horizontal velocity. Also, if the lumbar-thoracic spine is not flexed at this point of the skill, the center of mass will be excessively lifted during flight. This is

Figure 12.5 The high hurdling form of Willie Davenport, USA Olympic team. Preparatory phase: frames 1–4. Lift-off: frames 5–11. Flight phase: frames 12–28. Recovery: frames 29–32. *Courtesy Visual Track and Field Techniques, 292 So. La Cienaga Blvd., Beverly Hills, CA 90211.*

Theory of Practice: Techniques of Sport Motion Analysis

Figure 12.5—*Continued*

Noncinematographic and Basic Cinematographic Analyses

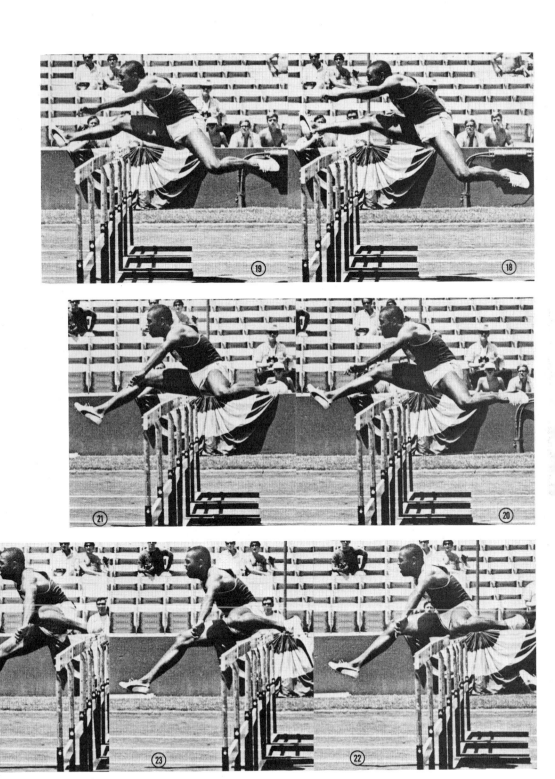

Theory of Practice: Techniques of Sport Motion Analysis 343

Figure 12.5—*Continued*

Noncinematographic and Basic Cinematographic Analyses

undesirable in hurdling, although it would be desirable in high jumping. At the time the athlete lands after clearing the hurdle, is the lumbar-thoracic spine moving into an extended position? This is highly desirable from the standpoint of maintaining a dynamic equilibrium, and it provides the performer with a good position to move down the track toward the next hurdle.

If the length of the running stride of the performer is adequate for his sprinting style, the two major checkpoints for **base of support observations** in hurdling are at lift-off in the flight phase and immediately upon landing during the recovery phase. In both cases, the athlete is in a unilateral weight-bearing position. At the time of takeoff, does the ankle of the supporting leg move through a complete range of motion for plantar flexion? This is desirable to help determine the optimal angle of projection. Upon landing, do the landing ankle and foot reach the ground in a position of slight plantar flexion (frame 28)? This is essential, because a considerable amount of energy and momentum must be absorbed as the ankle is dorsiflexed eccentrically. At this point, a good hurdler should have full extension in both knee and hip joints. If the base of support foot were in the normal anatomic, weight-bearing position at the time contact was made with the track, excessive forces would be transferred through the ankle, knee, and hip joints into the pelvis and spinal column. As a result, there would be some likelihood of trauma occurring. Another observation that can be made noncinematographically by the coach is to check on foot position during the running portion of the skill, as well as during takeoff and landing. Does the athlete "toe in" or "toe out"? Both of these movements, which may occur either at the knee joint or hip joint, are contraindicated. These rotary movements, which occur elsewhere but are observed in the foot, will cause excessive lateral forces that interfere with the desired linear motion.

There should be absolutely no motion within the **cervical spine and head** during hurdling. Although there are rather extreme or ballistic actions occurring within hip and shoulder joints, musculature around the cervical spine should keep this area stabilized. If the coach observes lateral plane or transverse rotations within the cervical spine, they will usually indicate that there are undesired rotations elsewhere in the spinal column.

The importance of the **arm movements** during hurdling is often overlooked. Arm movements contribute only minimally to vertical lift during this skill; their main purpose in hurdling is to counterbalance the ballistic actions necessary within hip joints. The action of the arms helps maintain rotational equilibrium during the process of hurdling. Just prior to reaching the hurdle, does the left arm diagonally adduct at the same time the right hip joint is moving into full flexion? This is a very important torque-countertorque aspect of hurdling. During the latter part of the flight phase, does the left shoulder move back into diagonal abduction as the left leg is being diagonally adducted at the left hip joint? Again, this is a torque-countertorque series of movements between the two body segments that is vitally important for maintaining dynamic equilibrium. During the flight phase, does the right shoulder joint remain in an extended position while the elbow is flexed almost to the midposition? At the beginning of the recovery phase, are the arms in good position for linear movement to the next hurdle? If arm adjustments have to be made at this point, the athlete has probably landed in a position of disequilibrium.

Leg motion is of vital importance in hurdling. One of the first movements which should be critically observed noncinematographically by the coach is the degree of flexion in the right knee joint immediately following the preparatory phase and during the first portion of the lift phase. At this point, does the right knee flex through a great enough range of motion to use effectively the conservation of angular momentum in the right lower limb? This knee action contributes significantly to the ultimate lifting force of the body in hurdle clearance. Also, this conservation of angular momentum contributes substantially to the velocity of the right leg moving from extension at the right hip joint into extreme flexion of the right hip joint. This must be a high-velocity, ballistic action of the right hip joint in order to lift the body into the proper trajectory for hurdle clearance. Does the right knee joint remain relatively extended throughout the flight and landing phases? Upon landing, does the right knee remain extended or flexed? Many inexperienced hurdlers will utilize knee and hip flexions upon landing to absorb shock instead of making the shock-absorption adjustment within the ankle movement. Upon contact with the ground, the extended right leg serves as a lever for subsequent linear motion. In regard to observations of the left leg, are the left hip and knee joints completely extended at takeoff? This extension in the left hip is important in view of the fact that it allows a maximum range of motion at takeoff. Are excessive left knee and hip flexions seen immediately prior to takeoff? If so, this is an indication that the athlete is trying to jump instead of hurdle. During the flight phase, does the left hip joint move from extension into diagonal adduction as the athlete crosses over the hurdle? This diagonal movement at the left hip joint, together with concurrent flexion of the left knee, constitute the optimal series of motions needed for the left leg to clear the hurdle effectively. At the point of impact with the ground, is the left hip joint moved into a position of full flexion? Is the left knee fully flexed at the time of impact? These two motions within the left knee and left hip joint are essential to provide a smooth transition from an effective rotational component to the linear force desired in running, i.e., subsequent motions within both lower limbs at this point should be primarily through the anteroposterior plane. If the coach observes any rotations through the transverse plane in hip and knee joints during the running or lift-off phase, these motions would interrupt the kinetic chain and cause an inefficient "flow of motion" for the hurdler.

The noncinematographic technique involves critical observations of joint angles and movement within body segments, as indicated above. In addition, the individual making the analysis must ask and answer relevant questions based on the accepted techniques for executing the skill. Finally, the implications of having the performer adhere to the governing biomechanic principles must be fully understood as they relate to the outcome of the performance. Ultimately, any joint motion adjustments made by the performer must be based on these principles in order to obtain the best results.

As stated previously, the only difference between noncinematographic and basic cinematographic analyses is the fact that the coach can study the performance at his or her leisure by viewing film. There is no time pressure to make decisions while doing basic cinematographic analysis. The film can and should be viewed a number of times at various projector speeds, and critical aspects of the performance should be viewed frame by frame. This procedure can objectify many of the observations of the coach who has viewed the performance noncinematographically. In other words, basic cinematographic analyses are really an extension of numerous observations made without the advantage of utilizing film.

During practice sessions and games, the coach has the opportunity to observe the same performance by an athlete many times. Through this accumulation of observations of the same performance, the coach has an opportunity to discover positive and negative "motion tendencies" of the performer. These multiple observations during noncinematographic analysis are a positive aspect of this analytic procedure. In contrast, the coach observing a film loop of a performer, for example, is given the opportunity to view only one performance. The performance on film may or may not be indicative of the athlete's most commonly used performance style. This aspect of basic cinematographic analysis is a disadvantage in comparison to noncinematographic analysis. However, the performance on film can be observed more critically and sequentially without any pressure of time restricting the coach. This is a definite advantage of basic cinematographic over noncinematographic analysis. Because of the advantages and disadvantages of these two analytic styles or techniques, in the best coaching the two types of analyses will be used concurrently to objectify the nature of movements by performers.

Film for purposes of basic cinematographic analysis should be taken during athletic contests. This principle should be followed as much as possible in order to determine true movement trends. If film is to be taken of a given skill during a practice session, the coach should do everything possible to place the athlete under gamelike conditions. If not, the athlete is very likely to act out what she or he considers to be the stereotype of perfect mechanics. From an analytic standpoint, this type of performance on film is not indicative of what the athlete would actually do in a game situation. Another factor to consider in filming practice situations is that some athletes simply do not practice in the same way they perform, no matter how much extrinsic motivation is applied by the coach.

For purposes of basic cinematographic analysis, filming speeds of 32 frames/second or 64 frames/second are recommended. The determining factor for filming speed is the nature of the skill being filmed. For a ballistic type of action such as a golf swing, the filming speed should be at least 64 frames/second. This filming speed will be fast enough to determine the nature and characteristics of the golfer's motions. It will not clearly define what is happening during the angular motion of the golf club. For most body motions, 32 frames/second is entirely adequate. For example, 32 frames/second is fast enough for the study of such activities as hurdling, wrestling, movement of football players, and basketball shooting. From the viewpoint of expense, there may be times during analyses of sports where 16 frames/second would be adequate for basic cinematographic analysis. In school situations, this is one of the economic factors in filming.

Figure 12.6 shows a sophisticated camera unit commonly used for filming interscholastic and intercollegiate athletic events. It allows the choice of optimal film speeds for the activity being recorded on film. The external power pack decreases the weight and facilitates usage of the system.

Figure 12.7 shows an excellent 16mm analysis projector designed for conducting basic cinematographic analyses of game film. This projector has complete stop motion designed to protect the film from burning, and it has constant focus. It also has a frame-by-frame, flickerless projection mode that protects against eyestrain and fatigue of the physical educator. The hand-held frame counter and remote control device allow convenient selection and repetition of the desired action sequences both in forward and reverse directions. These are all desired features of an analysis projector of any make or model.

The development of portable videotape units has brought this medium into extensive use in many school systems for recording athletic events. Videotape has the advantage of nearly instant playback so the performer can be observed immediately following the performance. The disadvantage of most systems which are currently being used in educational institutions is that they are difficult to use as compared to frame-by-frame analysis of film. However, in many cases television can be used to objectify the noncinematographic analysis that has been

Figure 12.6 A commonly used 16mm game film camera system, Bolex Rex-5 with Lafayette power pack and camera drive. *Compliments of Lafayette Instrument Company, Lafayette, IN.*

Noncinematographic and Basic Cinematographic Analyses

Figure 12.7 Basic cinematographic analysis 16mm projector, Lafayette Analyzer with digital frame counter. *Compliments of Lafayette Instrument Company, Lafayette, IN.*

made previously, since the coach can view the same actions several times. The videotape can often be used for a positive reinforcement of a skill improvement. From the economic point of view, the fact that videotape can be reused many times makes it an important tool for analysis in daily teaching and coaching situations.

Modern computer technology has developed systems which can directly process the information in the videotape signal to recognize patterns or calculate changes from one frame to the next. These are proving useful in research situations to reduce videotape data. Unfortunately, the cost and complexity of such systems preclude their use in the usual teaching-coaching situation.

In viewing a performance at the basic cinematographic level, it is essential to use a segmental approach to the analysis. The frames in which specific body-segment motions are decisive can thus be isolated by the coach and studied with specific reference to that body segment. One advantage of the cinematographic record is that it is possible to observe synchronization of various body-segment movements much more precisely than in a noncinematographic observation. To return to figure 12.5 of the hurdler, the first observations focus on the **pelvic girdle and lumbar-thoracic spine.** The two motions to be observed most closely in the analysis of the frames during this action are the transverse rotation of these segments of the body and the angle of lean for the lumbar-thoracic flexion during passage over the hurdle.

Hurdling

Observation of transverse rotation can be made by scanning the frames and observing specific points on the athlete's jersey. It will be observed in figure 12.5 that there is a slight left rotation in frames one through five; this is a normal occurrence of the running sequence between hurdles. Starting at frame six, the lumbar-thoracic region is returned to the midposition facing down the track, and will be observed to remain within a few degrees of that position throughout the entire takeoff phase and flight over the hurdle. No further rotation occurs during these phases. In the observation of a film of an unskilled performer, rotations often occur due to improper motions of arms or legs. These rotations provide excessive torque on the lumbar-thoracic region; and the extraneous motions will be quite evident by observing these points on the athlete's jersey. Such motions represent a disequilibrium during the flight phase and indicate energy outputs not coordinated with the desired motion of the athlete down the track.

Flexion of the lumbar spine during the flight phase can be observed in frames fourteen through twenty. The purpose of this flexion is twofold: (1) to reduce the height to which the center of mass must be lifted in the passage of the various body segments over the hurdle, and (2) to provide a more streamlined effect and thus reduce the drag coefficient of the performer. It should be remembered that for velocities attained by a rapidly moving athlete, the effect of the retarding force of air can be several pounds. This force opposes the vector of motion and tends to reduce horizontal velocity while the athlete is airborne. Since the purpose of a hurdling event is to maintain maximum possible horizontal velocity, any action which minimizes these retarding forces helps to decrease the total time for the race. Minimizing the total vertical lift of the performer's center of gravity also minimizes trajectory time in air. In this case an excessive lift not only reduces horizontal velocity but also greatly increases the amount of time the athlete is airborne and unable to exert further forces to move down the track. It should be observed in frames twenty-two and twenty-three that the athlete is clearing the hurdle with a minimum excess space. This also helps minimize the lifting of the center of mass.

With respect to the **base of support** in this action, the two critical regions are the takeoff during frames seven through eleven, and the recovery phase in frame thirty-two. In the sequence involving the takeoff, the important point to watch is the high degree of plantar flexion that helps lift the center of mass and determine the angle of projection. This is seen in frame twelve. Several points should be noted with reference to the base of support during the time of landing. It is important that the landing be made with the foot pointed straight forward in the direction of motion. This precludes any dissipation of force in an unintended direction. Secondly, the landing is also made with motion from plantar flexion to dorsiflexion in order to absorb downward momentum, with minimum shock to the remainder of the body. Finally, it should be noted that as the body reaches the full weight-bearing position, the center of mass is almost directly over the point of support. Thus, the forces from that point of support have no component in the backward direction to reduce horizontal velocity. The subsequent plantar flexion supplies a force to accelerate the runner horizontally as he moves into the running phase to approach the next hurdle. This is seen in frames thirty-one and thirty-two.

With respect to the **head and cervical spine,** stabilization or complete lack of motion throughout this sequence is of importance. As was indicated earlier, any excess motion would indicate a case of disequilibrium or a situation in which energy is being placed into nonproductive movement.

Observation of the **right arm** of this performer indicates one of the shortcomings of cinematographic analysis of a single event. Since the event can be seen from only one view, there are likely to be hidden segments of the body. In this case, it will be noted that from frames twelve through twenty-nine, the arm is virtually hidden. In this case, the fact that it cannot be seen is significant, since it means the athlete is dynamically stabilizing the arm with a shoulder extension and slight elbow flexion during the flight phase of this performance. This helps stabilize the lumbar-thoracic region in the midposition avoiding extraneous torques due to motion at the right shoulder.

Motion of the **left arm,** on the other hand, is used to counter the strong torque being generated by motions of the legs. Left-arm motion must begin before the athlete becomes airborne. A strong diagonal adduction of the **left shoulder** is observed in frames one through twelve, with the shoulder remaining stabilized thereafter until frame eighteen. The purpose of this initial motion is to provide a torque which counters the torque produced by motion of the right leg and also to furnish a small amount of lift during the termination of the upward component of diagonal adduction. Utilization of an extreme range of motion in this diagonal adduction will result in an excessive raising of the center of gravity, contributing to a prolonged time of flight.

In frames eighteen through thirty-two, the **left shoulder** undergoes a ballistic diagonal abduction which serves two purposes. First, the torque generated by this diagonal abduction tends to balance the countertorque being generated by the forward motion of the left leg passing over the hurdle. Second, the downward vertical component of the diagonal abduction is tending to provide a slight lift to maintain the center of mass over the hurdle during the final part of the trajectory. It will be noted that this combination of moves leaves the arms in the proper position for running immediately upon impact with the track following the flight portion of the hurdle. This can be observed in frames twenty-nine and thirty. This decreases the time necessary to begin linear acceleration during the running phase, and again decreases the total time of the race.

Another observation which can be made with respect to the **hands** is the position of the fingers throughout the various frames recorded in this motion. A tightly clenched fist is quite often an indication of excessive internal tension within the muscular structure. This leads to a decreased performance. Another indication of this condition can often be found in the position of the mouth and jaw muscles. This performer is remarkably free of internal tensions for an event such as the 110-meter high hurdles.

Cinematographic data are particularly helpful in the motions of the **knee and hip joints,** since these are quite complex and ballistic in this event. Frames five through twelve are especially significant in showing the contribution of the right leg to the lifting phase of the takeoff. It should be noted in frames seven through eleven that the knee joint is extremely flexed in order to obtain the maximum possible angular velocity through conservation of angular momentum and accelerating torque at the hip joint. This extreme angular motion of the right hip joint is necessary to enable the performer to become airborne as soon as possible, and to furnish the major portion of vertical lift necessary for this event. This high rotational velocity is also an aid to extreme hip flexion observed during the initial portion of the flight phase toward the hurdle. This extreme flexion can be seen in frames thirteen and fourteen. During these frames, knee extension necessary to bring the right foot over the hurdle can be observed. The observations of the left leg in frames nine and ten during the lift phase reveals a very small flexion at the knee and hip. This is significant, since it indicates the performer is hurdling and not jumping during takeoff. Frames eleven to thirteen indicate the plantar flexion governing the angle of projection into the air for the trajectory. Frames fifteen through twenty-nine indicate the ballistic diagonal adduction of the left hip joint, which serves the purpose of bringing the left foot over the hurdle following the pelvic region. It will be noted that this performer stops the diagonal adduction as the motion intersects the anteroposterior plane and immediately begins the extension movement of the hip and knee joints associated with the running phase. This can be observed in frames thirty-one and thirty-two.

The right knee is maintained in a nearly stabilized extension from frame sixteen through the culmination of this performance, and the right hip, approaching the landing point, moves into extension so the leg can serve as a lever for rotation as the performer moves into the running phase.

Examination of frames eight, fifteen, twenty-two, and twenty-eight can be used to indicate the synchronization of the beginning and terminating phases of arm and leg motions. This is an advantage which the basic cinematographic technique has over noncinematographic analysis. It is impossible even for the most experienced physical educator to observe precise synchronization of widely separated body segments. However, this can be done quite easily on film.

Frames twenty-nine, thirty, and thirty-one indicate the motion of a second performer who is trailing behind the principal performer. The reader should compare these with frames twenty, twenty-one, and twenty-two. What are the differences in the styles of the two performers? At 32 frames/second, for example, how much difference in time is indicated that one might attribute to these differences in styles?

High Jump
Earlier in this chapter the high jump was examined noncinematographically, using the Fosbury Flop as shown in figure 12.4. In actions such as the high jump, cinematographic analysis is necessary because of the large number of motions which take place within a very limited time span. In the successful execution of the high jump, these motions need to be almost perfectly synchronized. This synchronization is difficult to determine noncinematographically. Thus, observation made during practice or meets should be supplemented by cinematographic analysis, at least at the basic level.

The preparatory run to the bar as shown in frames one through fifteen of figure 12.4 serves several purposes. Basically, the athlete at this point develops kinetic energy which will be partially converted into potential energy in lifting the center of mass for passage over the bar. In addition, he is developing horizontal velocity which will carry him across the bar during the actual jumping phase. It should be noted that horizontal velocity developed during the run, as illustrated in frames one through seven, is greater than the optimal velocity for sequencing of motions to clear the bar. In frames eight to fifteen the athlete will actually suffer a deceleration as a result of the motions necessary to place him in the best position for jumping.

Also important in this run phase is the lowering of the **center of mass** as observed in frames eight, nine, ten, and eleven. The purpose of this strategy, which is more pronounced in some jumpers than in others, is to provide a greater range of vertical motion while still in contact with the ground. It also places the athlete in a more stable equilibrium throughout the last stride onto the jumping foot. The backward lean of the lumbar-thoracic region, as shown in frames thirteen, fourteen, and fifteen, is actually a compensatory movement for ballistic anteroposterior motion of the arms.

During the initial part of the movement phase, the **lumbar-thoracic and pelvic regions** are stabilized but rotate through an angle of approximately 90°, as seen in frames nineteen to twenty-five. This is the time of maximum rise of the center of mass of the performer. The angular motion is arrested by compensatory torques provided by the left arm and right leg. These are completed by frame twenty-six, with the performer's chest facing upward in the correct position for passing over the bar. At this point, as seen in frame twenty-six, the very important hyperextension of the lumbar spine is initiated. This hyperextension is continued through frame thirty, as the lumbar-thoracic portion of the body passes over the bar. Because of the low position of the legs during this phase and the extreme closeness of the remainder of the performer's body to the height of the bar, his center of gravity will lie near or below the bar level. One of the advantages of this form in the high jump is the fact that the performer can sometimes clear the bar without his center of gravity ever reaching that height. The hyperextension of the lumbar spine serves the purpose of forcing the lumbar-thoracic segment, the shoulders, and the cervical segment downward, with resulting counterforce upward on the pelvic segment. As the athlete's pelvic area passes over the bar, he returns to an extension of the lumbar-thoracic spine as a preparation for joint motions within the lower limbs. The synchronization of the lumbar hyperextension with the horizontal velocity as the athlete is passing over the bar enables him to clear it with a minimal raising of the center of mass. Throughout the movement phase the horizontal motions of the performer include a component passing across the bar together with an almost equal component sliding along the bar. During the preparatory motions it is essential that both optimal magnitude and angle of horizontal motion be maintained in order to supply these components.

The critical frames with reference to the **base of support** for this high jumper are frames twelve through twenty. Frames twelve through fifteen illustrate the planting of the jumping foot. Frames fourteen and fifteen illustrate placement of the foot in a normal weight-bearing position, in such manner that horizontal momentum carries him into an optimal position for moving vertically. It should be observed at this point that the vertical lift arises from three sources: (1) actions of the arms upward, (2) action of the right leg moving upward, and (3) forceful extension of the left hip, knee, and ankle in raising the entire body. The counterforce necessary to the execution of all three of these segment movements is delivered by the ground onto the left foot. It is for this reason that the final placement of that foot with respect to the position of the other body segments is so vitally important. It will be noted in frames twenty and twenty-one that for this athlete the range of motion for plantar flexion seems to be somewhat limited in the left ankle. This decreases the total lift that he is obtaining in this jump. With an increased range of motion, he would be able to remain in contact with the ground longer and supply still greater upward velocities to all three of the aforementioned segmental motions. At this point in the skill, he needs a greater range of left plantar flexion and subsequent drills should be designed to reinforce that motion (force-time relationship).

A sequence of photographs illustrates the stabilizing of the **cervical spine and the head** throughout the entire preparatory phase and initial portion of the movement phase. In frame nineteen there is the beginning of a lateral flexion to the right at the cervical spine. Beyond frame twenty-seven the angle of viewing completely hides this portion of the body. As discussed earlier in this chapter, it might be advantageous to combine a hyperextension of the cervical spine with the hyperextension of the lumbar spine, and likewise a return to extension in the cervical spine at the time of return to extension in the lumbar spine. This combination would enhance the force-counterforce actions discussed in connection with lumbar hyperextension. The athlete's attention should be focused on these motion adjustments during practice.

During the preparatory phase, motion of the **arms** into shoulder hyperextension allows use of the arms as a lifting device. In frames one to four the arms are being held in an abducted position, rather than participating in the natural running motion. The critical points for observation in this phase of the movement are frames twelve and thirteen, where the shoulders are in an extreme bilateral hyperextension. The symmetrical nature of this motion is necessary to obtain maximum lift without introducing an unwanted torque by a stronger motion of one arm than the other. The maximum extension should occur just before planting the jumping foot, as seen in frame thirteen. In frames thirteen to nineteen the arms are brought forward and upward into a bilateral shoulder-joint flexion. This provides a large amount of upward-directed momentum used by the body throughout the jumping process. As the athlete exerts a force with the shoulder musculature to stop this upward motion of the arms, the resulting counterforce provides a lift to the remainder of the body. During the flight over the bar, diagonal abduction of the left shoulder, together with extension of the left elbow joint, is used to provide the countertorque to stop rotation of the body. This is seen in frames twenty-seven, twenty-eight, and twenty-nine. This action, plus the diagonal abduction of the right shoulder, adds to flight stability during passage over the bar. A horizontal abduction of both shoulder joints accompanies the return to extension in the spine in preparation for landing. This is seen in frames thirty-one through thirty-four. This lifting of the arms is forcing the total body downward before the legs have cleared the bar, and it detracts from the total performance. It is possible that the use of a total body "tuck position" in this latter phase could contribute slightly to the maximum possible height of the jump. Such a "tuck" would have to be initiated about frame thirty.

Frame seventeen indicates the **hip and knee flexions** which represent an optimal compromise between the range of upward motion available, power to be obtained from concentric contractions of the left knee extensors, and maintenance of an optimal horizontal velocity. Although a greater range of motion and muscular force might be obtained through a higher degree of flexion at the left knee and hip joints, it would interrupt the summation of motions in such a way as almost to negate the preparatory phase.

The prime component of vertical lift in the Fosbury Flop comes from the ballistic flexion of the right hip. The rapid upward motion of the right leg, interrupted in frame twenty, provides the major portion of the upward component of momentum during the initial part of the movement phase. In the example shown, the ability to move this hip rapidly into flexion from the extension shown in frame thirteen is partially diminished by the failure to flex the knee joint completely during this motion. In frames fourteen through eighteen this joint is carried at about a 90° flexion, whereas an angle of about 130° should be possible for this type of motion. The reader should compare the motion of the right leg in this illustration with that of the hurdler discussed previously, in which a more optimal use of this conservation of angular momentum was made. A higher degree of flexion reduces the total moment of inertia of the leg, thereby allowing a much greater acceleration with the same torque being produced at the hip. This, in turn, produces a larger upward velocity before the interruption, as indicated previously in frame twenty.

Once airborne, the right leg returns to a partial extension at the hip in order to become symmetrical with the left leg, as in frame twenty-eight. As the lumbar-thoracic spine returns to extension in frame thirty-one, the hips begin the flexion necessary to allow the heels to clear the bar. This combination of hip flexion with partial knee extension completes the final phase of passage over the bar.

Figure 12.8 shows a sequence of film frames of a javelin thrower following his approach run through the crossover step into the culmination of the throw and follow-through.

Javelin

In this sequence, the *preparatory phase* is extended from frames one through twenty. This begins before frame one, in the preliminary run, which is not shown in these photographs. This extended preparatory phase is a necessary part of the javelin throw because of the complex relationship between the various segment motions that must be synchronized in order to achieve optimal velocity and release angle of the javelin. Because of the length of the javelin, it is awkward to place the body in the correct position with the proper muscles on stretch in order to achieve a final optimal velocity. Also, this velocity must be directed at an angle between 30° and 45° above the horizontal, and the javelin should be tilted slightly above the line of motion in order to achieve the flight interaction with the air. This requires a complex combination of rotational movements involving the pelvic girdle, spinal column, and right shoulder joint.

Following the extended preparation, the *motion phase* is short. This occurs in frames twenty-one through twenty-five, a total time of about one-eighth of a second. After release following frame twenty-five, frames twenty-six to thirty show the beginning of the *follow-through*. This strategy is designed to prevent the javelin thrower from fouling by crossing the line. No formal recovery phase is required, since the performance is completed with this one action.

The motions of the pelvic region and lumbar-thoracic spine are extremely important in the javelin throw during the final aspect of the preparatory phase, and provide an excellent example of the Serape Effect as described in Chapter 7.

Figure 12.8 The javelin throw as executed by Janis Lusis, USSR. Preparatory phase: frames 1–20. Motion phase: frames 21–25. Follow-through: frames 26–30. *Courtesy Visual Track and Field Techniques, 292 So. La Cienaga Blvd., Beverly Hills, CA 90211.*

Figure 12.8— *Continued*

Noncinematographic and Basic Cinematographic Analyses

During the crossover-step portion of the preparatory phase, the coach should observe the magnitude of the vertical lift of the body. If this is excessive, a portion of the kinetic energy gained during the initial run is being converted into vertical motion rather than being retained as horizontal motion. In the example, the maximum vertical position occurs near frame six but can be seen to be only four to six inches above the average vertical position throughout the sequence. Throughout the preparation phase, the **lumbar-thoracic spine** is held dynamically stable in lateral flexion combined with a slight rotation to the right. This stability exists through frame sixteen at which time both lateral flexion and rotation are increased as the muscles are placed on stretch in preparation for the motion phase. This lateral flexion of the lumbar-thoracic spine is necessary in order to correctly establish the subsequent angle of release of the javelin.

Frames seventeen through twenty illustrate the Serape Effect as used in this event. Left transverse pelvic rotation combined with right rotation of the lumbar-thoracic spine places the large, left spinal column rotators on stretch to optimize the subsequent lumbar-thoracic rotation, as seen in frames twenty-one through twenty-six.

The role of the left transverse pelvic rotation needs to be emphasized. Omission of this movement at the initiation of the motion phase will have an adverse effect on the velocity of the throwing limb as well as on the subsequent velocity and horizontal distance traversed by the javelin. This left transverse rotation of the pelvic girdle is the initial movement in the Serape Effect. The torque produced by this motion results in a countertorque by the lumbar-thoracic spine which provides optimal rotation of the upper portion of the body. These **pelvic and spinal column motions** not only place large spinal-rotator musculature on stretch, but also subsequently increase the total range of motion through which force can be placed against a javelin. Frames seventeen through twenty of figure 12.8 show a classic example of the use of the Serape Effect in the javelin throw.

The placement of the **base of support,** especially during the final step, is necessary in order to allow the proper sequence of rotations to take place without loss of either linear or angular momenta. The final placement of the right foot in this sequence is observed in frame thirteen. During this performance, placement was made with the right foot forward with respect to the center of gravity of the performer. As observed in frame thirteen, the backward lean of the performer causes forces being transferred from the ground to include a horizontal component somewhat impairing the horizontal velocity gained during the run. It would be better if the landing could occur with the body position observed in frame fourteen.

Placement of the left foot during the final weight-shifting process, as observed in frame twenty, is also critical. The foot must be placed to the left of the midline of the body and pointed in the intended direction of the throw to provide stability and allow pelvic and spinal rotations to occur. Foot placements should be timed to allow a smooth shifting of weight from one to the other, with minimal loss of horizontal velocity.

The **head and cervical spine** are maintained in a relatively stable position through most of the preparatory phase, as indicated in frames one through twenty.

In frames twenty-one through twenty-five, an extreme cervical rotation is observed. This adds to the rotation of the lumbar-thoracic region and also aids in diagonal adduction of the right shoulder joint. This is another example of the use of the torque and countertorque principle in order ultimately to increase the total velocity of the throwing limb. At the point of release seen in frame twenty-six of this sequence, the performer is not looking in the direction of flight of the javelin. If he attempts to look in this direction of motion too soon, the summation of motions critical to the event will be interrupted. The head does not return to the position which allows the performer to observe javelin flight until frame twenty-eight, approximately one-tenth of a second *after* release.

The prime function of the **left arm** throughout this performance is to provide balance and add to the torque-countertorque sequence of motions within the final throwing phase.

In frames one through ten, the left shoulder is in a position of diagonal adduction as an aid to equilibrium of the body. In frames eleven through twenty-one, the left shoulder is diagonally abducted, and the left elbow moves into extension. This places muscles on stretch and aids in the final production of torque within the spinal column for rotation at lumbar-thoracic spine, and the final medial rotation and diagonal adduction of the right shoulder. In frames twenty-two through thirty, although the left arm is hidden by the body, the left shoulder is adducted through the lateral plane to provide a balance mechanism to offset ballistic motions of other body segments. Flexion occurs at the left elbow as a result of the ballistic motion of the right arm through the crossed-extensor reflex.

Through frame sixteen, the **right arm** is diagonally abducted at the shoulder in a stable position. The wrist is also stabilized in slight hyperextension in order to maintain the desired angle for the javelin with respect to the horizontal. Any wobbling motion of the javelin during the preparatory phase not only results in inefficient use of energy by the performer, but also may contribute to poor flight characteristics of the javelin after release. Since flight orientation of the javelin is so critical and a spin about the longitudinal axis must be given in the final phases, it is highly important that the orientation of the javelin during the preparatory phase be a fixed part of the performance.

Frames seventeen through twenty-five provide an example of development of angular momentum through diagonal adduction of the right shoulder joint through the high diagonal plane. There is a subsequent increase in limb velocity by a decrease of the moment of inertia through flexion at the right elbow. This flexion is observed in frames twenty-three and twenty-four of the sequence. Also occurring in frames twenty-four and twenty-five is the medial rotation of the right shoulder joint. This provides a positive acceleration prior to release of the javelin. As in previously described throwing motions, it is necessary to observe that this medial rotation takes place *before the release of the object*. If it is delayed until the follow-through phase, it cannot contribute any forces to increase the velocity of the sport object. Following release, medial rotation of the throwing limb can only contribute to the aesthetic aspect of the performance.

During the crossover or diagonal action of the **legs,** the performer in this sequence is minimizing vertical lift while maintaining a horizontal velocity by an inversion of the left transverse tarsal and subtalar joints. A complete plantar flexion of the left ankle at this point would provide an undesired vertical component of velocity. This would result in a loss of horizontal velocity. During the landing on the right foot, as observed in frames thirteen through fifteen, it is necessary that the right knee joint undergo flexion. The motion is eccentrically controlled by the right quadriceps femoris muscle group. This places these muscles on stretch for the subsequent powerful extension of the right knee as seen in frames eighteen through twenty-one. This knee extension provides the push against the ground and precedes the pelvic rotation so vital to the Serape Effect. Frame twenty shows the leg position which provides the optimal horizontal force component for initiating this rotation. At the same time, there is a minimizing effect on any vertical component of motion at this stage of the performance. If the performer simply pivots through this step, it will be extremely difficult to obtain the necessary pelvic rotation to achieve an optimal Serape Effect and subsequent ballistic rotations.

In frames eighteen through twenty, the left hip is placed in slight flexion, with the knee stabilized in an extended position. This provides the base on which the final rotation takes place. As the body moves forward, the hip gradually returns to an extended position.

The javelin throw is an excellent example of a sequence of motions timed to produce a skilled performance as described in earlier chapters. An optimal performance requires a force generated from the right leg, which provides a rotation for the pelvic region. In turn, this produces rotations at lumbar-thoracic spine followed by diagonal-plane motions at the shoulder and flexion-extension at the elbow. Each of these motions must be performed with proper timing to achieve a maximum performance.

Following the basic cinematographic analysis, teaching-coaching recommendations are made, based on the observations. The most serious problem is defined, and this should receive the immediate attention of the coach and athlete at the next practice session(s).

Kicking   Figure 12.9 shows the phases for a soccer-style kick as used in American football. The pictures were processed from actual frames of 16mm "drill film" taken at 32 frames/second by a typical, nonprofessional school photographer. *One purpose of including this figure is to give the reader some realistic insight into the type of film one must work with when making analyses from game or drill film.* Videotape is worse. It can be seen that data extraction for intermediate cinematographic analysis would be difficult from this type of film; nevertheless, film of this quality can be utilized for making relevant observations leading to better instruction based on biomechanic principles.

In figure 12.9 there are four selected frames to depict the preparatory, motion, impact, and follow-through phases of the skill. Figure 12.9*(a)* illustrates the preparatory approach to the ball, with the athlete moving linearly at an approximate 45° angle to the goal line. Figure 12.9(b) shows the left lower limb ("plant leg") stabilized to initiate rapid acceleration of the kicking limb at the start of the motion phase. Figure 12.9(c) is the point of impact, or the conclusion of the motion phase. Figure 12.9(d) shows the follow-through and the results of the torque-countertorque inherent in this skill. The observations made below were from the total (frame-by-frame) film sequence.

Soccer-style kicking incorporates strong rotational motions in the **pelvic girdle and lumbar-thoracic segments.** These rotations during the preparatory and motion phases are necessary in order to place muscle groups on stretch. This myologic component observes the biomechanic principle of lengthening the interval of time over which the force is effective. As can be seen in figures 12.9(a) and 12.9(b), the performer has right lateral and left transverse rotations of the pelvis, and slight right transverse rotation of the lumbar-thoracic spine. The degree of these rotations does not conform to the stereotype of perfect mechanics for this skill. The athlete is not utilizing the Serape Effect efficiently, and the pelvic girdle remains dynamically stable throughout the remainder of the performance. The ultimate result of the torque-countertorque can be seen in the follow-through, figure 12.9(d).

Two factors related to the **base of support** must be considered in this skill: (1) the position of the support foot in relation to the body mass is important in preserving equilibrium during the strong ballistic motions involved, and (2) the position of the foot with respect to the ball is critical in attaining the most efficient plane for the kicking limb to apply force to the ball. There appear to be no problems for the athlete in figure 12.9 in maintaining equilibrium throughout the skill. However, the placement of the left foot is too far from the ball; this does not allow the impact between the ball and kicking limb to take place at the point of maximum velocity of the foot. This reduces effectiveness when greater distances are required for field goals. In reaching for the ball, there is a dissipation of the rotational motion of the kicking limb.

The **cervical spine and head** remain in a flexed and stabilized position throughout the performance. The eyes remain in contact with the ball and with the initial location of the ball after impact. If the cervical spine were rotated through a transverse plane, there would be a negative, compensatory effect on the rotation within the lumbar-thoracic region.

The **arm motions** contribute to preserving equilibrium and to maximizing the moment of inertia of the lumbar-thoracic region providing greater torque development. A more rapid diagonal abduction of the right shoulder joint would facilitate a greater right transverse rotation of the lumbar-thoracic spine. This athlete abducts the right shoulder joint a few degrees, then stabilizes it; therefore, he does not obtain any reciprocal benefit from the motion.

The **hip and knee motions** appear to be adequate through all phases of the skill. Figure 12.9(b) shows optimal knee flexion for kicking limb velocity through conservation of angular momentum. The kicker uses a forceful diagonal adduction of the right hip joint, as seen in figures 12.9(c) and 12.9(d).

Figure 12.9 The soccer-style kick: (a) preparatory phase, (b) motion phase, (c) impact, and (d) follow-through. (An example of non-professional film.)

(a)

(b)

Noncinematographic and Basic Cinematographic Analyses

(c)

(d)

Figure 12.10 Lateral
and anterior camera
angles of a softball pitch
performed by Linda
Schulz of Southwest
Missouri State University.

(a)

(b)

Noncinematographic and Basic Cinematographic Analyses

The athlete shown in these photographs has three interrelated problems in his execution of soccer-style kicking. In rank order of importance they are—

1. Improper placement of the left (support) foot with respect to the ball in the motion phase;

2. Inadequate right transverse lumbar-thoracic rotation in the preparatory phase; and

3. Passive use of the right arm (shoulder joint) during the preparatory phase.

Coaching must be based on analyses, and the suggestions should be positive and designed not to confuse the student-athlete. Three critical problems have been defined for the kicker, but only one adjustment in kicking should be made initially. For example, drills should be designed for the kicker to reinforce the conclusion that a change in left foot placement is needed and desired. It is entirely possible that the correction of this fault may also alleviate to some degree the other two problems noted above, without burdening the athlete with too much detail initially. It is essential, however, that the coach know *WHY* adjustments are made in the skill of the athlete.

As mentioned earlier in this chapter, performers should be viewed from several angles to analyze at the noncinematographic level. This is even more critical for basic cinematographic analysis, and must be a consideration in giving the filming directions to people who operate cameras during contests or practice sessions.

Camera and Observation Angles for Analysis

Ideally, performances should be viewed anteriorly, posteriorly, and laterally. For some skills, diagonal views may be substituted, due to unusual angles of release or other critical motions. Camera angles are limited at times, due to safety considerations or psychological factors related to the performer being filmed. As an example, a low anterior view of a hammer thrower utilizing a regular lens might be desirable from an analytic standpoint, but the camera operator would be in jeopardy! Psychologically, high jumpers do not like to be filmed from the line of their approach. Practical considerations regarding observation angles must also be taken into account.

Figure 12.10 shows lateral and anterior views of a softball fast pitch. It can be seen that the lateral view fails to clearly depict a major fault in this performer. The placement of her left foot while striding in the preparation phase is inappropriate. The nature of the stride problem can been seen more readily in the anterior view rather than the lateral view. By striding across the midline of the body with the left lower limb, she negates rotations within pelvic and lumbar-thoracic segments necessary for force development. This also alters the plane of motion to be traversed by the throwing limb. For an effective pitching motion, the stride with her left foot should be adjusted several inches to her left. This problem could not be identified and ameliorated if one observed the performance from a lateral perspective only.

In this chapter there was a discussion of the two most commonly used types of biomechanic analysis, noncinematographic analysis and basic cinematographic analysis. These two levels of biomechanic analysis will be utilized most frequently by teachers of physical education and by athletic coaches during their

Summary

day-to-day relationships with students. The intermediate cinematographic analysis technique discussed in chapter 14 will not be used as extensively in the daily teaching-coaching process.

Four examples of noncinematographic analyses were presented: (1) sprint start, (2) shot put, (3) high jump, and (4) hurdling. The primary purpose for presenting the sprint start as a noncinematographic analysis was to accentuate a performance where linear acceleration is the critical factor. In shot-putting, it is essential that rotational body motions be controlled to provide linear motion for the shot, with maximum force application through the time of release. High jumping involves a smooth transition or transfer of horizontal to vertical motion. In contrast, hurdling involves maintenance of a maximum horizontal velocity during performance of vertical lift. In each of these examples a segmental analysis approach was stressed.

The examples of hurdling and high jumping were continued through the basic cinematographic level. In addition, examples of javelin throwing and soccer-style kicking were added. A short discussion of the importance of observation and camera angles was also included.

13

Extracting Data Manually from Film

The purpose of this chapter is to assist the student to gain an understanding of the procedures and principles underlying the manual extraction of biomechanic data from film. *These are used for intermediate level analyses.* This chapter is an introduction to intermediate cinematographic analysis of sport film.

In the previous chapter we discussed the qualitative analysis of performance data, either during live action or from film or video records. Most of the analyses made by active teacher-coaches will obviously be at this level, since the amount of time available for an individual analysis is necessarily limited.

However, at some point in the analysis of sport motion, techniques in which numerical values are developed will have definite advantages. The purpose of biomechanic analysis rests on the ability to describe and compare motions precisely on a numerical basis. The entire concept of scientific measurement involves establishing a numerical expression of some property of a system; and the function of biomechanic measuring instruments is to supply these data.

A prime advantage of numerical expression is that it is formulated as an objective quantity, and therefore can be communicated with a high degree of precision to other people in biomechanics or related fields. The establishment of measuring standards has resulted in rapid gains in the correlation of sport studies with other scientific areas of medical and physiologic research. The field of biomechanics is now closely associated with other areas of human biology. As a result, physical educators can now benefit greatly from the results of such research, and vice versa.

There are also coaching advantages to be gained by quantitative analysis of data. Permanent numerical records make possible the adjustment of present performances in the light of past efforts. For example, a coach who finds that a football backfield is taking a few tenths of a second longer to perform play execution than did previous groups may want to modify the blocking strategy in the line on such plays. Or, a baseball coach may want numerical data on pitching speeds to determine more accurately the state of anatomic readiness of various staff members.

The value of such data in sport is attested to by the vast amount of statistics now compiled and maintained by professional and international teams for use in their coaching, evaluation, and player selection procedures. While it is true that individual performances are influenced greatly by current environmental and psychological factors, individual performance trends (both positive and negative) which are developing gradually are often overlooked in the absence of numerical records.

A second advantage of numerical or quantitative analysis of performance is in the psychological motivation it can furnish to some athletes. Efforts to improve one's "sport statistics" can sometimes provide extra incentive, for example, in those periods of a team sport season where the championship has been won or lost. The use of bar or line graphs and other techniques can display clearly the analysis data of individual performance. These provide motivational effect upon some athletes.

Although numerical data are sometimes directly displayed by the measuring equipment, as in a radar or infrared speed gun, it often is necessary for the teacher-coach to extract manually the data from film or other records. Even with the use of sophisticated equipment systems it is necessary for the physical educator to understand the theoretical bases and procedures by which the numbers are assigned in order to perform calibration and checking procedures.

Probably the best method to attain this learning objective is to perform the task of manually extracting performance data from film during one's first exposure to biomechanic subject matter. This should be done prior to actual experiences with more sophisticated scientific hardware such as computerized automatic film reading digital analysis systems.

The manual data extraction process is viewed by the authors as a logical learning progression from simple to complex procedures. The student should have a greater understanding and appreciation of more sophisticated data extraction theory, procedures, and equipment if he or she has learned the more cumbersome manual techniques initially. The simple manual extraction of data, once learned and understood, can lead to better comprehension of the more complex procedures encountered later in the professional preparation process or in the field.

Pragmatically, for many physical educators, it is imperative that they learn how to extract data from film manually, because their secondary schools and universities may never provide them with the budget to purchase the more expensive hardware needed to obtain and analyze performance data. This is an economic fact of life; however, it does not mean that one cannot use one's knowledge of biomechanics and quantitative analysis skills to improve the performance levels of the students.

Numerical data are normally obtained from film or video tape by projecting the images onto a flat surface. Simple segment drawings or direct measurements of the coordinates of interest can be made easily and quickly. (See figs. 13.3 and 13.4.) Correct filming and measuring procedures can reduce the time and effort required, while assuring consistency and accuracy in manually extracting the numerical data from film.

This chapter is designed to assist the student in learning these basic procedures. Once these are learned, the progression should be made to performing the same or similar functions by utilizing computerized data analysis systems. Undergraduate students taking a biomechanics course should be assigned projects to manually extract data from film.

Quantitative analysis can be made from most film taken during instructional or athletic situations. It is fully recognized that this type of film often has limitations, especially when compared to the filming criteria used for research purposes. *Observations and calculations must take these limitations into account, but this does not negate the use of film of this type for instruction utilizing biomechanic subject matter.*

Numerical data can be extracted from game film if plans and preparations are made for the filming process in advance:

1. Place the camera operator in a fixed position at a right angle and known distance to the photographic field. The most difficult quantitative measurements are those where the performer is moving either toward or away from the camera. The distance of the camera from the photographic field should be great enough to minimize the rotational components of the camera during the filming process. If possible, distances from the camera to the photographed subjects should be known. Also, the camera should be on a tripod whenever possible, to minimize vertical and horizontal errors (fig. 13.1).

2. Filming speeds in frames per second must be known. Most sport performances can be filmed for intermediate analysis at filming speeds ranging from sixteen to eighty frames per second. Sport skills involving high-velocity ballistic actions of limbs should be filmed at sixty-four to eighty frames per second. Game film for team sports such as basketball and football can be filmed at speeds of sixteen to thirty-two frames per second. However, as one example, if a football coach wants to make detailed observations of a passer during a game, filming speed on the passing downs should be increased to sixty-four or eighty frames per second.

3. Known scales in feet or meters should be included in film backgrounds when possible. As an example, vertical and horizontal distance lines can be painted on the face of spectator stands next to a track. Film of runners or hurdlers would include the runner as well as these known distance marks on the stands. This would facilitate analyses of vertical and horizontal components of velocities and accelerations.

4. If possible, include a large clock with a sweep hand at least once per filming within the photographic field. Most swimming teams have large interval training clocks with sweep hands which can be included in the film background for swimming or any other sport. This serves as a check on the calibration of film speed within the camera. Spring-driven cameras tend to lose frame speed as they wind down; consequently, they should be wound frequently. Electric-powered cameras are recommended, but not all athletic areas or fields have electric outlets readily available. Battery-powered cameras are available, which do not limit the mobility of the operator (fig. 13.2).

The use of these standardized filming procedures makes the subsequent steps in data extraction more efficient and useful. With fixed background material, scale factors and calibration standards, film can be used many times, and the

Figure 13.1 A tripod mounted Photo-Sonics camera utilized to film a field event at the Montreal Olympics. *Courtesy Instrumentation Marketing Corporation, Burbank, CA.*

Figure 13.2 A 16-millimeter Photo-Sonics camera being utilized during a National Football League game. The belt battery pack around the camera operator's waist allows access to any part of the stadium or sidelines during the contest. *Courtesy Instrumentation Marketing Corporation, Burbank, CA.*

Extracting Data Manually from Film

comparison of performances has added validity. Problems or errors which have developed can be compensated for by limited changes in background or technique.

The easiest numerical calculations to be made on film are time measurements. Modern camera framing rates usually vary less than five percent except under the most stressful environmental conditions. Therefore, a simple counting of the frames indicates the time intervals accurately, with the smallest interval depending on the framing rate.

Since many performance parameters occur in tenths of seconds, nominal framing rates will allow these to develop over an interval of a few frames. Higher framing rates give added accuracy, but are more costly in terms of film used.

A rate of 32 frames per second means that each frame represents a time interval of .03125 seconds. Multiplying this by the number of frames counted gives the performance interval with a probable error less than .03 seconds. For long intervals it is useful to have a projector or film editor with a frame counter attached, but this is not absolutely necessary.

As mentioned earlier, time measurement by this procedure can be calibrated and checked by including a clock with a sweep second hand in some filming fields.

As an example of such an analysis, the film of the high jump by Sapka (figure 12.4) in the previous chapter shows the takeoff foot in contact with the ground for 6 to 7 frames. That represents a time interval of about 0.2 seconds. His head crosses the bar in frame 25 and the knees in frame 39. This indicates a time interval of about 0.4 seconds. The fact that three major muscular changes (spinal hyperextension, lumbar flexion, and hip flexion) must be made within this interval indicates the coordination required for this performance.

Numerical extraction of positional data is performed by super-imposing simplified drawings of specific information or points from several frames onto a single background. Measurements against this background then furnish a record of the changes of position as a function of time.

In order accurately to transfer information from the picture on the film into a set of numbers, it is usually necessary to enlarge the image. This may be accomplished through projection on a screen by use of such instruments as Vanguard and Kail projectors. Any projector used for this purpose must be capable of *sustained single-frame projection*. This is necessary because of the time it takes for making measurements or drawings from film. If only a few measurements are needed from each frame, it is usually more efficient and easier to make direct ruler measurements on the projected image. For example, in the sequence on hurdling, figure 13.3, measurements can be made directly from the pictures to indicate the distance from the takeoff point to the hurdle and from the hurdle to the landing point. Using the measured height of the hurdle as a scaling factor, it is possible to calculate the actual flight distance during this performance. In this example, a measurement will show that the horizontal distance from the

Figure 13.3 Hurdling trajectory.

Extracting Data Manually from Film

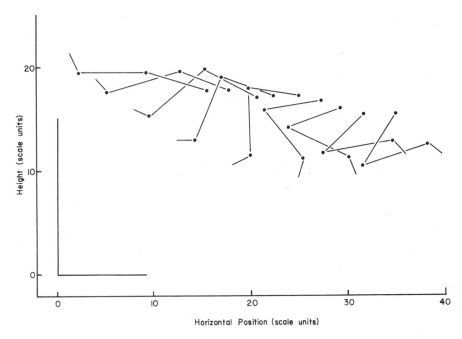

Height (scale units)

Horizontal Position (scale units)

Figure 13.4 Lower limb segments of hurdler in figure 13.3.

takeoff point to the base of the hurdle is almost exactly twice the measured distance of the hurdle height. Since a high hurdle is 42 inches in height; this means that the takeoff point was almost exactly 7 feet in front of the hurdle. Similar measurements will indicate a distance to the landing point of approximately 4.45 feet. Of course, these distances should not be expected from every high hurdler, but it is possible to determine rather quickly from film this type of performance parameter for any hurdler.

If, however, a series of measurements is needed on each frame or if comparative measurements between successive frames are to be used, it is usually more efficient to redraw the essential parts of the image in a more permanent form. In order to do this efficiently, a person must be familiar with the basic subject matter of anatomic kinesiology. Rather than make a detailed contour drawing of the performer on a frame-by-frame basis, it is only necessary to mark positions of specific landmarks and/or joints. These can be used for subsequent analysis. Such a drawing is shown in figure 13.4.

Figure 13.4 represents a portion of the right lower limb movement of the hurdler as seen in figure 13.3. *Approximate positions* of the center of the right hip and knee joints and the medial malleolus of the right tibia were recorded from frames eight through sixteen of figure 13.3. The recording of information from successive frames on a single drawing is extremely useful for calculation of rotation and linear velocities and accelerations.

It is useful to show the projected image either on a screen with a graphic scale or on a piece of graphing paper. If care is taken to adjust the size of the image to the graphing scale, positions can be read most conveniently. It is easier to construct a set of numbers by comparison to a graphic scale than it is to make actual measurements with a ruler. It should be noted that in such a drawing, angles can be measured directly by use of a protractor. Angles measured in this manner have some built-in error. This will be discussed later in this chapter.

It is also necessary to include in the drawing some part of the known scale factors from the film in order to scale the physical motion. In figure 13.4, the primary scale factor was the forty-two-inch height of the hurdle. This scale factor was chosen because the hurdle is located at the same distance from the camera as the hurdler. This provides a known linear dimension.

Quantitative analysis is often avoided because of the mathematical calculations involved. In most cases, the mathematical computations are not of a complex variety. They require only a few steps of multiplication or division, but they usually make use of a large amount of data. As a consequence, it becomes an extremely time-consuming procedure if performed manually. Most school districts or universities will have at least a desk-type calculator which can perform these multiplications and divisions. In fact, it is becoming quite common in many school systems to have either a computer terminal facility or a small self-contained desk-top computer capable of repetitive calculation. Use of such instruments can assure saving of time for the type of calculation involved at this level of analysis. It is only necessary to tell the computer once what actions are to be taken with the data, and from then on the results of any performance are simply entered as new data. This elimination of time as an inhibiting factor should ultimately lead to a greater use of biomechanic analysis at this level, provided the physical educator has a background in computer operation.

Once the superimposed drawing is completed, the measurements should be made directly against the scale of the drawing. A table of such measurements is easy to check and correct or to augment with new measurements when desired. Scale measurements should be made both of the actual motion involved and of the background objects. These latter numbers are used to establish the scaling constants for converting scale measurements to actual distances in performance space.

Figure 13.5 shows a table of numbers used in the transformation from the scale drawing given in figure 13.4 to the actual spatial coordinates of the approximate joint centers of the hip and knee as well as the medial malleolus at the ankle. In this case, the information from the various frames was projected and recorded on a scaled paper with magnification to the point that the height of the hurdle represented fifteen scale units. This gave a scaling factor of fifteen units equal to 42 inches—3.5 feet. The numbers in the upper portion of the figure represent values as read from the graph paper. The lower portion of figure 13.5 presents corresponding spatial coordinates. The horizontal coordinate represents distance from the hurdle, while the vertical coordinate is the height of the body segment above the ground. Thus, conversion between upper and lower tables was simply a matter of multiplying each number by the factor 3.5 divided by 15.

Extracting Data Manually from Film

Frame Number	8	9	10	11	12	13	14	15	16
Scale position (scale units)									
Hip									
Horizontal	33.5	30.4	28.0	26.2	24.2	21.3	19.8	17.0	14.8
Vertical	15.5	15.5	16.8	16.7	17.1	17.0	17.0	17.7	17.6
Knee									
Horizontal	30.3	26.3	23.0	20.5	19.0	16.3	14.7	11.8	8.8
Vertical	10.5	11.7	14.0	15.8	17.8	18.8	19.8	19.5	19.3
Ankle									
Horizontal	36.7	33.3	29.0	24.4	19.4	13.7	9.2	5.0	2.2
Vertical	12.6	12.8	11.2	11.0	11.3	12.8	15.2	17.3	19.2
Space position (feet)									
Hip									
Horizontal	7.82	7.09	6.53	6.11	5.65	4.97	4.62	3.97	3.45
Vertical	3.62	3.62	3.92	3.90	3.99	3.97	3.97	4.13	4.11
Knee									
Horizontal	7.07	6.14	5.37	4.78	4.43	3.80	3.43	2.75	2.05
Vertical	2.45	2.73	3.27	3.69	4.15	4.39	4.62	4.55	4.50
Ankle									
Horizontal	8.56	7.77	6.77	5.69	4.53	3.20	2.15	1.17	0.51
Vertical	2.94	2.99	2.61	2.57	2.64	2.99	3.55	4.04	4.48

Figure 13.5
Changes from scale to space coordinates of the hurdler.

Determination of Angles from Film

The determination of angular information on film is often less accurate and reliable than the determination of spatial positions. One reason for this lies in the fact that the planes in which rotation is taking place are often tilted in relation to the line of sight. This means that the reference circle for angular measurement contains components within or outside of the film plane; therefore, angles cannot be objectively determined.

Careful planning of the camera position relative to the motion to be studied helps in angular determination. If the major motions occur perpendicular to the line of sight of the camera, all angles will be more precisely viewed. But, in many cases errors of several percent will exist in the individual items of data.

Angles are determined by imagining a reference circle on the ground, on walls, or on some other visible surface. A direction within that circle must be chosen to represent zero degrees, and the angles are measured with reference to that landmark direction. Often there will be objects or planes within the filmed image to establish an easily maintained set of directions.

It is standard practice to measure angles counterclockwise as the positive direction and clockwise as the negative direction. Orientations are commonly expressed as being between zero and 360 degrees around the reference circle. Practice is necessary to place angles in the proper quadrant and to estimate the fraction of that 90 degree sector to describe the total angle.

Figure 13.6 shows a film sequence of a championship discus performer. This differs from the hurdle example because the action occurs within a relatively small spatial environment and is primarily rotational in nature. Techniques for the discus throw utilize (1) a combination of body and upper-limb rotations to produce a high velocity of the discus, (2) release of the discus at the proper flight angle, and (3) the proper flight orientation of the discus to achieve optimal interaction with air. If these technique factors are followed, maximum distance of discus flight will be attained. In order to perform this type of discus technique successfully, a sequence of rotational motions must be timed to produce an optimal velocity and angle of release of the discus.

Quantitative analysis of the frames in figure 13.6 provides the reader with examples which differ from the parameters analyzed previously on the hurdler. Measurement of angles and positions in this performance is much more difficult and therefore much less precise than in the previous example. In the case of the hurdler, motions of interest were all perpendicular to the line of observation by the camera. For this reason, the critical changes of position could be accurately measured from one frame to another. In the discus example, the essential changes are those in the direction in which certain segments of the body are oriented. The determination of these directions is less precise than determinations of spatial positions. These filming limitations must be recognized, and they have to be accepted during intermediate cinematographic analysis. In order to achieve precision in the measurement of joint angles, triaxial procedures must be utilized. Actual performances in competitive situations and film taken during games differ from film taken of a performance in a laboratory. Although these limitations are evident, there is still a possibility of gaining considerable information about a performer's actions, data which can be used for instructional or coaching purposes. *This realistic objective of using subject matter from biomechanics is very relevant to enhancing the quality of instruction in physical education and athletic programs.*

To begin the analysis of this performance, a set of data was extracted from the frames shown in figure 13.6. Directions of orientation were determined for (1) the pelvic girdle, (2) lumbar-thoracic spine, (3) cervical spine and head, and (4) the right upper limb. The latter were oriented with respect to the earth. The numbers within figure 13.7 were assigned with reference to the discus circle, using the following criteria: (1) 0° represents the direction toward the observer or camera, (2) 90° is to the right of the frame, (3) 180° represents the direction away from the observer, and (4) 270° represents the direction to the left of the frame. In addition, the vertical height of the discus in each frame was measured with reference to the scale of the chain link fence in the background of the frame. Because the fence is farther away from the camera than the performer, the two-inch mesh cannot be used directly for scaling. However, the size of the discus

Figure 13.6 The discus throw as performed by Jay Silvester, USA Olympic team. *Courtesy Visual* *Track and Field Techniques, 292 So. La Cienaga Blvd., Beverly Hills, CA 90211.*

Theory of Practice: Techniques of Sport Motion Analysis

Figure 13.6—*Continued*

Extracting Data Manually from Film

Frame Number	7	8	9	10	11	12	13	14	15	16	17	18	19	20
Pelvic direction (°)	135	150	165	190	230	250	270	300	320	350	0	20	50	70
Lumbar-thoracic direction (°)	90	120	150	170	210	230	245	260	290	320	340	0	20	40
Head direction (°)	150	165	180	190	230	260	280	300	330	10	30	60	75	90
Right upper limb direction (°)	320	340	350	30	100	120	135	150	180	200	220	250	260	270
Discus height (scale units)	20.5	19	19	16	13	12	12	11.5	—	13	13	15	16	18

Frame Number	21	22	23	24	25	26	27	28	29	30	31	32	33	34
Pelvic direction (°)	90	120	150	180	200	220	240	260	290	320	350	0	0	10
Lumbar-thoracic direction (°)	60	80	110	140	150	160	180	210	250	280	320	0	10	30
Head direction (°)	120	150	170	180	200	220	240	250	270	310	0	30	50	70
Right upper limb direction (°)	290	310	340	0	20	40	70	90	130	180	220	250	280	310
Discus height (scale units)	20	23	24	25	24	23	20	17	13	—	14	19	25	31

Note: In estimating the directions the following system was used: 0° toward observer, 90° to right of picture, 180° away from observer, 270° to left of picture.

Scale for height of discus determined from chain link fence in background.

Figure 13.7 Table of basic angular and height data for the discus performance.

(Diameter = 8.75 inches) can be used to establish the true scaling factor between motions in the planes of the performer and the fence. In the case of this sequence, the scaling factor was such that the diagonal distance across one of the mesh squares represents 2.9 inches. Figure 13.7 provides angular and height data which were determined from frames seven through thirty-four of figure 13.6.

By calculating the difference between the observed orientations of various body parts it is possible to measure the angle of rotation of various joints. This is illustrated for the above data in figure 13.8.

The position of the **cervical spine and head** in the discus throw is important from biomechanic and psychologic points of view. Rotations in the cervical spine can be used to supply a torque-countertorque effect to increase the angular ve-

Theory of Practice: Techniques of Sport Motion Analysis

Frame Number		7	8	9	10	11	12	13	14	15	16	17	18	19	20
Angle															
Cervical	(°)	60	45	30	20	20	30	35	40	40	50	50	60	55	50
Lumbar-thoracic	(°)	45	30	15	20	20	20	35	40	40	30	20	20	30	30
Shoulder	(°)	40	50	70	50	20	30	20	20	20	30	30	20	30	40

Frame Number		21	22	23	24	25	26	27	28	29	30	31	32	33	34
Angle															
Cervical	(°)	60	70	60	40	50	60	60	40	20	30	40	30	40	40
Lumbar-thoracic	(°)	30	40	40	40	50	60	60	50	40	40	30	0	−10	−20
Shoulder	(°)	40	40	40	50	40	30	20	30	30	10	10	20	0	−10

For explanation of angles, see text.

Figure 13.8 Table of joint angles and ranges of motion for the discus performance.

locity of the lumbar-thoracic region. Therefore, a large angle of rotation of the cervical spine is required preceding the final phase of the performance. When this is suddenly decreased, the countertorque follows. From a psychologic point of view, there is a strong tendency to want to see the direction in which one is moving. This human trait can be observed in the discus thrower. He not only rotates the cervical spine; his eyes are also rotated. This is observed in frames thirteen through eighteen of figure 13.6.

The cervical angle listed in figure 13.8 is the difference in orientation between head and lumbar-thoracic spine directions as given in figure 13.7. In essence, figure 13.8 provides ranges of motion associated with cervical rotation for each frame of the sequence.

Likewise, the difference between the orientation of the **lumbar-thoracic body segment and the pelvic girdle** is presented in figure 13.8 as the lumbar-thoracic angle. This rotation is especially important because it places the critical muscles on stretch for the final execution of the Serape Effect. This strong muscle action provides rotational acceleration of the lumbar-thoracic spine, and this motion is primarily responsible for the ultimate velocity of the discus.

The shoulder angle given in figure 13.8 describes the relative orientation between the lumbar-thoracic segment and the right upper limb. The numbers were chosen so a 0° angle was arbitrarily established as any position of the right upper limb lying in the lateral plane. The number of degrees shown in each frame in the table, therefore, represents the angle in which the right upper limb is *deflected posteriorward* through either diagonal or transverse planes. The negative number given for the shoulder in frame thirty-four represents a 10° anterior deflection, while the negative numbers for the lumbar-thoracic angle in frames thirty-three and thirty-four represent a rotation of the lumbar-thoracic spine beyond the stabilized position of the pelvic girdle at that point in the performance.

Extracting Data Manually from Film

Frame Number	8	9	10	11	12	13	14	15	16
Segment orientation									
Femur									
Length (ft)	1.39	1.30	1.33	1.35	1.23	1.24	1.36	1.29	1.40
Angle (°)	32.7	46.9	60.7	81.0	97.5	109.7	118.6	109.0	106.1
Tibia-fibula									
Length (ft)	1.57	1.65	1.55	1.44	1.51	1.52	1.67	1.66	1.54
Angle (°)	−108.2	−99.0	−64.8	−37.0	−5.0	22.2	50.1	72.1	89.3
Knee flexion (°)	39.1	34.1	54.5	62.0	77.5	93.5	111.5	143.1	163.2
Angular velocity (rev/sec)									
Femur	1.26	1.23	1.80	1.47	1.08	0.79	−0.85	−0.26	
Tibia-fibula	0.82	3.04	2.47	2.95	2.51	2.39	1.96	1.53	
Knee	−0.44	1.81	0.67	1.38	1.42	1.60	2.84	1.78	

Femur and tibia-fibular angles were measured clockwise with the vertical downward position representing zero degrees.

The knee flexion angle represents the angle between the femur and the tibia-fibula.

Body segment orientations and angles can also be calculated by finding the space coordinates of each end of the segment and determining the differences between the horizontal and vertical coordinates of these segments. The orientation and length of each segment can then be calculated by the following:

Figure 13.9 Angular positions and segment lengths of the hurdler shown in figure 13.3.

$$\triangle x = x_b - x_a$$

$$\triangle y = y_b - y_a$$

$$\text{Orientation angle with horizontal} = \tan^{-1} \frac{\triangle y}{\triangle x}$$

$$\text{Length} = \sqrt{\triangle x^2 + \triangle y^2}$$

Figure 13.9 presents data taken from the film of the hurdler shown in figure 13.3. Figure 13.9 indicates angular positions and lengths of the femur and tibia-fibula segments of the right lower limb and the angle formed at the hurdler's knee. These can be measured directly from figure 13.4.

The second part of figure 13.9 indicates the apparent length of the femur and tibia-fibular segments as calculated from the recorded landmark positions. This calculation was made primarily to check for errors in the recording of the landmark positions. Since these two lengths should remain relatively fixed, a calculation indicating a large discrepancy provides a check against any errors made in the earlier portion of the analysis. For example, it should be noted that the lengths of the tibia-fibular segment vary between 1.51 and 1.67 feet in all frames except frame eleven. The lower figure of 1.44 feet in that frame probably indicates about a three-fourth-inch error in the location of the malleolus. A progressive error in the measurement of such fixed lengths may indicate corrections necessary due to the camera angle during filming.

In the film sequence used for quantitative analysis in this chapter, an attempt is made to show what type of numerical data could be obtained from typical film taken of athletes in action. The user of this type of film must accept the fact that there is an accuracy loss one would not have when extracting data from research film; but this loss of accuracy is not great enough to negate its use for *instructional purposes*. For example, the accuracy of our input data for the performance film analyzed in this chapter was estimated to be ±5 percent when compared to film meeting research criteria. With this small loss of accuracy of the input data, no intermediate results should be interpreted as being more accurate from a truly scientific point of view. Thus, the rounding of numbers, approximations, and other mathematic and cinematographic factors are not particularly significant unless they introduce an error of greater than ±5 percent. This degree of accuracy loss is probably typical of what the physical educator working with game film would expect in most situations. The accuracy loss, however, could be greater or lower depending upon the control during filming. The numerical results of these analyses can be used to support or refute prior qualitative analyses. Such results, with their loss of scientific precision, limit the validity of their use to produce research findings.

Care should also be taken as to the methods used to present the results of quantitative analyses. Units should be selected to keep the sizes of the numbers in an understandable range, and the numbers should be grouped in a fashion which shows the comparisons desired in the study. In many cases it is better to show graphical plots of the results than to use long tables of numerical data.

Figure 13.10 illustrates a table of the numerical coordinates of the positions of the head, shoulder and hip of the hurdler discussed previously. Below that is a plot of the same data. While for some applications the actual numbers are needed, most people will understand the charcteristics of the hurdling motion more clearly from the graphical presentation. Presentation of quantitative analysis findings will depend upon the group you expect to communicate with during the dissemination of ideas. One's presentation or discussion with professional peers would be very different graphically and in content from the presentation of the same material to students or student-athletes. *Teaching-coaching is an art based on science. The art lies in how effectively the coach can communicate his or her knowledge to students. Therein lies a challenge, especially as related to the exercise sciences.*

It is essential to note that in the numeric calculation of the vertical position of the head, the total variation throughout this part of the hurdling process was from 5.90 feet in frame eight to 6.14 feet in frame fourteen. This represents a total rise of the head of less than three inches throughout the clearance of the

Frame Number	8	9	10	11	12	13	14	15	16	17	18	19	20	21	22	23	24
Height above ground (feet)																	
Head	5.9	5.95	5.97	6.02	6.07	6.11	6.14	6.11	6.07	6.02	6.02	6.02	5.97	6.07	6.02	6.02	
Shoulder	5.25	5.25	5.44	5.53	5.55	5.60	5.67	5.76	5.67	5.60	5.53	5.53	5.44	5.46	5.43	5.41	5.30
Hip	3.22	3.31	3.34	3.55	3.55	3.66	3.80	4.04	4.04	4.06	4.20	4.27	4.29	4.32	4.27	4.26	4.25

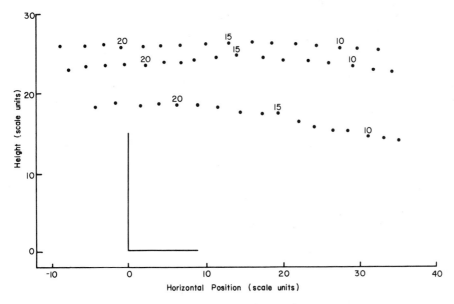

Figure 13.10
Tabular (above) and graphic (below) behavior of anatomic landmarks indicating the hurdler's trajectory as seen in figure 13.3. The top line of dots is the head, center line is the acromion process of the shoulder, and the bottom line is the hip joint. The numbers are the film frames for these anatomic landmarks.

forty-two-inch-high hurdle! While the shoulder shows a slightly larger variation throughout the process, part of this vertical motion can be attributed to the diagonal adduction of the upper limb as a portion of the torque-countertorque process. The vertical position of the center of the hip joint during passage over the hurdle was calculated to be between fifty and fifty-two inches. This is only eight to ten inches above the hurdle. When the diameter of the thigh musculature is considered, this is excellent hurdle clearance with minimal vertical lift.

Figure 13.11 represents a similar presentation of the data presented for the discus thrower in figure 13.8. The reader is also referred to figure 13.6 for the actual film of the discus performance from which these data were taken.

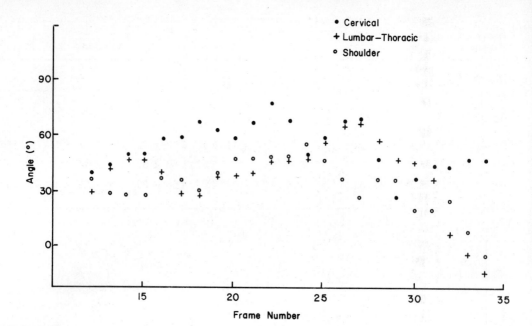

Figure 13.11
Sequential joint angles during the discus performance seen in figure 13.6. The reader is also referred to the data presented in figure 13.8.

Extracting Data Manually from Film

14

Intermediate Cinematographic Analysis and Biomechanic Research

Extracting numerical data for objectifying performance parameters is an integral function of intermediate cinematographic analysis. Therefore, the examples presented in chapter 13 should be considered as an introduction to intermediate level analysis procedures. One purpose of this chapter is to continue with the discussion of intermediate cinematographic analysis and its procedures by using additional examples from the hurdling (figure 13.3) and discus (figure 13.6) performances. An additional objective is to supply the beginning student of the analysis of sport motion with a brief overview of biomechanic research. A detailed discussion of research techniques and procedures is beyond the scope of an introductory text in biomechanics or the analysis of sport motion. For the student interested in learning more about research in this field, several excellent sources are cited in the Bibliography.

As the name implies, intermediate cinematographic analysis involves transitional elements from the first two levels on the analysis hierarchy and an initial progressive step of an academic nature into the biomechanic research level. Intermediate cinematographic analysis can also be used extensively in teaching-coaching situations to objectify those observations from the first two levels which tend to be more subjective. Unlike the first two analytic levels, mathematical computations are used during intermediate analysis to enhance precision and clarify observations made by the teacher or coach. As a result of these computations, considerably more time is required to undertake an intermediate-level analysis as compared to a basic cinematographic analysis. It is this time element which discourages extensive use of intermediate cinematographic analysis by busy teachers and coaches. However, when time permits, it is a functional and practical technique for potential use as an instructional adjunct in class and/or athletic situations.

Intermediate Cinematographic Analysis

The undergraduate student taking the first course(s) in biomechanics or kinesiology should be given the opportunity to extract numerical data manually from sport film. When students are fully acquainted with basic cinematographic techniques and have worked with quantitative problems involving calculations of such parameters as velocities and accelerations, they should progress to intermediate level projects for total sport motion analysis. The hardware for doing this need not be ultrasophisticated. Figure 14.1 shows two students working with a simple projector and a common computer terminal. Basic analyses and manual extraction of data from film can be accomplished by using any 8 or 16 millimeter

projector and a good calculator (Appendix C). Computer systems are desired, but not necessary to conduct detailed intermediate level analyses. Most college and secondary school physical educators have ready access to projectors and calculators, and these tools can be utilized to enhance instruction quality from information gained by quantitatively analyzing sport performance film. Most universities do not have sophisticated biomechanics laboratories.

Intermediate cinematographic analysis has two functions. First, it offers the practical objective of using biomechanics in the teaching-coaching situation. Second, it provides the interested student an introduction to some of the methods and procedures of biomechanic research.

How should findings of an intermediate level analysis be utilized? The immediate objective for a physical educator working with students in class situations, or with student-athletes, should be to utilize the findings of the biomechanic analysis to improve sport performances of those students and help them reach their optimum potential as performers. This objective is not of immediate concern to the research investigator. This is another major difference between what is described here as intermediate cinematographic analysis and biomechanic research. Subject matter from biomechanics can no longer be considered the exclusive property of the research investigator. The relationship of biomechanics subject matter to skill development is too important not to be utilized in day-to-day teaching and coaching situations. It can and should be utilized regularly by professional physical educators working with students of both sexes, all ages, and at every level of the performance continuum, from the adapted physical education student to the Olympic athlete preparing for international competition.

In the hurdling sequence described in the previous chapter (figure 13.3), a first step in intermediate analysis would be to calculate the velocities of selected points and segments from the positional data derived and presented there. By using the changes of position in successive frames and the time interval given by the framing rate, such velocities are calculated as the simple ratio of change of position per unit time.

Figure 14.2 presents the calculated numbers for linear velocities of the three landmark positions. These velocities require the use of the film speed of thirty-two frames per second to determine the time interval over which linear changes of position occur. They are expressed, in this case, in terms of horizontal and vertical components, as well as the total speed of the landmark position.

One interesting factor in these calculations is the extremely low values for the vertical component of velocity for the hip joint throughout the motion. Most of these values can be attributed to slight variations in the location of landmark positions during the graphic analysis phase. At this film speed, an error of one-half inch in location of a landmark position produces a velocity error of 1.3 ft/sec (½ inch × 32 = 16 inches = 1.3 feet). This problem of errors in velocities and accelerations is increased immensely for analyses made at higher film speeds. This is one reason why accurate studies of higher order accelerations require the techniques of biomechanic research.

In a similar manner, angular velocities can be calculated as the change of orientation per unit time. Figure 14.3 indicates the angular velocities of the leg segments of the hurdler taken from the data in figure 13.3. Again a film speed of 32 frames/second was used for purposes of calculating time intervals.

Velocity is then calculated by dividing the change of angle between frames by the time necessary for change to occur. It can be noted from these figures that the highest angular velocities are reached by the tibia-fibular segment. This portion of the lower limb benefits from hip flexion and knee extension forces. It can

Figure 14.2 Linear velocities of the hurdler seen in figure 13.3.

Frame Number	8	9	10	11	12	13	14	15	16
Hip velocity									
Horizontal	23.4	17.9	13.4	14.7	21.8	21.2	20.8	20.2	
Vertical	0	9.6	−0.6	2.9	−0.6	0	5.1	−0.6	
Total speed	23.4	20.3	13.5	15	21.8	21.2	21.4	20.3	
Knee velocity									
Horizontal	29.8	24.6	18.9	11.2	20.2	21.8	21.8	22.4	
Vertical	9.0	17.3	13.4	14.7	7.7	7.4	−2.2	−1.6	
Total speed	31.1	30.1	23.2	18.5	21.6	23.9	21.9	22.5	
Ankle velocity									
Horizontal	25.3	32.0	34.6	37.1	42.6	33.6	31.4	21.2	
Vertical	1.6	−12.2	−1.3	2.2	11.2	17.9	15.7	14.1	
Total speed	25.3	34.2	34.6	37.2	44.0	38.1	35.1	25.4	

All velocities in feet/second.

Theory of Practice: Techniques of Sport Motion Analysis

Frame Number	8	9	10	11	12	13	14	15	16
Angular velocity (rev/sec)									
Femur	1.26	1.23	1.80	1.47	1.08	0.79	−0.85	−0.26	
Tibia-fibula	0.82	3.04	2.47	2.95	2.51	2.39	1.96	1.53	
Knee	−0.44	1.81	0.67	1.38	1.42	1.60	2.84	1.78	

Figure 14.3 Angular velocities of the hurdler seen in figure 13.3.

also be noted that maximum elevation of the femur segment (hip flexion) is achieved in frame fourteen, and in the final two frames the femur segment is actually returning to the horizontal.

Similar calculations can be presented for the example of the discus throw.

Figure 14.4 represents the angular velocities of the pelvic girdle, lumbar-thoracic spine, and right upper limb. These were calculated by taking the difference in orientation between successive frames and utilizing the 32 frame/sec filming speed to transform the differences in angular position into angular velocity. In order to smooth out the "scatter" due to errors in determining precise angle positions, the average motion over a three-frame interval centered at the indicated frame was used. For example, in frames seven through nine there was a total change of orientation by the pelvic girdle of 55°. This is an average of 18° per frame. In other words, a total of 55° of rotation occurred between frames seven and ten. Multiplying this average of 18° per frame by 32 (the filming speed of the camera) and dividing by 360 provides the average velocity at this interval in revolutions per second. The calculation gives the value of 1.60, which is seen in the line for the pelvic velocity between frames eight and nine in figure 14.4. In this way, average velocities during each interval of the pelvic girdle, lumbar-thoracic spine, and right upper limb were determined.

Information about a performance can be absorbed by most individuals more efficiently if it is presented in the form of a graph rather than a table of numbers. Therefore, in figures 13.11 and 14.5, information from the previous tables has been plotted in graphic fashion to show the correlations between joint angles and the angular velocities at various times in the performance. It can be seen that both the cervical and lumbar-thoracic angles progressively develop to a maximum about frame twenty-eight. This decreases rapidly as the lumbar-thoracic spine is accelerated rotationally to its final peak velocity as noted in figure 13.11.

The graph of the velocity data shows certain items of interest. During the very early part of the performance, the rotational velocity of the throwing limb reaches a peak between frames ten and eleven. Paradoxically, this represents a rotational velocity nearly as high as the final velocity seen near the time of release. However, as can be observed in the frames of figure 13.6, the discus is not being held as far from the center of rotation of the body as it is in the final rotation, and the angular position at this point is not one in which the elevated trajectory could be given to the discus. Discus throwers, especially those at the

Frame Number	7	8	9	10	11	12	13	14	15	16	17	18	19	20
Velocity (rev/sec)														
Pelvic	1.96	1.60	2.40	2.49	2.40	2.04	2.04	2.40	1.78	1.78	1.78	2.04	2.04	2.04
Lumbar-thoracic	2.67	2.40	2.67	2.40	2.22	1.51	1.78	2.22	2.40	2.04	1.78	1.78	1.78	1.78
Right upper limb	1.78	2.40	3.40	3.40	3.11	1.51	1.78	1.96	2.04	2.04	1.78	1.51	1.16	1.51

Frame Number	21	22	23	24	25	26	27	28	29	30	31	32	33	34
Velocity (rev/sec)														
Pelvic	2.40	2.67	2.40	2.04	1.78	1.78	2.04	2.40	2.67	2.67	0.89	0	0.89	
Lumbar-thoracic	2.04	2.40	2.04	1.51	1.16	1.78	2.67	2.93	3.29	3.29	3.56	0.89	1.78	
Right upper limb	2.04	2.04	2.04	1.78	2.04	2.04	2.67	3.29	3.82	3.82	2.93	2.67	2.67	
Right upper limb (Corrected)									3.82	3.82	3.89	3.94	3.94	

Figure 14.4 Table of angular velocities for the discus performance seen in figure 13.6.

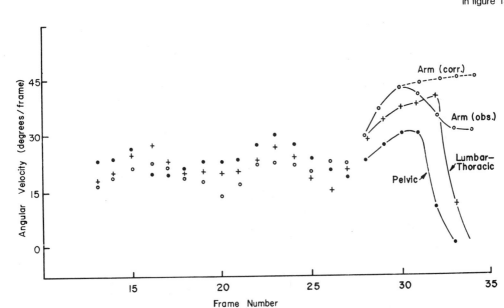

Figure 14.5 Sequential angular velocities during the discus performance seen in figure 13.6.

secondary-school level of education, who release the discus during the early stages of the throw at this early peak of velocity of the throwing limb may obtain fair results. However, they will never reach their full potential as discus throwers. One psychological problem for the athlete and the coach at the secondary-school level seems to be that a large and strong performer can win points for the team in meets without ever developing to his full potential. This is where the coach must interject his own philosophic view regarding the role of athletics in education. Points now or excellence later?

From the table in figure 14.4 and the graph in figure 14.5, there is an apparent decrease of angular velocity by the right upper limb in the final frames. This apparent decrease comes from the fact that the angles were measured with reference to the horizontal discus circle. In these final frames, the right upper limb is moving through a diagonal plane. Therefore, it has only a component of its motion apparent relative to the circle. The corrected numbers given in the final line for these frames represent the change from the horizontal component to the total motion in that frame. The dotted lines given for the motion of the right upper limb in the graph on figure 14.5 indicates the same set of numbers.

It will be noted in the final portion of figure 14.5 that there is a succession of maximal rotational velocities reached by the various body segments. The maximum velocity for the pelvic region occurs about frame thirty-one. The lumbar-thoracic spine achieves maximum velocity about frame thirty-three, and maximum velocity of the throwing limb occurs between frames thirty-three and thirty-four at the time of release. It is this excellent sequencing of motion that must be observed in a skill of this nature. This is a classic example of summation of motion by an elite athlete.

It will be noticed in frame thirty-three of figure 13.6 that this performer leaves the ground prior to release of the discus. This is allowable as a result of his excellent throwing style. A man of this size and moment of inertia is capable of accomplishing this without any loss to either body stability or final velocity of the discus. In contrast to this, such an action might not be advisable for smaller or younger performers who are not as tall, strong, and skilled in this very difficult event. They should have both feet in contact with the ground when the discus is released to take advantage of the force-counterforce principle.

Biomechanic Research	Biomechanic research is the ultimate level of the analysis hierarchy in studying human body motion. It is at this level that new knowledge is discovered and recorded in the literature for use by others. It often represents more abstract and theoretical studies, as contrasted with the more pragmatic approaches involved in coaching and the formal approaches of teaching. Many students may, as a result of their teaching-coaching experiences, decide that their interest lies at the research level.

Conducting biomechanic research requires extensive multidisciplinary professional preparation at both undergraduate and graduate levels. This includes study of mathematics through calculus, advanced physics, computer sci-

Figure 14.6 An automatic film reading digital analysis system consists of a front projection film reader console with digitizer table, control panel, cursor, floating keyboard and display, foot switch, x, y, frame LED display, film transport, light source and optics, plus PDS Digitizing Camera relay optics, and other features for extracting data. Photo-Sonics/PDS Model 80 *Courtesy Instrumentation Marketing Corporation, Burbank, CA.*

ence, human engineering, and other related science course work. As a consequence, the most advanced analytic step in the hierarchy, biomechanic research, is beyond the scope of this book, except to inform the reader that this level exists and follows logically after the student has attained an understanding of the first three levels of biomechanic analysis. *Most professional physical educators will utilize the first three levels of biomechanic analysis in the hierarchy, and be consumers as opposed to producers of biomechanic research.*

The purpose of this section is to present the biomechanic research level in a very brief and introductory manner. A few basic concepts about research as well as some of the instrumentation utilized in biomechanic research are presented to give the reader some insight into the potential scope and depth of biomechanic research projects. The interested and academically qualified student is encouraged to pursue this area of biomechanics.

Research involves controlled observations in which systematic investigations are conducted to seek new truths in an academic discipline. The results of such investigations serve to establish facts from which generalized principles can be developed. These principles are often described in terms of models or "laws" which organize, integrate, and express the known data about a given phenomenon. The models can then be used to design new or critical studies which lead to more data and a better understanding of the phenomenon. In the field of biomechanics, the principles and models produced by such research then serve as guides for action and training in the conduct of programs for sport, exercise, and dance.

Research usually involves descriptive analyses performed at several levels of sophistication. *Biomechanic research is conducted to solve delimited motion problems, accurately describe motion, and discover new knowledge about neuromuscular performances.* The most commonly researched parameter is the relationship between space and time relative to human performances. Recordings of motions of various body segments moving through space in a variety of paths and at different velocities concurrently require complex research techniques. Facilities for obtaining data from which descriptive analyses can be conducted require excellent equipment and a large financial outlay. Extensive labor-hour involvement is also an important consideration. As a result, only a relatively small number of biomechanic research facilities are in operation in colleges and universities throughout the United States.

Techniques of biomechanic research are changing rapidly due to the impact of developments in modern technology and instrumentation. The manned space flight program led to many new instruments and techniques for monitoring the actions occurring during human performance. Smaller, lighter detection equipment and advances in electronic transmission now allow measurements to be made without the bulky instrumentation and wiring so often used in the past. And developments in optics and computers now allow data analysis to be done by sophisticated systems (figure 14.6) instead of by tedious reading and calculation by hand.

Cinematography

Cinematographic procedures serve as the most common "tool" of biomechanic research for physical educators. In general, the cinematographic procedure outlined previously in this chapter under the heading "Intermediate Cinematographic Analysis" has many of the same components as biomechanic research. The principal difference lies in the precision of measurement to obtain accurate information for analysis. Another difference is that a researcher tends to delimit her or his study to a considerable extent in the laboratory situation; whereas analyses on the lower levels of the hierarchy are concerned with the totality of the skill in the game situation. Furthermore, the research worker is not concerned about immediate utilization of the findings of the investigation for instructional purposes.

Cinematographic data obtained during biomechanic research situations are more precise than data obtained during intermediate cinematographic analysis of game film. Control and precision are characteristics of research. These factors must be considered strengths; however, they are also weaknesses when athletic skills are studied. The reader should consider the following question in this respect: Is a performance of a shot-putter, as one example, filmed in a sterile laboratory situation in front of mirrors, on unusual surfaces, and with the performer wired for determination of joint angles and muscle-action potentials the same type of performance as that which is seen in competition? The superficiality of the testing situation would appear to be a logically significant limitation of biomechanic research techniques applied to athletic skills when compared with the other levels of cinematographic analysis discussed earlier in this chapter and in chapter 13. When possible and feasible, cinematographic data for research purposes should be obtained when performers are in competitive situations.

Specialized cameras which have precisely maintained filming speeds are available for use in biomechanic research laboratories. Some of these have very high framing rates to allow study of ballistic motions. In certain instances these are modifications of more common cameras, with electronic speed control to allow accurate framing to 500 frames per second (fig. 14.7). In other designs, the film advance utilizes fixed pins to engage the holes in the film, for precise positioning and holding of each frame. These pin-registered cameras (fig. 14.8) are capable of sharp, precise pictures controllable at speeds to 500 frames per second. Since their pin action is controlled by electronic pulses, they can be used under a wide range of conditions where synchronization or resonance is important.

The high speed camera shown in figure 14.9 was designed for filming with the specific objective of analyzing sport motion. It has film speeds continuously variable from 10 to 500 frames per second, and it can accomodate 1200 feet of film in its film magazine. This camera with its accessories is designed to acquire film data efficiently in sport settings, which can then be analyzed effectively through the data processing systems shown in figures 14.6 and 14.12.

The cameras shown in this chapter are examples of the type which should be utilized when filming sports. The cameras in figures 14.7 and 14.9 are versatile enough to be used in game filming situations for subsequent basic cinematographic analyses. They all can be used in laboratory settings, but are also used to film for research and other purposes in more dramatic situations (figure 14.10).

Figure 14.7 Photo-Sonics Actionmaster/500 high-speed camera with pistol grip shoulder pad and orienting viewfinder. *Courtesy of Instrumentation Marketing Corporation, Burbank, CA.*

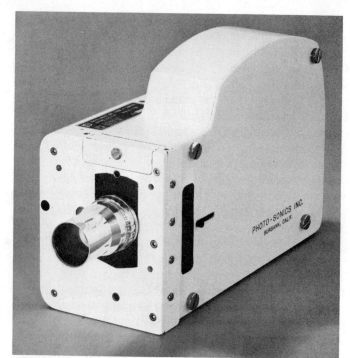

Figure 14.8 Photo-Sonics Model 16mm-1PL high-speed intermittent pin-registered camera, capable of internal control to 500 fps or external electronic pulse control. *Courtesy of Instrumentation Marketing Corporation, Burbank, CA.*

Figure 14.9 The Biomechanics/500 camera was designed specifically to facilitate the analysis of sport motion. Film speeds are continuously variable from 10 to 500 frames per second. *Courtesy of Instrumentation Marketing Corporation, Burbank, CA.*

Intermediate Cinematographic Analysis and Biomechanic Research

The Headingley Library

Figure 14.10
Research filming is not limited to the biomechanics laboratory. The Olympic Games serve as a "dynamic laboratory" for taking film for biomechanic research purposes. *Courtesy Instrumentation Marketing Corporation, Burbank, CA.*

Recording three-dimensional action on two-dimensional film constitutes a major problem of cinematography. This is due to parallax. *Parallax is the apparent displacement of an observed object due to a change in the position of the observer.* Because most biomechanic analyses involve measurement of angles formed by limbs and body segments moving at joints, parallax must often be overcome in planes of motion not parallel to the film, if accurate joint angles are to be determined. Unless the angle formed by two levers is in a plane of motion perpendicular to the line of observation by the camera, the *actual angle* of that joint will be different from that observed on film.

In research situations, parallax must be considered. To this end, three views of any given angle must be photographed to ascertain the actual angle of any given joint motion. This may be accomplished by using three synchronized cameras to film the performance. Or, it can be accomplished with one camera utilizing several mirrors set at strategic angles within the laboratory. The three-camera or the multiple-mirror technique using one camera must provide photographs of a given angle in which the visual axis *(the visual axis being a straight line perpendicular to the film surface)* of each photograph is perpendicular to the others. The measured angles from these three views are combined to yield the actual angle. This method of determining accurate joint angles during biomechanic research is known as *triaxial analysis*. There are several methods to attain this.

When filming a performer in the standing position, the three-camera technique involves placement of two cameras approximately at waist height of the performer, with their visual axes in a perpendicular relationship to each other. The third camera is placed above the performer with its visual axis perpendicular to the other two axes.

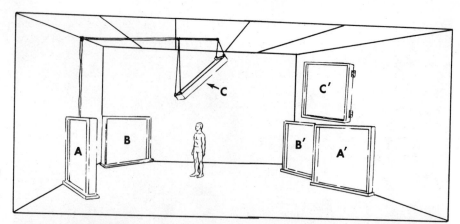

Figure 14.11 Triaxial cinematographic analysis mirror setting—a tool for biomechanic research.

Figure 14.12 Utilization of an automatic film reading digital analysis system— Model 78. *Courtesy Instrumentation Marketing Corporation, Burbank, CA.*

Figure 14.11 illustrates mirror placement in a laboratory setting in which only one motion-picture camera is used for triaxial analysis. The photographic area shown involves six mirrors ranging in size from five feet by five feet to five feet by eight feet. Mirrors *A* and *B* are positioned in such a way that the images produced in mirrors *A'* and *B'* are representative of the two visual axes that would be produced by using two cameras whose visual axes lie perpendicular to each other. Mirror *C*, the overhead mirror, produces an image in mirror *C'*. This is representative of the visual axis that would be produced by an overhead camera.

Intermediate Cinematographic Analysis and Biomechanic Research

The image in mirror C' is perpendicular to the visual axes of the other two cameras. It can be seen through this very limited discussion of triaxial analysis that it would be difficult to utilize this procedure in the competitive situation while taking game film. Other filming techniques are emerging which modify and simplify the procedures considerably.

With modern electronic data processing systems and computational techniques, it is possible to make corrections for filming angle and parallax that would have been too involved and time-consuming in the past. Computer programming is available to allow data to be recovered from an action filmed by two cameras at any two different viewing angles. This requires reading the film with an analyzer capable of measuring the horizontal and vertical components of position and the angles of various coordinate landmarks in the field of view (fig. 14.12).

Through the use of a dichroic mirror placed in front of the motion-picture camera lens, a clockface and grid may be superimposed over the image of the performer being photographed (fig. 14.13). Camera speeds may be checked accurately by placing a clock with a sweep-second hand in the photographic field. A grid may be used to provide reference points for measurements. In addition, placement of a carpenter's level in the scene provides a reference to the horizontal. It also allows for the presence of an object of known length.

Electrogoniometry provides accurate measurement of joint angles and ranges of motion through the use of recorded signals produced by an external electric source. Two lever arms hinged to form a joint are attached to a performer across joints of the limbs or body segments as shown in figure 14.14. The recorder produces a visual record of the number of degrees through which the levers move. The utilization of the electrogonimeter in research situations is a useful procedure to verify or determine ranges of motion and joint angles.

Modern electrogoniometric equipment uses solid state devices to make the detection equipment smaller and lighter; this also permits the changing of the measured angle into a series of coded pulses. These can be transmitted to a data processing system. This allows a continuous pattern of joint angles to be recorded during the performance of a maneuver by an athlete or dancer.

Electrogoniometry

Figure 14.13 Use of a dichroic mirror.

Theory of Practice: Techniques of Sport Motion Analysis

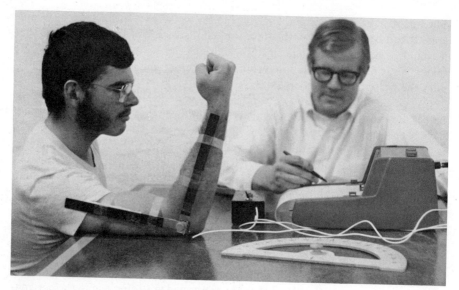

Figure 14.14
Utilization of the electrogoniometer to determine ranges of motion of joints during biomechanic research.

Force and Acceleration Measurement

Forces occurring in various performances can be measured and recorded by the use of strain gauges and piezoelectric devices. Strain gauges are devices whose electrical resistivity changes when they are deformed. Piezoelectric materials develop an electrical voltage when under stress. In either case, the device can be used to transform mechanical forces into electrical signals. These signals may be voltage or current levels that can be used to operate electronic recorders, or they may be in the form of pulse codes that can be processed by a computer.

Combinations of these devices are often incorporated into force plates. When three strain gauges or transducers are utilized, force can be measured through the three cardinal planes of the body. No appreciable movement of the platform on which the performer applies force is necessitated during the measurement. Through the use of multiple strain gauges and subsequent mathematical calculations, force in any direction may be determined. The research implications in sport are unlimited for force and counterforce determinations.

These devices may also be utilized in conjunction with a specially mounted mass to detect and measure accelerations. When the system is accelerated, the inertia of the mass causes a force that can be measured by the strain gauge or transducer. Again, these may be mounted in a system to determine the directional components of the acceleration.

Electromyography

The type and magnitude of muscle-action potentials being produced during a muscle contraction may be measured by electromyography. These data can be translated into the approximate amount of internal force being exerted during any motion. Correlations between force-time analyses and electromyographic information add to the depth of study of total performance. It is recommended that cinematographic studies of athletic skills also include synchronized electromyographic extraction of data. Figure 14.15 shows the physiograph being uti-

lized to obtain electromygraphic data from a subject performing work on a bicycle
ergometer during a research project to determine the efficiency of various bi-
cycle-seat settings for the sport of cycling. The electrodes are on the rectus femoris muscle.

It is sometimes more convenient to record the original data for biomechanic analysis on a single frame of film rather than on a series of frames. This can be
accomplished by using a series of light flashes to record multiple images as a
function of time. Very sophisticated stroboscopic flashes have been developed for
this purpose. These can provide accurately calibrated flash rates over a wide range
of frequencies, or they may be triggered by electronic pulses from other measuring devices. The resultant photograph then serves as primary data for measurements of position as a function of time for any part of the image.

**Stroboscopic
Analysis**

Because biomechanic research relies mainly on cinematographic techniques,
electronic data reduction methods become necessary to handle the large amount
of information produced from high-speed motion-picture photography. Quantitative information used to calculate such factors as velocities and accelerations
may be obtained through the use of motion analyzers with automatic readout
systems (figures 14.6, 14.12 and 14.16). The film image is projected in such a
way that measurements of X and Y coordinates, angles, time, and other information may be obtained from the film. This information is then fed into a computer system for an analysis of the data.

Computer Analysis

Biomechanic research techniques generate data at a prodigious rate. These
data, combined with the more complex analyses required for such items as parallax corrections and accurate scaling, require a vast amount of computation.
For these reasons and others, the use of computer techniques in biomechanic
research becomes a necessity rather than a convenience. Most research groups
in universities have access to a large computer facility, either through time-shar-

Theory of Practice: Techniques of Sport Motion Analysis

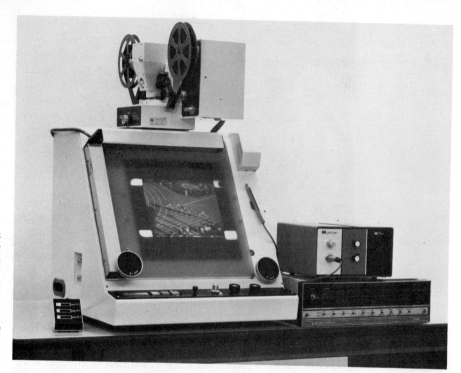

ing terminals or card-punch machines within their laboratories. Furthermore, computer-center personnel, especially in universities, have data reduction and computer skills to help analyze accumulated data. These services must be utilized during biomechanic research. Many new uses are being designed for computers in biomechanic research; consequently, the student planning to work in this field should arrange to take considerable course work related to cybernetics.

The Future for Biomechanic Research

The technological developments that have caused changes in the techniques of biomechanic research during the past few years are continuing at an accelerated pace. The miniaturization of detection and broadcasting equipment should soon allow a performer to be monitored in all of the areas described on the past few pages with little discomfort or added weight. This means that concurrent data on motion, forces, and muscle potentials could be routinely gathered during practice or the performance itself. This would do much to bring the theoretical and pragmatic fields of biomechanics closer together.

Lasers, infrared detection equipment, and microwave devices now make it possible for measurements of position and speed to be made directly during athletic performances. During the last two Olympics, laser-ranging equipment has been used for accurate measurement of distances in the field events. Infrared and microwave (radar) guns have become common for the measurement of pitching speed in professional baseball. The development of techniques and the more widespread availability of sophisticated equipment promise great strides in analysis, both in the teaching-coaching area and in biomechanic research.

Perhaps the most significant changes in the near future will be ascribed to the rapid development and use of microcomputer chips. These devices, which incorporate complete computer and control circuitry into a space smaller than a single letter in this text, promise extremely sophisticated measuring devices at a reasonable cost. In the near future there will be instruments in the field which will be able to measure a wide range of parameters concurrently, reduce the data, and report the performance characteristics of a performer almost immediately. Such capabilities should have a profound impact on the present methods of utilizing the exercise sciences in teaching physical education classes and coaching athletic teams.

Appendix A

Conversion Tables for Mechanical Units

This appendix has been compiled to provide conversion factors between the most commonly used sets of units for the description of human body motion. All possible conversion factors are not presented.

Although the pound is technically a unit of force and the kilogram a unit of mass, these units are both in common usage as force and mass units. Therefore, the conversions for both have been included. It should be remembered, however, that when using many formulae for the interactions of biomechanics, a correction factor is needed if mass is expressed in pounds, or force is expressed in kilograms.

Interconversions within the metric system have not been given because they are always expressed in terms of powers of 10. The following list of prefixes indicates the most commonly used factors:

Mega	1,000,000	10^6
Kilo	1,000	10^3
Centi	0.01	10^{-2}
Milli	0.001	10^{-3}
Micro	0.000001	10^{-6}

Distance

To Change From	To	Multiply By
feet	meters	0.3048
yards	meters	0.9144
meters	feet	3.28
meters	yards	1.094
miles	kilometers	1.609
kilometers	miles	0.621
inches	centimeters	2.54
centimeters	inches	0.3937

Velocity

In all conversions using only the form distance/sec, the conversion factors are the same as in the "Distance" section above.

miles/hour	feet/second	1.467
miles/hour	meters/second	0.447
miles/hour	kilometers/hour	1.609
feet/second	miles/hour	0.682
meters/second	miles/hour	2.237
kilometers/hour	miles/hour	0.621

Acceleration

feet/second²	meters/second²	0.3048
meters/second²	feet/second²	3.28
miles/hour per second	feet/second²	1.467
miles/hour per second	meters/second²	0.447
meters/second²	miles/hour per second	2.237
feet/second²	miles/hour per second	0.682

To Change From	To	Multiply By	Force
pounds	newtons	4.45	
pounds	kilograms	0.454	
kilograms	newtons	9.8	
kilograms	pounds	2.2	
newtons	kilograms	0.102	
newtons	pounds	0.225	
pounds	kilograms	0.454	**Mass**
pounds	slugs	0.0311	
kilograms	pounds	2.2	
kilograms	slugs	0.0685	
slugs	pounds	32.174	
slugs	kilograms	14.59	
pounds/inch²	kilograms/cm²	0.0703	**Pressure**
pounds/inch²	kilograms/meter²	703.0	
pounds/inch²	newtons/cm²	0.689	
pounds/inch²	dynes/centimeter²	68947.0	
kilograms/cm²	pounds/inch²	14.22	
kilograms/meter²	pounds/inch²	0.00142	
foot-pounds	Joules	1.356	**Energy**
foot-pounds	kilocalories	0.00032	
Joules	foot-pounds	0.738	
Joules	kilocalories	0.000239	
kilocalories	foot-pounds	3087.4	
kilocalories	Joules	4186.0	
foot-pounds/second	watt	1.356	**Power**
foot-pounds/second	horsepower	0.0018	
watt	foot-pounds/sec	0.738	
watt	horsepower	0.00134	
watt	kilocalories/hr	0.860	
horsepower	foot-pound/sec	550	
horsepower	watts	745.2	
horsepower	kilocalories/hr	641.3	
kilocalories/hr	foot-pounds/sec	0.858	
kilocalories/hr	watt	1.163	
kilocalories/hr	horsepower	0.00156	
slug-feet/sec	pound-feet/sec	32.174	**Momentum**
slug-feet/sec	kilogram-meters/sec	4.45	
pound-feet/sec	slug-feet/sec	0.0311	
pound-feet/sec	kilogram-meters/sec	0.138	
kilogram-meters/sec	slug-feet/sec	0.225	
kilogram-meters/sec	pound-feet/sec	7.227	

To convert between the pound-second, newton-second, and the kilogram-second, see the **Impulse** conversion factors under "Force" for the pound, newton, and kilogram.

One pound-second of impulse produces a change of momentum of 1 slug-foot/sec.
One newton-second of impulse produces a change of momentum of 1 kilogram-meter/sec.

Angle	To Change From	To	Multiply By
	radians	degrees	57.296
	radians	revolutions	0.159
	degrees	radians	0.0175
	degrees	revolutions	0.00278
	revolutions	radians	6.28
	revolutions	degrees	360.0
Moment of Inertia	slug-feet^2	pound-feet^2	32.174
	slug-feet^2	kilogram-meter^2	1.3558
	pound-feet^2	slug-feet^2	0.0311
	pound-feet^2	kilogram-meter^2	0.0421
	kilogram-meter^2	slug-feet^2	0.738
	kilogram-meter^2	pound-feet^2	23.70
Torque	pound-feet	meter-newtons	1.356
	pound-feet	meter-kilograms	0.138
	meter-newtons	pound-feet	0.737
	meter-newtons	meter-kilograms	0.102
	meter-kilograms	pound-feet	7.23
	meter-newtons	meter-kilograms	9.8

Angular Momentum The units of angular momentum are slug-feet^2-radians/sec, pound-feet^2-radians/sec, and kilogram-meter^2-radians/sec. The conversion factors are in the same as given under "Moment of Inertia" for slug-feet^2, pound-feet^2, and kilogram-meter^2.

Appendix B

I. Linear Relationships

Basic Mathematical Relationships

If data when plotted form a relatively straight line, this can be expressed by the equation:

$$y = mx + b$$

where b is the value of the line at $x = o$ and $m = \dfrac{y - y_o}{x - x_o}$

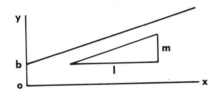

If the plot is position vs time, m is velocity. If the plot is velocity vs time, m is acceleration.

II. Quadratic Equation

If a quantity is expressed by the equation $ax^2 + bx + c = o$ then

$$x = \dfrac{-b \pm \sqrt{b^2 - 4ac}}{2a}$$

This equation is often encountered in problems involving position as a function of time during constant acceleration.

III. Circular Relationships

The circumference of a circle is $2\pi r$. The area of a circle is πr^2. The length of an arc is rA where A is expressed in radians.

1 radian = 57.3 degrees.

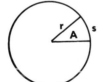

IV. Triangular (trigonometric) Relationships

sine of $A = \dfrac{a}{c}$

cosine of $A = \dfrac{b}{c}$

tangent of $A = \dfrac{a}{b}$

$A = $ arc tan $\dfrac{a}{b}$ or $\tan^{-1}\dfrac{a}{b}$

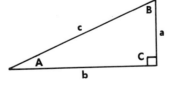

Pythagorean Relationship $c^2 = a^2 + b^2$

For any triangle

Law of sines $\dfrac{a}{\text{sine A}} = \dfrac{b}{\text{sine B}} = \dfrac{c}{\text{sine C}}$

Law of cosines $a^2 = b^2 + c^2 - 2bc \, \text{Cos A}$

V. Addition of Vectors

Triangle construction

$C = A + B$

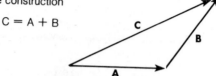

Parallelogram construction

$C = A + B$

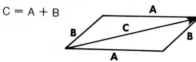

Use of components

$A_x = A \cos a$ $A_y = A \sin a$

$B_x = B \cos b$ $B_y = B \sin b$

$C_x = A_x + B_x$ $C = \sqrt{C_x^2 + C_y^2}$

$C_y = A_y + B_y$ $c = \text{arc tan} \dfrac{C_x}{C_y}$

Subtraction of vectors

$A - B = A + (-B)$ where $-B$ indicates a reversal of direction.

VI. Multiplication of vectors

A. Scalar multiplication

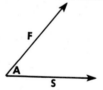

$F \bullet S = FS \cos A$

This is used for calculating the work done by force F moving a distance S.

B. Vector or cross multiplication

$r \times F = rF \sin \theta$

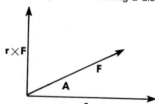

and is a vector perpendicular to both r and F.
This is used with rotational quantities

Torque $= r \times F$

Angular momentum $= m(r \times v)$

$(m = \text{mass}, v = \text{linear velocity})$

ANGLE		Sine	Cosine	Tangent
Degree	Radian			
0°	0.000	.0000	1.0000	.0000
1°	0.017	.0175	.9998	.0175
2°	0.035	.0349	.9994	.0349
3°	0.052	.0523	.9986	.0524
4°	0.070	.0698	.9976	.0699
5°	0.087	.0872	.9962	.0875
6°	0.105	.1045	.9945	.1051
7°	0.122	.1219	.9925	.1228
8°	0.140	.1392	.9903	.1405
9°	0.157	.1564	.9877	.1584
10°	0.175	.1736	.9848	.1763
11°	0.192	.1908	.9816	.1944
12°	0.209	.2079	.9781	.2126
13°	0.227	.2250	.9744	.2309
14°	0.244	.2419	.9703	.2493
15°	0.262	.2588	.9659	.2679
16°	0.279	.2756	.9613	.2867
17°	0.297	.2924	.9563	.3057
18°	0.314	.3090	.9511	.3249
19°	0.332	.3256	.9455	.3443
20°	0.349	.3420	.9397	.3640
21°	0.367	.3584	.9336	.3839
22°	0.384	.3746	.9272	.4040
23°	0.401	.3907	.9205	.4245
24°	0.419	.4067	.9135	.4452
25°	0.436	.4226	.9063	.4663
26°	0.454	.4384	.8988	.4877
27°	0.471	.4540	.8910	.5095
28°	0.489	.4695	.8829	.5317
29°	0.506	.4848	.8746	.5543
30°	0.524	.5000	.8660	.5774
31°	0.541	.5150	.8572	.6009
32°	0.559	.5299	.8480	.6249
33°	0.576	.5446	.8387	.6494
34°	0.593	.5592	.8290	.6745
35°	0.611	.5736	.8192	.7002
36°	0.628	.5878	.8090	.7265
37°	0.646	.6018	.7986	.7536
38°	0.663	.6157	.7880	.7813
39°	0.681	.6293	.7771	.8098
40°	0.698	.6428	.7660	.8391
41°	0.716	.6561	.7547	.8693
42°	0.733	.6691	.7431	.9004
43°	0.750	.6820	.7314	.9325
44°	0.768	.6947	.7193	.9657
45°	0.785	.7071	.7071	1.0000

ANGLE		Sine	Cosine	Tangent
Degree	Radian			
46°	0.803	.7193	.6947	1.0355
47°	0.820	.7314	.6820	1.0724
48°	0.838	.7431	.6691	1.1106
49°	0.855	.7547	.6561	1.1504
50°	0.873	.7660	.6428	1.1918
51°	0.890	.7771	.6293	1.2349
52°	0.908	.7880	.6157	1.2799
53°	0.925	.7986	.6018	1.3270
54°	0.942	.8090	.5878	1.3764
55°	0.960	.8192	.5736	1.4281
56°	0.977	.8290	.5592	1.4826
57°	0.995	.8387	.5446	1.5399
58°	1.012	.8480	.5299	1.6003
59°	1.030	.8572	.5150	1.6643
60°	1.047	.8660	.5000	1.7321
61°	1.065	.8746	.4848	1.8040
62°	1.082	.8829	.4695	1.8807
63°	1.100	.8910	.4540	1.9626
64°	1.117	.8988	.4384	2.0503
65°	1.134	.9063	.4226	2.1445
66°	1.152	.9135	.4067	2.2460
67°	1.169	.9205	.3907	2.3559
68°	1.187	.9272	.3746	2.4751
69°	1.204	.9336	.3584	2.6051
70°	1.222	.9397	.3420	2.7475
71°	1.239	.9455	.3256	2.9042
72°	1.257	.9511	.3090	3.0777
73°	1.274	.9563	.2924	3.2709
74°	1.292	.9613	.2756	3.4874
75°	1.309	.9659	.2588	3.7321
76°	1.326	.9703	.2419	4.0108
77°	1.344	.9744	.2250	4.3315
78°	1.361	.9781	.2079	4.7046
79°	1.379	.9816	.1908	5.1446
80°	1.396	.9848	.1736	5.6713
81°	1.414	.9877	.1564	6.3138
82°	1.431	.9903	.1392	7.1154
83°	1.449	.9925	.1219	8.1443
84°	1.466	.9945	.1045	9.5144
85°	1.484	.9962	.0872	11.430
86°	1.501	.9976	.0698	14.301
87°	1.518	.9986	.0523	19.081
88°	1.536	.9994	.0349	28.636
89°	1.553	.9998	.0175	57.290
90°	1.571	1.000	.0000	

Appendix B

The easy availability of manual and desk top calculators has eliminated much of the uncertainty and tedium formerly associated with analysis calculations. Electronic arithmetic assures speed and accuracy unknown to earlier generations of students. Thus, these devices should be used wherever possible to facilitate concentration on concepts and procedures, rather than on the numerical manipulations.

Examples of Electronic Calculator Use in Intermediate Analysis

This appendix gives three examples of analysis calculations which can be easily made with small electronic devices. This will serve as an introduction to the variety of possible uses. The data used was taken from analyses given earlier in the text, and the examples use an algebraic type language. Different calculators will show a slight variation in keys or in the order of punching for certain operations; but these examples concentrate on the procedure. They are not intended to provide a comprehensive introduction to the details of programming.

A. SCALING EXAMPLE (see Figure 13.5)

Operational keys needed: X, ÷, Store, Recall

The first step is to calculate and store the scale factor for the photographs. The data in this exercise were plotted so the hurdle height of 3.5 feet measured 15 scale units. This scaling factor can be calculated and stored by the following set of commands:

$$15$$
$$÷$$
$$3.5$$
$$=$$
$$\text{Store}$$

This procedure stores the fact that 4.286 scale units equals one foot. The data in the first column are then scaled by the following set of commands:

33.5	15.5	30.3	10.5	36.7	12.6
÷	÷	÷	÷	÷	÷
Recall	Recall	Recall	Recall	Recall	Recall
=	=	=	=	=	=

The recall command gives the scale factor each time it is needed. The equal command gives the space positions in the first column of the lower table. The remainder of the table is generated by repeating the procedure.

B. VECTOR ADDITION (see Figure 8.3)

Operational keys needed: Sine, Cosine, Tangent, Inverse, +, (,), X, $\sqrt{\quad}$

Vector addition can be done with an electronic calculator by finding the horizontal and vertical components of each vector and then combining these to give the direction and magnitude of the resultant.

In the figure representing the volleyball spike, one should consider the basic vectors to have the following characteristics:

	Magnitude	Direction
B	12 ft/sec	+45°
S	18 ft/sec	0° (horizontal)
A	40 ft/sec	−30°
W	6 ft/sec	−37°

The following set of instructions will provide the sum of the horizontal components of these vectors:

$$12 \times \text{Cos } 45 + 18 + \quad (\,40 \times \text{Cos } 30\,) + \quad (\,6 \times \text{Cos } 37\,) =$$

In these instructions, the parentheses are used to allow the calculation of a complete vector component before adding or subtracting it from the sum. On some calculators there is an Accumulate Key which will do the same thing without the use of parentheses. Also, on some calculators it is necessary to enter the angle size before hitting the cosine key.

Performing the above set of instructions will give the sum of the horizontal components of these vectors to be 65.9 feet per second. The vertical components are found by:

$$12 \times \text{Sin } 45 - \quad (\,40 \times \text{Sin } 30\,) - (\,6 \times \text{Sin } 37\,) =$$

This summation yields the vertical components to be − 15.1 feet/sec. The total vector can then be found by:

$$\sqrt{(\,65.9 \times \quad 65.9 + 15.1 \times \quad 15.1\,)} =$$

The magnitude of the final hand velocity obtained is 67.6 feet/second. The direction is found by:

$$\text{Inverse Tangent} (\,15.1 \div 69.5\,) =$$

giving an angle of 12.26° below the horizontal.

C. CALCULATION OF ROTATIONAL VELOCITIES (see Figures 13.9 and 14.2)

While the hip, knee, and ankle velocities in figure 14.2 were actually calculated from positional data on the film, it is possible to check these figures by calculating the speed induced by the rotations and lever arms in figure 13.9. For example, the motion of the knee will be a combination of hip velocity and the velocity induced by femur rotation.

Appendix C

This last component is given by the following key order:

$$\begin{array}{cc} 1.39 & X \\ X & 1.26 \\ 6.283 & = \end{array}$$

giving a velocity of 11 ft/sec at an angle 32.7° from the vertical. The 6.283 converts rev/sec to radians/sec. This must be added to the hip velocity to obtain calculated knee velocity:

$$\begin{array}{c} 11 \\ X \\ \text{Sin } 32.7 \\ + \\ 23.4 \text{ for the horizontal,} \end{array}$$

and,

$$\begin{array}{c} 11 \\ X \\ \text{Cos } 32.7 \text{ for the vertical} \end{array}$$

giving values of 29.3 ft/sec for the horizontal and 9.26 ft/sec for the vertical in close agreement with the previous calculations.

This technique can be extended to more complex rotational motion but would require precise film data for better accuracy.

Bibliography

Adrian, M.; Tipton, C. M.; and Karpovich, P. *Electrogoniometry Manual*. Springfield, Mass.: Springfield College, 1965.

Alhabashi, Z. *Kinesiology in the Field of Physical Education*. Maktabet: Al-Kahira Al-Haditha, 1964.

Amar, Jules. *The Human Motor*. New York: E. P. Dutton and Co., 1920.

American Academy of Orthopedic Surgeons. *Measuring and Recording of Joint Motion*. Chicago: The Academy, 1965.

American College of Sports Medicine (ed.) *Guidelines for Graded Exercise Testing and Exercise Prescription* (Second Edition). Philadelphia: Lea and Febiger, 1980.

Anderson, T. McClurg. *Human Kinetics and Analyzing Body Movements*. London: William Heinemann Medical Books, Ltd., 1951.

Aristotle. *Progressions of Animals*. Translated by E. S. Forster. Cambridge: Harvard University Press, 1945.

Asmussen, Erling and Jorgensen, Kurt (eds.) *International Congress of Biomechanics*. Baltimore: University Park Press, 1978.

Bade, Edwin. *The Mechanics of Sport*. Kingswood: Glade House, 1962.

Barham, Jerry N. *Mechanical Kinesiology*. St. Louis: C. V. Mosby Co., 1978.

Barham, Jerry N., and Thomas, William L. *Anatomical Kinesiology*. New York: Macmillan Co., 1969.

Barham, Jerry N., and Wooten, Edna P. *Structural Kinesiology*. New York: Macmillan Co., 1973.

Barnett, C. H.: Davies, D. V.; and MacConaill, M. A. *Synovial Joints, Their Structure and Mechanics*. Springfield, Ill.: Charles C Thomas, Publisher, 1961.

Barrow, Harold M. *Man and Movement: Principles of Physical Education*. Philadelphia: Lea and Febiger, 1977.

Basmajian, J. V. *Muscles Alive. Their Functions Revealed by Electromyography*. Baltimore: The Williams and Wilkins Co., 1962.

Beevor, C. *The Croonian Lectures on Muscular Movements*. London: Macmillan, Ltd., 1903.

Bernstein, N. A. *Investigations on the Biodynamics of Walking, Running, and Jumping*. Moscow: Central Scientific Institute of Physical Culture, 1940.

——. *The Co-ordination and Regulation of Movements*. New York: Pergamon Press, 1967.

Birdshistell, R. L. *Introduction to Kinesics: An Annotation System of Analysis of Body Motion and Gesture*. Washington: D.C.: U.S. Foreign Service, 1952.

Bleustein, J., ed. *Mechanics and Sport*. New York: The American Society of Mechanical Engineering, 1976.

Bootzin, David, and Muffley, Harry C. *Rock Island Arsenal Biomechanics Symposium*. New York: Plenum Press, 1969.

Borelli, G. A. *De Motu Animalium*. Ludguni Batavorum, 1680–81.

Bourne, G. H. B., ed. *The Structure and Function of Muscle*. New York: Academic Press, 1960.

Broer, M. R. *An Introduction to Kinesiology*. Englewood Cliffs, N.J.: Prentice-Hall, 1968.

——. *Efficiency of Human Movement*. Philadelphia: W. B. Saunders Company, 1973.

——. *Laboratory Experiences: Exploring Efficiency of Human Movement*. Philadelphia: W. B. Saunders Company, 1973.

——, and Houtz, Sara J. *Patterns of Muscular Activity in Selected Sport Skills: An Electromyographic Study*. Springfield, Ill.: Charles C Thomas, Publisher, 1967.

Broer, Marion R. and Zernicke, Ronald F. *Efficiency of Human Movement* (Fourth Edition). Philadelphia: Saunders Company, 1979.

Brown, Roscoe C., and Kenyon, Gerald S., eds. *Classical Studies on Physical Activity*. Englewood Cliffs, N.J.: Prentice-Hall, 1968.

Brunnstrom, Signe. *Clinical Kinesiology*. Philadelphia: F. A. Davis Co., 1972.

Bunn, John W. *Scientific Principles of Coaching*. Englewood Cliffs, N.J.: Prentice-Hall, 1972.

Campbell, E. J. M. *The Respiratory Muscles and the Mechanics of Breathing*. London: Lloyd-Luke (Medical Books) Ltd., 1958.

Canna, D. J., and Loring, E. *Kinesiography*. Fresno, Calif.: The Academy Guild Press, 1955.

Cerquiglini, S.; Venerando, A.; and Wartenweiler, J., eds. *Biomechanics III*. Basel, Switzerland: S. Krager AG, 1973.

Chesterman, W. D. *The Photographic Study of Rapid Events*. London: Oxford University Press, 1951.

Clarke, Harrison H. *Muscular Strength and Endurance in Man: International Research Monograph*. Englewood Cliffs, N.J.: Prentice-Hall, 1966.

Clarys, J., and Lewillie, L. *Swimming II*. Baltimore: University Park Press, 1975.

Close, J. R. *Motor Function in the Lower Extremity: Analyses by Electronic Instrumentation*. Springfield, Ill.: Charles C Thomas, Publisher, 1964.

——. *Functional Anatomy of the Extremities*. Springfield, Ill.: Charles C Thomas, Publisher, 1973.

Cochran, Alastair, and Stobbs, John. *The Search for the Perfect Swing*. New York: J. B. Lippincott Company, 1968.

Cooper, John M., ed. *C.I.C. Symposium on Biomechanics*. Chicago: Athletic Institute, 1971.

——, Adrian, Marlene, and Glassow, Ruth B. *Kinesiology*. St. Louis: The C. V. Mosby Company, 1982.

——, and Glassow, Ruth B. *Kinesiology*. St. Louis: The C. V. Mosby Company, 1976.

——, and Siedentrop, D. *The Theory and Science of Basketball*. Philadelphia: Lea and Febiger, 1969.

Counsilman, James E. *The Science of Swimming.* Englewood Cliffs, N.J.: Prentice-Hall, 1968.

Crouch, J. E. *Functional Human Anatomy.* Philadelphia: Lea and Febiger, 1972.

Cureton, T. K. *Physics Applied to Health and Physical Education.* Springfield, Mass.: International YMCA College, 1936.

Daish, C. B. *The Physics of Ball Games.* London: The English Universities Press Ltd., 1972.

Daniels, L., and Worthingham, C. *Muscle Testing Techniques of Manual Examination.* Philadelphia: W. B. Saunders Co., 1972.

Donskoi, D.D. *Biomechanik der Korperubüngen.* Berlin: Sportverlag, 1961.

DuBois, J., and Santschi, W. R. *The Determination of the Moment of Inertia of Living Human Organisms.* New York: John Wiley & Sons, 1963.

Duchenne, G. B. A. *Physiology of Motion.* Translated by E. B. Kaplan. Philadelphia: W. B. Saunders Company, 1959.

Duvall, Ellen Neall. *Kinesiology: The Anatomy of Motion.* Englewood Cliffs, N.J.: Prentice-Hall, 1959.

Dyson, Geoffrey H. G. *The Mechanics of Athletics.* London: University of London Press, 1973.

Elftman, H. *Skeletal and Muscular Systems: Structure and Function in Medical Physics.* Chicago: Year Book Publications, 1944.

Esch, D., and Lepley, M. *Evaluation of Joint Motion: Methods of Measuring and Recording.* Minneapolis: University of Minnesota Press, 1974.

———. *Musculoskeletal Function: An Anatomy and Kinesiology Laboratory Manual.* Minneapolis: University of Minnesota Press, 1974.

Eshkol, N., and Wachman, A. *Movement Notation.* London: Weidenfeld and Nicolson, 1958.

Evans, F. Gaynor, ed. *Biomechanical Studies of the Musculoskeletal System.* Springfield, Ill.: Charles C Thomas, Publisher, 1961.

———. *Studies on the Anatomy and Function of Bone and Joints.* New York: Springer-Verlag, 1966.

———. *Mechanical Properties of Bone.* Springfield, Ill.: Charles C Thomas, Publisher, 1973.

Feather, N. *An Introduction to the Physics of Mass, Length, and Time.* Edinburgh: University Press, 1959.

Fetz, F. *Bewegungslehre der Leibesubungen.* Frankfurt a.M.: Wilhelm Limpert-Verlag GmbH, 1972.

Finley, F. Ray. *Kinesiological Analysis of Human Locomotion.* Eugene: University of Oregon Press, 1961.

Fischer, O. *Theoretische Grundlagen fuer eine Mechanik der lebenden Koerper.* Berlin: B. G. Teubner, 1906.

———. *Kinematik organischer Gelenke.* Braunschweig: F. Vieweg & Sohn GmbH, 1907.

Frankel, Victor H., and Burstein, Albert H. *Orthopaedic Biomechanics.* Philadelphia: Lea and Febiger, 1970.

Frankel, Victor H. and Nordin, Margareta. *Basic Biomechanics of the Skeletal System.* Philadelphia: Lea and Febiger, 1980.

Frost, Harold M. *An Introduction to Biomechanics.* Springfield, Ill.: Charles C Thomas, Publisher, 1967.

———. *Orthopaedic Biomechanics.* Springfield, Ill.: Charles C Thomas, Publisher, 1973.

Fung, Y. C.; Perrone, N.; and Anliker, M., eds. *Biomechanics: Its Foundations and Objectives.* Englewood Cliffs, N.J.: Prentice-Hall Inc., 1972.

Gage, Howard (ed.). *Biomechanical and Human Factors Conference.* New York: American Society of Mechanical Engineers, 1967.

Gans, Carl. *Biomechanics: An Approach to Vertebrate Biology.* Philadelphia: J. B. Lippincott Company, 1974.

Ganslen, R. V., and Hall, K.G. *The Aerodynamics of Javelin Flight.* Fayetteville: University of Arkansas Press, 1960.

Ganslen, R. V. *Mechanics of the Pole Vault.* St. Louis: John S. Swift Co., 1963.

George, Gerald S. *Biomechanics of Women's Gymnastics.* Englewood, N.J.: Prentice-Hall, Incorporated, 1980.

Goldsmith, Werner. *Impact.* New York: St. Martin's Press, 1961.

Govaerts, A. *Biomechanics: A New Method of Analyzing Motion.* Brussels: Brussels University Press, 1962.

Gowitzke, Barbara A. and Milner, Morris. *Understanding the Scientific Bases of Human Movement* (Second edition). Baltimore: The Williams and Wilkins Company, 1980.

Gray, Henry. *Anatomy of the Human Body.* Philadelphia: Lea and Febiger, 1982.

Gray, J. *How Animals Move.* London: Cambridge University Press, 1960.

Grieve, D. W., et al. *Techniques for the Analysis of Human Movement.* London: Lepus Books, 1975.

Groves, Richard, and Camaione, David N. *Concepts in Kinesiology.* Philadelphia: W. B. Saunders Company, 1975.

Hall, Michael C. *The Locomotor System: Functional Anatomy.* Springfield, Ill.: Charles C Thomas, Publisher, 1965.

Halliday, D., and Resnick, R. *Fundamentals of Physics,* Second Edition. New York: John Wiley & Sons, 1981.

Harris, Ruth W. *Kinesiology Workbook and Laboratory Manual.* Boston: Houghton and Mifflin Co., 1977.

Hawley, Gertrude. *An Anatomical Analysis of Sport.* Cranbury, N.J.: A. S. Barnes & Co., 1940.

———. *The Kinesiology of Corrective Exercise.* Philadelphia: Lea and Febiger, 1949.

Hay, James G. *The Biomechanics of Sports Techniques.* Englewood Cliffs, N.J.: Prentice-Hall, 1978.

———. *A Bibliography of Biomechanics Literature.* Iowa City: Department of Physical Education for Men. University of Iowa, 1981.

Hays, Joan F. *Modern Dance: A Biomechanical Approach to Teaching.* St. Louis: The C. V. Mosby Company, 1981.

Hemming, George W. *Billiards Mathematically Treated.* London: Macmillan, 1899.

Hertel, Heinrich. *Structure-Form-Movement.* New York: Reinhold Publishing Corp., 1966.

Hickman, C. N.; Nagler, F.; and Klopsteg, Paul E. *Archery: The Technical Side.* Redlands, Calif.: National Field Archery Association, 1947.

Higgins, Joseph R. *Human Movement.* St. Louis: C. V. Mosby Co., 1977.

Hill, A. V. *Muscular Movements in Man.* New York: McGraw-Hill Book Co., 1927.

———. *Living Machinery.* New York: McGraw-Hill Book Co., 1927.

———. *First and Last Experiments in Muscle Mechanics.* Cambridge: University Press, 1970.

Hinson, Marilyn M. *Kinesiology.* Dubuque, Iowa: Wm. C. Brown Company, Publishers, 1977.

Hochmuth, G. *Biomechanik sportlicher Bewegungene.* Frankfurt a.M.: Wilhelm Limpert-Verlag GmbH, 1967.

Hollinshead, H. W. *Functional Anatomy of the Limbs and Back.* Philadelphia: W. B. Saunders Company, 1960.

Hopper, B. J. *Notes on the Dynamical Basis of Physical Movement*. Twickenham, Middlesex. England: St. Mary's College, 1959.

————. *The Mechanics of Human Movement*. New York: American Elsevier Publishing Co., Inc. 1973.

Howard, I. P., and Templeton, W. B. *Human Spatial Orientation*. New York: John Wiley & Sons, 1966.

Howell, A. Brazieir. *Speed in Animals: Their Specialization for Running and Leaping*. New York: Hafner Publishers, 1965.

Hoyle, G. *The Nervous Control of Muscular Contraction*. New York: Cambridge University Press, 1958.

Hyzer, William G. *Engineering and Scientific High Speed Photography*. New York: Macmillan Co., 1963.

International Society of Electrophysiological Kinesiology. *Proceedings: Fourth Congress of the International Society of Electrophysiological Kinesiology*. Boston: C. J. DeLuca, 1979.

Jensen, Clayne R., and Schultz, Gordon W. *Applied Kinesiology*. New York: McGraw-Hill Book Co., 1977.

Johnson, Warren R., and Buskirk, E., eds. *Science and Medicine of Exercise and Sports*. New York: Harper & Row, Publishers, 1974.

Jones, F. W. *Structure and Function as Seen in the Foot*. London: Bailliere, Tindall and Cox, Ltd., 1949.

Joseph, J. *Man's Posture: Electromyographic Studies*. Springfield, Ill.: Charles C Thomas, Publisher, 1960.

Karas, V., and Stapleton, A. *Application of the Theory of the Motion System in the Analysis of Gymnastic Motions*. New York: S. Karger, 1968.

Katz, B. *Nerve, Muscle, and Synapse*. New York: McGraw-Hill Book Co., 1966.

Kelley, David L. *Kinesiology: Fundamentals of Motion Description*. Englewood Cliffs, N.J.: Prentice-Hall, 1971.

Kendall, M. B. *Muscles: Testing and Function*. Baltimore: Wiliams and Wilkins, 1971.

Kenedi, R. M., ed. *Symposium on Biomechanics and Related Bioengineering Topics*. New York: Pergamon Press, 1955.

Klopsteg, P. E., and Wilson, P. V., eds. *Human Limbs and Their Substitutes*. New York: McGraw-Hill Book Co., 1954.

Knuttgen, H. G. *Neuromuscular Mechanisms*. Baltimore: University Park Press, 1976.

Komi, P. V., ed. *Biomechanics V-A and V-B*. Baltimore: University Park Press, 1976.

Krause, J. V., and Barham, Jerry. *The Mechanical Foundations of Human Motion: A Programmed Text*. St. Louis: The C. V. Mosby Company, 1975.

Kreighbaum, Ellen and Barthels, Katherine. *Biomechanics: A Qualitative Approach for Studying Human Movement*. Minneapolis: Burgess Publishing Company, 1981.

Krogman, W. M., and Johnston, F. E. *Human Mechanics (Four Monographs Abridged)*. AMRL Technical Documentary Report 63–123. Wright-Patterson Air Force Base, Ohio: 6570th Aerospace Medical Research Laboratories, December, 1963.

Kuel, J., et al. *Energy Metabolism of Human Muscle*. Baltimore: University Park Press, 1972.

Lam, C. R. *An Introduction to Biomechanics*. Springfield, Ill.: Charles C Thomas Publishing Co., 1967.

Landry, Fernand and Orban, William (eds.). *Biomechanics of Sports and Kinanthropometry*. Miami: Symposia Specialists, 1978.

Larson, L. *Foundations of Physical Activity*. New York: Macmillan Publishing Co., 1976.

LeVeau, Barney. *Biomechanics of Human Motion*. Philadelphia: W. B. Saunders Co., 1977.

Levens, Alexander. *Graphical Methods in Research*. New York: John Wiley & Sons, 1965.

Lewille, L., and Clarys, J. P., eds. *International Symposium on Biomechanics in Swimming, Water Polo, and Diving*. Brussels, 1971.

Lipovetz. F. J. *Basic Kinesiology*. Minneapolis: Burgess Publishing Co., 1952.

Logan, Gene A. *Adapted Physical Education*. Dubuque, Iowa: Wm. C. Brown Company Publishers, 1972.

————, and McKinney, Wayne C. *Kinesiology*. Dubuque, Iowa: Wm. C. Brown Company Publishers, 1970.

————, and McKinney, Wayne C. *Anatomic Kinesiology*. Dubuque, Iowa: Wm. C. Brown Company Publishers, 1982.

Long, C. *Normal and Abnormal Motor Control in the Upper Extremities*. Washington: Social and Rehabilitation Services, 1970.

Lucas, D. B., and Inman, V. T. *Functional Anatomy of the Shoulder Joint*. Berkeley: University of California Medical School, 1963.

Luttgens, Kathryn, and Wells, Katherine. *Kinesiology: Scientific Basis of Human Motion*. Lavallette, N.J.: Saunders College Publishing, 1982.

MacConaill, M. A., and Basmajian, J. V. *Muscles and Movement: A Basis for Human Kinesiology*. Baltimore: The Wiliams and Wilkins Co., 1969.

McCormick, E. J. *Human Factor Engineering*. New York: McGraw-Hill Book Co., 1970.

Maquet, Paul G. J. *Biomechanics of the Knee*. New York: Springer-Verlag, 1976.

Marey, Etienne J. *Movement*. Translated by Eric Pritchard. London: William Heinemann, Limited, 1895.

Margaria, Rodolfo. *Biomechanics and Energetics of Muscular Exercise*. Oxford: Clarendon Press, 1976.

Martin, Thomas P., ed. *Biomechanics of Sport, Selected Readings*. Brockport, N.Y.: T. P. Martin, 15 Beverly Dr., 1976.

Mascelli, Joseph V., and Miller, Arthur. *American Cinematographer Manual*. Hollywood: American Society of Cinematographers Holding Corporation, 1966.

Massey, Benjamin H., et al. *The Kinesiology of Weight Lifting*. Dubuque, Iowa: Wm. C. Brown Company Publishers, 1959.

Metheny, Eleanor. *Body Dynamics*. New York: McGraw-Hill, Inc., 1952.

Miller, Doris, and Nelson, Richard C. *The Biomechanics of Sport*. Philadelphia: Lea and Febiger, 1973.

Montagu, M. F. A. *A Handbook of Anthropometry*. Springfield, Ill.: Charles C Thomas, Publisher, 1960.

Morehouse, L. E., and Cooper, J. M. *Kinesiology*. St. Louis: The C. V. Mosby Company, 1950.

Morris, Roxie. *Correlation of Basic Sciences with Kinesiology*. New York: American Physical Therapy Association, 1955.

Morton, D. J., and Fuller, D. D. *Human Locomotion and Body Form*. Baltimore: The Williams and Wilkins Co., 1952.

Muybridge, Eadweard. *The Human Figure in Motion*. New York: Dover Publications, Inc., 1955.

Naylor, T. H. *Computer Simulation Techniques*. New York: John Wiley & Sons, 1966.

Nelson, Richard, and Morehouse, Chauncey, eds. *International Seminar on Biomechanics*. Baltimore: University Park Press, 1974.

Nemessuri, M. *Funktionelle Sportanatomie*. Berlin: Sportverlag, 1963.

Northrip, John W.; Logan, Gene A.; and McKinney, Wayne C. *Introduction to Biomechanic Analysis of Sport*. Dubuque, Iowa: Wm. C. Brown Company Publishers, 1979.

O'Connell, Alice L., and Gardner, Elizabeth B. *Understanding the Scientific Bases of Human Movement*. Baltimore: The Williams and Wilkins Co., 1972.

Ostyn, Michael; Beunen, Gaston; and Simons, Jan (eds.). *International Seminar on Kinanthropometry*. Baltimore: University Park Press, 1979.

Peterson, A. P. G., and Gross, E. E. *Handbook of Morse Measurements*. Concord, Mass.: General Radio Co., 1974.

Piscopo, John and Baley, James. *Kinesiology: The Science of Movement*. New York: John Wiley and Sons, 1981.

Plagenhoef, Stanley. *Fundamentals of Tennis*. Englewood Cliffs, N.J.: Prentice-Hall, 1970.

————. *Patterns of Human Motion: A Cinematographic Analysis*. Englewood Cliffs, N.J.: Prentice-Hall, 1971.

Posse, N. *The Special Kinesiology of Educational Gymnastics*. Boston: Lothrop, Lee and Shepard Co., Inc., 1890.

Rasch, Philip J., and Burke, Roger K. *Kinesiology and Applied Anatomy*. 5th ed. Philadelphia: Lea and Febiger, 1978.

Rebikoff, Dimitri, and Chernery, Paul. *A Guide to Underwater Photography*. New York: Greenberg, 1955.

Rodahl, K., and Horvath, S. M. *Muscle as a Tissue*. New York: McGraw-Hill, Inc., 1962.

Ruch, T. C., and Patton, H. D., eds. *Physiology and Biophysics*. Philadelphia: W. B. Saunders Company, 1965.

Scott, M. Gladys. *Analysis of Human Motion*. New York: Appleton-Century-Crofts, 1963.

Sharpley, F. *Biomechanics for Beginners*. Ardmore, New Zealand: Ardmore Teacher's College, 1968.

Simons, E. N. *Mechanics for the Home Student*. London: Iliffe and Sons Ltd., 1950.

Skalak, Richard and Nerem, Robert M. (eds.). *Biomechanics Symposium*. New York: American Society of Mechanical Engineers, 1975.

Skalak, Richard and Schultz, Albert (eds.). *Biomechanics Symposium*. New York: American Society of Mechanical Engineers, 1977.

Skarstrom, W. *Gymnastic Kinesiology*. Springfield, Mass.: F. A. Bassette Co., 1909.

————. *Kinesiology of Trunk, Shoulders and Hip*. Springfield, Ill.: Charles C Thomas, Publisher, 1946.

Slocum, D. B., and Bowerman, William. *The Biomechanics of Running*. Clinical Orthopaedics No. 23. Philadelphia: J. B. Lippincott Co., 1962.

Spence, D. *Essentials of Kinesiology*. Philadelphia: Lea and Febiger, 1975.

Squire, P. J. *Biomechanics of Sport and Human Movement: A Reference Bibliography*. Edinburgh: Dufermline College of Physical Education, 1977.

Stapp, J. P. *Jolt Effects of Impact on Man*. San Antonio, Tex.: Brooks Air Force Base, 1961.

Steindler, Arthur. *Mechanics of Normal and Pathological Locomotion in Man*. Springfield, Ill.: Charles C Thomas, Publisher, 1935.

————. *Kinesiology of the Human Body under Normal and Pathological Conditions*. Springfield, Ill.: Charles C Thomas, Publisher, 1973.

Stoddard, J. T. *The Science of Billiards*. Boston: Butterfield, 1913.

Strasser, H. *Lehrbuch der Muskel and Gelenkmechanik*. Berlin: J. Springer, 1913.

Streeter, V. L. *Fluid Mechanics*. New York: McGraw-Hill, 1966.

Terauds, Juris (ed.). *International Symposium of Science in Weightlifting*. Del Mar, CA: Academic Publishers, 1979.

————. *Science in Biomechanics Cinematography*. Del Mar, CA: Academic Publishers, 1979.

Terauds, Juris and Dales, George G. (eds.). *International Symposium of Science in Athletics*. Del Mar, CA: Academic Publishers, 1979.

Terauds, Juris and Danials, Dayna (eds.). *International Symposium of Science in Gymnastics*. Del Mar, CA: Academic Publishers, 1979.

Thompson, Clem W. *Kranz Manual of Kinesiology* (Eighth edition). St. Louis: The C. V. Mosby Company, 1977.

Tichauer, E. R. *The Biomechanical Basis of Ergonomics: Anatomy Applied to the Design of Work Situations*. New York: Wiley and Sons, 1978.

Tricker, R. A. R. and Tricker, B. J. K. *The Science of Movement*. New York: American Elsevier Publishing Company, 1967.

VanVeen, F. *Handbook of Stroboscopy*. Concord, Mass.: General Radio Co., 1977.

Vredenbregt, J., and Wartenweiler, J., eds. *International Seminar on Biomechanics*. Baltimore: University Park Press, 1971.

Waddell, J. H., and Waddell, J. W. *Photographic Motion Analysis*. Chicago: Indust. Lab. Publ., 1955.

Wartenweiler, J.; Jokl, E.; and Hebbelnick, M., eds. *International Seminar on Biomechanics*. New York: S. Krager, 1968.

Webster, F. A. M. *Why? The Science of Athletics*. London: John F. Shaw and Co., Ltd., 1936.

Wells, Katharine F., and Luttgens, K. *Kinesiology*. Philadelphia: W. B. Saunders Co., 1976.

Wickstrom, Ralph L. *Fundamental Motor Patterns*. Philadelphia: Lea and Febiger, 1977.

Widule, Carol J. *Analysis of Human Motion*. Lafayette: Balt Publishers, 1977.

Williams, M., and Lissner, H. R. *Biomechanics of Human Motion*. Philadelphia: W. B. Saunders Company, 1962.

Wilt, Fred. *Mechanics Without Tears*. Tucson, Ariz.: United States Track and Field Federation, 1970.

Winter, David. *Biomechanics of Human Movement*. New York: John Wiley and Sons, 1980.

Woolridge, D. E. *Mechanical Man: The Physical Bases of Intelligent Life*. New York: McGraw-Hill Book Co., 1968.

Wright, W. *Muscle Function*. New York: Paul B. Hoeber, Inc., 1928.

Yamada, Hiroshi. *Strength of Biological Materials*. Baltimore: The Williams and Wilkins Co., 1970.